Innovations in Biomolecular Modeling and Simulations
Volume 1

RSC Biomolecular Sciences

Titles in the Series:

1: Biophysical and Structural Aspects of Bioenergetics
2: Exploiting Chemical Diversity for Drug Discovery
3: Structure-based Drug Discovery: An Overview
4: Structural Biology of Membrane Proteins
5: Protein–Carbohydrate Interactions in Infectious Disease
6: Sequence-specific DNA Binding Agents
7: Quadruplex Nucleic Acids
8: Computational and Structural Approaches to Drug Discovery: Ligand–Protein Interactions
9: Metabolomics, Metabonomics and Metabolite Profiling
10: Ribozymes and RNA Catalysis
11: Protein–Nucleic Acid Interactions: Structural Biology
12: Therapeutic Oligonucleotides
13: Protein Folding, Misfolding and Aggregation: Classical Themes and Novel Approaches
14: Nucleic Acid–Metal Ion Interactions
15: Oxidative Folding of Peptides and Proteins
16: RNA Polymerases as Molecular Motors
17: Quantum Tunnelling in Enzyme-Catalysed Reactions
18: Natural Product Chemistry for Drug Discovery
19: RNA Helicases
20: Molecular Simulations and Biomembranes: From Biophysics to Function
21: Structural Virology
22: Biophysical Approaches Determining Ligand Binding to Biomolecular Targets: Detection, Measurement and Modelling
23: Innovations in Biomolecular Modeling and Simulations: Volume 1

How to obtain future titles on publication:
A standing order plan is available for this series. A standing order will bring delivery of each new volume immediately on publication.

For further information please contact:
Book Sales Department, Royal Society of Chemistry, Thomas Graham House, Science Park, Milton Road, Cambridge, CB4 0WF, UK
Telephone: +44 (0)1223 420066, Fax: +44 (0)1223 420247, Email: books@rsc.org
Visit our website at http://www.rsc.org/Shop/Books/

Innovations in Biomolecular Modeling and Simulations
Volume 1

Edited by

Tamar Schlick
Department of Chemistry, Courant Institute of Mathematical Sciences, New York University, New York, NY 10012, USA

RSC Publishing

RSC Biomolecular Sciences No. 23

ISBN: 978-1-84973-461-5
ISSN: 1757-7152

A catalogue record for this book is available from the British Library

Published by The Royal Society of Chemistry,
Thomas Graham House, Science Park, Milton Road,
Cambridge CB4 0WF, UK

Registered Charity Number 207890

For further information see our web site at www.rsc.org

Printed in the United Kingdom by Henry Ling Limited, at the Dorset Press, Dorchester, DT1 1HD

Preface

Modelers to the Rescue

"The purpose of models is not to fit the data but to sharpen the questions."
–Samuel Karlin, 1983.

We are in an era of data deluge. Merchants know all about what we eat, drink, wear, and read; doctors subject us to numerous tests and procedures to extract numbers and images that are intended to capture our state of health; and large-scale genome sequencing companies are producing so much data that computer disks via FedEx, rather than Internet downloads, are the only viable mode of data transfer![1] No wonder cartoonist Chris Wildt drew men and women of science pouring over paper files, computer screens, and endless diagrams before giving up: "Well, we've seen all the data and crunched the numbers... heads or tails?"

Sifting through biological and chemical data to extract knowledge, and eventually an understanding, is a task attributed today to bioinformaticians and other computational scientists through data mining approaches and computer modeling and simulation. As recently surveyed in a field perspective,[2] the trajectory of biomolecular modeling and simulation has evolved from initial euphoria and unrealistic expectations in the 1980s to a more mature, cautiously optimistic viewpoint bolstered by demonstrated successes (Figure P.1). Clearly, biomolecular modeling is not likely to supplant experimentation in the near future for predicting and understanding biomolecular structure and function; however, iterative program design between experiment and theory is becoming a reality for advancing the field. The problem of early unrealistic expectation and hype was not so much due to the inherent imperfections, approximations,

RSC Biomolecular Sciences No. 23
Innovations in Biomolecular Modeling and Simulations: Volume 1
Edited by Tamar Schlick

Published by the Royal Society of Chemistry, www.rsc.org

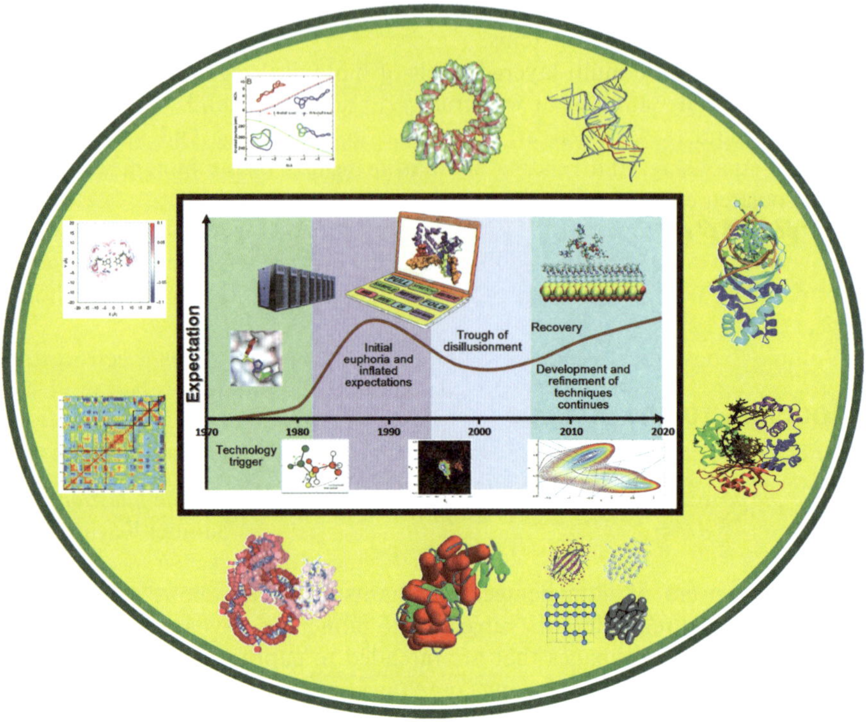

Figure P.1 Proposed expectation curve for the field of biomolecular modeling and simulation, with approximate timeline,[2] surrounded by representative images from work collected in these volumes. The field started when comprehensive molecular mechanics efforts began, and it took off with the increasing availability of fast workstations and later supercomputers. Following unrealistically high expectations and disappointments, the field is well on its fruitful trajectory of realistic progress and synergy between theory and experiment. The images taken from contributing chapters in this two-volume book are, *inside the box*: bottom right: contour plots connecting potential energy basins (A. E. Cardenas *et al.*); bottom middle: essential dynamics projection for HU protein/DNA junction complex (Wheatley *et al.*); top right: peptides on microarray surface (J. Feng *et al.*); and left middle: fragment grown for drug discovery applications (Z. Liu *et al.*). Outside the box, images clockwise from bottom right to top right are: proteins at various levels of resolution (N. Ceres *et al.*); coarse-grained model of the molecular motor myosin (E. R. Taylor); nucleosome chains (K. Rippe *et al.*); covariance matrix for protein/DNA/dNTP complex in DNA polymerases (M. Foley *et al.*); solvent charge density in an AT base pair plane (T. Luchko *et al.*); trefoil knot ensemble (G. Witz *et al.*); electrostatic potential of the nucleosome (R. C. Harris *et al.*); junction architecture in hairpin ribozyme (D. Lilley); structure of architectural protein HU bound to DNA (Wheatley *et al.*); and structure of DNA polymerase λ (M. Foley *et al.*).

and limitations of the models used, but rather an underestimation of the complexities of biological systems.

As we uncover the multi-layered levels of biological regulation, we are only beginning to understand what we actually do not understand. For this reason, modeling requires as much art as science in developing and appropriately applying, analyzing, and relating simulation results to the biological systems and associated experimental observations.

In this two-volume book, Innovations in Biomolecular Modeling and Simulations, twenty-six teams of leading scientists in biomolecular structure and modeling present the state-of-the-art in the field, from both a methodological and application perspective.

Volume 1 describes recent developments in force fields and electrostatics, conformational sampling and rate calculations, and coarse grained modeling and multiscale models.

Volume 2 describes atomistic simulations of nucleic acids and nucleic acid complexes; coarse-grained modeling approaches to study large-scale DNA folding, knotting, sliding and hopping; and the latest approaches and achievements in structure-based drug design.

In more detail, Volume 1 begins with Scheraga's personal perspective of the field; having worked in the field of protein structure and molecular biophysics for 65 years, Scheraga's trajectory from mainly experimental to mostly theoretical work on conformational energy calculations of polypeptides and proteins is an inspiration. The second historical perspective is contributed by writer Pollack, who tells the story of the emergence of the freely available molecular dynamics program NAMD from the dream of one immigrant, Klaus Schulten, who arrived to America in 1987 with a supercomputer in his backpack. Through interviews and studies of other pioneers, she weaves in history, science, and personal stories to present a fascinating historical account of the early days of molecular dynamics taking us to current-day applications.

Turning into the force-field advances section of Volume 1, MacKerell and co-workers describe one of the most important areas of current development, namely incorporating the polarizable component of electrostatic interactions into modern force fields; they also describe procedures for parameterizing these forms and testing them on model systems as well as highly-charged nucleic acids, where improvements can be demonstrated. Case and co-workers continue discussions of the fundamental electrostatics interactions by focusing on approaches using integral equations to model the interactions of biomolecules with water and ions and improve upon other implicit-solvation techniques such as Poisson Boltzmann approximations. Smith and co-workers share their related methodological advances using both classical molecular mechanics and quantum-mechanical techniques to study complex systems of interest to the energy bioscience field using supercomputing resources; implicit solvation and efficient scaling on large-scale machines are required to study behavior of multi-million-atom systems.

The second group of papers in Volume 1 describes the latest algorithms to survey conformational changes, associated pathways, and reactions rates from the molecular simulations. As reviewed recently,[3,4] various clever stochastic as well as deterministic approaches have been developed and applied in recent years to increase the variety of configurations captured, encourage transitions, and extract essential quantitative information on those rare events. The methods include Milestoning and Forward Flux methods based on trajectory fragments, as described by Elber and co-workers, and conformational dynamics and Markov state models as detailed by Darve and co-workers. The alternative Monte Carlo approach for large systems is described by Mezei, who discusses special implementational considerations for macromolecules.

While atomistic models are essential for describing detailed events at the fine level, coarse-grained models are required to describe events relevant to very large complex systems like the chromatin fiber or major rearrangements involving topological changes such as DNA knotting. For such problems, simplified representations are needed, and the challenge is not only to construct useful models that approximate many aspects of the systems, while detailing others, but to eventually telescope between the macroscopic and microscopic views. Examples of such approaches are described by Lavery and coworkers for proteins, by Taylor and co-workers for the molecular motor myosin, by Radhakrishnan and co-workers for protein-protein and protein-membrane networks, by Pettitt and co-workers for complexes at material surfaces, and by Fish and co-workers in the context of large-scale industrial applications.

Volume 2 begins with state-of-the-art examples of atomistic simulations of nucleic acids and nucleic acid complexes. Lankas describes the nuts and bolts of atomistic simulations of DNA in explicit solvent with a viewpoint into coarse-grained models and a focus on understanding sequence and anharmonic effects. Neidle and co-workers describe applications to quadruplex systems of nucleic acids, their complex hierarchical arrangements as studied by classical dynamics, enhanced sampling methods, and hybrid classical/quantum approaches, as well their relevance for supplementing current static structure resolution approaches. Rohs, Fenley and co-workers discuss electrostatic and structural aspects of DNA that have tailored DNA for interactions with proteins, as deciphered from atomistic studies of nucleic acids and their complexes. Schlick and co-workers describe insights into DNA polymerase repair mechanisms as deduced from atomistic dynamics and principal component analysis of several polymerase enzymes; the resulting correlated motions suggest mechanisms that are hybrid between traditional induced fit and conformational selections models. Mukerji, Beveridge and co-workers report their latest study for a structural prediction from molecular dynamics of a novel protein/DNA complex involving a four-way junction.

Continuing to RNA at atomic resolution, Sponer and co-workers describe the state of-the-art in simulating RNAs by all-atom molecular dynamics, a field that has lagged behind RNA's famous cousin due to force-field limitations, among others. Lilley then describes the fascinating and hierarchical folding of RNA junctions, important since junctions organize the global architecture of RNA.

Events that require different kinds of methods are then investigated in the second part of Volume 2. Witz and Stasiak describe studies of DNA knots and catenanes using topological frameworks explored by Monte Carlo and Brownian dynamics simulations. Rippe and co-workers describe Monte Carlo simulations of nucleosome chains using simple coarse-grained models that capture essential features of the systems. Levy and Marcovitz then describe a study of DNA sliding on the genomic level.

The book concludes with overviews of drug design approaches, from both industrial (Singh and co-workers) and academic perspectives (Amaro and co-workers, McCammon). McCammon's personal perspective focuses on the identification of two anti-viral drug targets for AIDS using a structure and dynamics-based approach.

The range of subjects and methods as reflected by these exciting contributions indicate not only the ingenuity of the methods and wide range of tools available to the molecular modeler, but also the endless questions that arise from the modeling. These questions, in turn, serve to advance the field toward an ultimate understanding of the complex interaction between biomolecular structure and function.

The data deluge is unlikely to cease in the near future. Nonetheless, our modeling approaches will undoubtedly become better and more diverse, allowing us to tackle yet more challenging problems and unravel complex relationships, patterns, and long-time events in biomolecules.

Tamar Schlick
New York

References

1. A. Pollack, DNA Sequencing Caught in Deluge of Data, 30 Nov., *The New York Times*, 2011, page C1.
2. T. Schlick, R. Collepardo-Guevara, L. A. Halvorsen, S. Jung and X. Xiao, Biomolecular Modeling and Simulation: A Field Coming of Age. *Quart. Rev. Biophys.*, 2011, **44**, 191–228.
3. T. Schlick, Monte Carlo, Harmonic Approximation, and Coarse-Graining Approaches for Enhanced Sampling of Biomolecular Structure. *F1000 Biol. Rep.*, 2009, **1**, 48.
4. T. Schlick, Molecular-Dynamics Based Approaches for Enhanced Sampling of Long-Time, Large-Scale Conformational Changes in Biomolecules. *F1000 Biol. Rep.*, 2009, **1**, 51.

Contents

Volume 1

Beginnings

RSC Biomolecular Sciences No. 23
Innovations in Biomolecular Modeling and Simulations: Volume 1
Edited by Tamar Schlick

Published by the Royal Society of Chemistry, www.rsc.org

Force Fields and Electrostatics

Coarse Graining and Multiscale Models

Volume 2

Atomistic Simulations of Nucleic Acids and Nucleic Acid Complexes

Beginnings

CHAPTER 1

Personal Perspective

HAROLD A. SCHERAGA

Baker Lab of Chemistry, Cornell University, Ithaca, NY 14853-1301, US
Email: has5@cornell.edu

My interest in biomolecular modeling and simulation has its origins in my graduate work at Duke University under the direction of Paul M. Gross and Marcus E. Hobbs, and in my year-long courses in quantum mechanics and statistical mechanics with Fritz London. Gross had previously spent a sabbatical leave with Peter Debye in Leipzig, and returned to Duke with an interest in the relation between molecular structure and dipole moments. Shortly before his arrival at Duke, London had formulated a quantum mechanical treatment of van der Waals forces, in which polarizability played an important role. In this atmosphere, I began graduate research using electrical birefringence (Kerr effect) to determine anisotropic polarizabilities of small organic molecules. This research was interrupted by the entry of the US into World War II, and my resulting participation in a war project at Duke.

One day I had a chance encounter in the chemistry library with a then new book by Edwin Cohn and John Edsall, titled *Proteins, Amino Acids and Peptides*, which contained chapters by several authors besides Cohn and Edsall, namely John Kirkwood, George Scatchard, and Larry Oncley. Edsall described flow birefringence and Oncley described dielectric dispersion of proteins. This appealed to me as a chance to take up the birefringence work that I had to drop at Duke and, as an ACS postdoctoral fellow at Harvard Medical School, I applied flow birefringence to proteins under Edsall's guidance in an atmosphere devoted to the physical chemistry of blood plasma proteins.

Then, at Cornell, I began experimental work on the mechanism of the action of thrombin on fibrinogen to produce the fibrin clot. In a limited proteolytic

RSC Biomolecular Sciences No. 23
Innovations in Biomolecular Modeling and Simulations: Volume 1
Edited by Tamar Schlick

Published by the Royal Society of Chemistry, www.rsc.org

reaction, thrombin releases peptides from fibrinogen, exposing a polymerization site on the resulting fibrin monomer. I used flow birefringence to elucidate the nature of the staggered-overlapped rod-like polymers formed from fibrin monomer on the pathway to the blood clot.

At the same time, Pauling and Corey had proposed the α and β structures of proteins, focusing on the backbone hydrogen bonds. With my first graduate student, Michael Laskowski, I examined the role of side-chain hydrogen bonds in proteins. Specifically, we demonstrated how side-chain hydrogen bonds are involved in the polymerization of fibrin monomer,[1] and also influence the pK's of ionizable groups[2] as well as limited proteolysis[3] in which it is necessary to break hydrogen bonds (during the hydrolysis of a peptide bond) to liberate a fragment which had been connected to the rest of the molecule by such hydrogen bonds.

This led to an attempt to determine protein structure by acquisition of distance constraints by location of side-chain hydrogen bonds experimentally. Charles Tanford had used UV titration of ribonuclease A (RNase A) in the pH region near the pK° of tyrosine, viz. ~10, to demonstrate that three of the six tyrosines had abnormally large pK's[4] and, with Jan Hermans, we used potentiometric titration to demonstrate that three of the eleven COOH groups had abnormally low pK's.[5] During my sabbatical leave with Kai Linderstrøm-Lang at the Carlsberg Laboratory in Copenhagen in 1956–57, with UV difference spectroscopy[6] (see Figure 1.1), I showed that the UV absorption spectrum of tyrosine varied with pH at low pH where COOH groups ionize, suggesting the proximity of COOH group(s) near tyrosine(s). Back at Cornell, I started a long series of experiments with graduate students and postdocs which ultimately located three Tyr–Asp interactions,[7] *viz.*, Tyr25 with Asp14, Tyr92 with Asp38, and Tyr97 with Asp83. These were subsequently verified by the crystal structure of RNase A.[8]

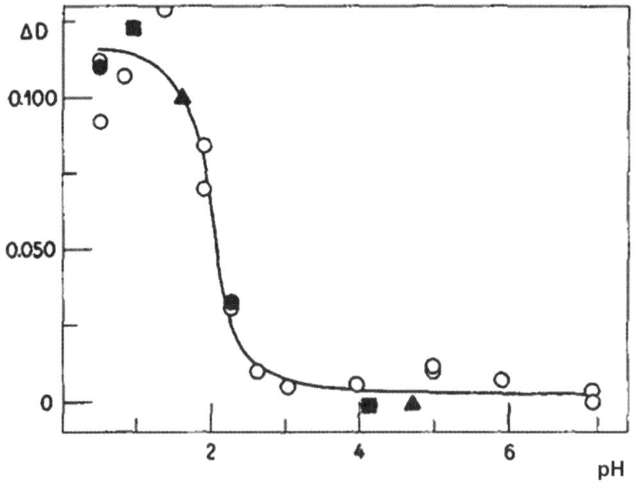

Figure 1.1 Difference in optical density at 287 mμ between a solution of ribonuclease at pH 6.95 and at the pH indicated. The protein concentration in all solutions was 1.94 mg/ml. From *Biochim. Biophys. Acta.*, 1957, **23**, 196.

Also, during my sabbatical leave at the Carlsberg Laboratory, Walter Kauzmann arrived in mid-year and, with Linderstrøm-Lang, we had many discussions about hydrophobic interactions. When Walter returned to Princeton, he wrote his famous article on hydrophobic interactions,[9] and upon my return to Cornell, I started a new graduate student, George Némethy, on a statistical mechanical theory of hydrophobic interactions.[10] Simultaneously, with Izchak Steinberg and George Némethy,[11] we discussed the interactions between hydrogen bonds and hydrophobic interactions, and pointed out how the nonpolar portions of so-called polar side chains can be involved in hydrophobic interactions with nearby nonpolar side chains, as shown in Figure 1.2, providing increased strength to the hydrogen bond, and its consequent influence on protein structure and stability.

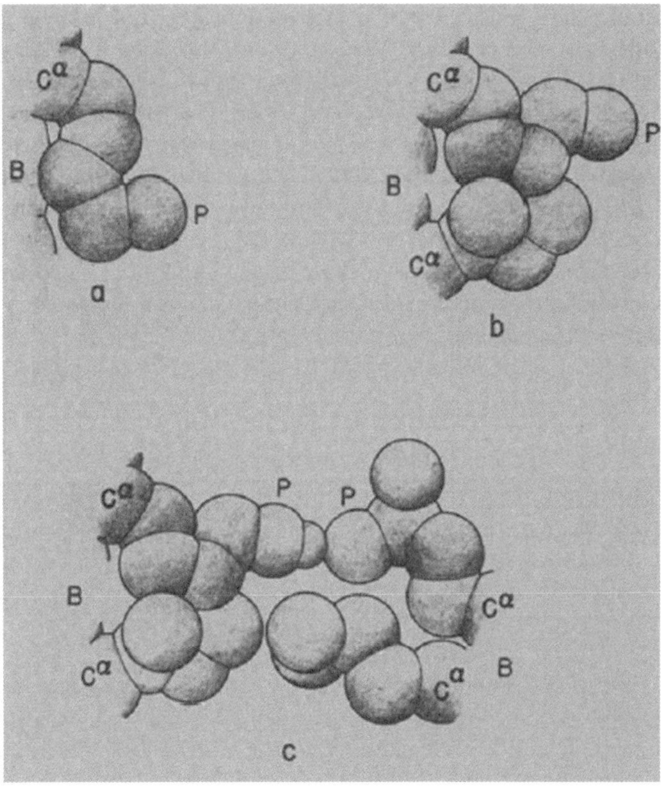

Figure 1.2 Schematic representation of various hydrophobic interactions of a polar side chain with its surroundings (B refers to the backbone, P to the polar head, and α to the α carbon): (a) interaction of a lysine side chain with the backbone; (b) interaction of a lysine side chain with a nearby isoleucine side chain; (c) interaction of two polar side chains (lysine and glutamic acid), engaged in hydrogen bonding, and with hydrophobic interactions with two nonpolar side chains (isoleucine and leucine, respectively). A hydrophobic interaction also appears between the two nonpolar side chains. From *Biopolymers*, 1963, **1**, 43.

With George Némethy, we decided to formulate a theoretical approach to determine protein structure[12,13] by making use of the distance constraints implied by the three Try–Asp interactions and the location of the four disulfide bonds in RNase A. This computational work evolved over the years, first by formulation of an all-atom force field, ECEPP,[14] an Empirical Conformational Energy Program for Peptides, and subsequently, by development of UNRES,[15] a united-residue coarse-grained force field.

In addition to the effort to determine the structure of RNase A, we also embarked on experimental work to determine its folding pathways by oxidation with two redox systems, GSSG/GSH[16] and DTT^{ox}/DTT^{red}.[17] Later, with UNRES, we expanded our computational approach to simulate folding pathways[18] and folding kinetics.[19]

The ultimate goal of all our research, beginning with our work on the thrombin–fibrinogen interaction, was to use physical chemistry to elucidate biological structure and function. With our coarse-grained force field, we have recently embarked on studies of protein–protein interactions, e.g., Aβ,[20] PICK1[21] and Hsp 70,[22] and have started to formulate a coarse-grained nucleic acid force field to be able to treat protein–nucleic acid interactions.

All of this work has been an excellent vehicle with which to train undergraduate, graduate, and postdoctoral students whose contributions to this research have been fundamental.

A more detailed description of some of the difficulties encountered and surmounted during the implementation of this research is provided in a prefatory chapter in *Annual Reviews of Biophysics*.[23]

References

1. J. M. Sturtevant, M. Laskowski, Jr, T. H. Donnelly and H. A. Scheraga, *J. Am. Chem. Soc.*, 1955, **77**, 6168.
2. M. Laskowski, Jr and H. A. Scheraga, *J. Am. Chem. Soc.*, 1954, **76**, 6305.
3. M. Laskowski, Jr and H. A. Scheraga, *J. Am. Chem. Soc.*, 1956, **78**, 5793.
4. C. Tanford, J. D. Hauenstein and D. G. Rands, *J. Am. Chem. Soc.*, 1955, **77**, 6409.
5. J. Hermans, Jr and H. A. Scheraga, *J. Am. Chem. Soc.*, 1961, **83**, 3293.
6. H. A. Scheraga, *Biochim. Biophys. Acta.*, 1957, **23**, 196.
7. H. A. Scheraga, *Fed. Proc.*, 1967, **26**, 1380.
8. A. Wlodawer and L. Sjölin, *Biochemistry*, 1983, **22**, 2720.
9. W. Kauzmann, *Adv. In Protein Chem.*, 1959, **14**, 1.
10. G. Némethy, *J. Phys. Chem.*, 1962, **66**, 1773; erratum, *J. Phys. Chem.*, 1963, **67**, 2888.
11. G. Némethy, I. Z. Steinberg and H. A. Scheraga, *Biopolymers*, 1963, **1**, 43.
12. G. Némethy and H. A. Scheraga, *Biopolymers*, 1965, **3**, 155.
13. H. A. Scheraga, *Adv. Phys. Org. Chem.*, 1968, **6**, 103.

14. F. A. Momany, R. F. McGuire, A. W. Burgess and H. A. Scheraga, *J. Phys. Chem.*, 1975, **79**, 2361.
15. A. Liwo, C. Czaplewski, J. Pillardy and H. A. Scheraga, *J. Chem. Phys.*, 2001, **115**, 2323.
16. Y. Konishi, T. Ooi and H. A. Scheraga, *Proc. Natl. Acad. Sci. U.S.A.*, 1982, **79**, 5734.
17. D. M. Rothwarf, Y.-J. Li and H. A. Scheraga, *Biochemistry*, 1998, **37**, 3760.
18. A. Liwo, M. Khalili and H. A. Scheraga, *Proc. Natl. Acad. Sci. U.S.A.*, 2005, **102**, 2362.
19. M. Khalili, A. Liwo and H. A. Scheraga, *J. Mol. Biol.*, 2006, **355**, 536.
20. A. Rojas, A. Liwo, D. Browne and H. A. Scheraga, *J. Mol. Biol.*, 2010, **404**, 537.
21. Y. He, A. Liwo, H. Weinstein and H. A. Scheraga, *J. Mol. Biol.*, 2010, **405**, 298.
22. E. Golas, G. G. Maisuradze, P. Senet, S. Ołdziej, C. Czaplewski, H. A. Scheraga and A. Liwo, *J. Chem. Theory and Computation*, submitted.
23. H. A. Scheraga, *Annual Reviews of Biophysics*, 2011, **40**, 1.

CHAPTER 2

Fashioning NAMD, a History of Risk and Reward: Klaus Schulten Reminisces

LISA POLLACK

Beckman Institute for Advanced Science and Technology, University of Illinois at Urbana-Champaign, Urbana, Illinois 61801, US
Email: lpollack@illinois.edu

2.1 Introduction

With a homemade parallel supercomputer in his backpack, Klaus Schulten waited patiently in the Chicago O'Hare airport, hoping for no trouble getting through customs after his flight from Germany. It was the summer of 1988 and Schulten was about to start a new job at the University of Illinois. It was also the height of the Cold War, the ultimate peak in US–Soviet tensions, and supercomputers were causing the Reagan administration much consternation. Although Reagan had escalated the arms race and all its accompanying technological advances while in office, he wanted to keep burgeoning supercomputer developments out of the hands of the Soviets, for fear they would build better weapons. In 1986, the government announced a proposed policy to restrict Soviet scientists from using supercomputers housed in US universities.[1] The policy objective was clear: no access to scientists from 20 Communist bloc countries. The implementation was not so clear, however. The universities were incensed about this policy, worrying about infringements on their academic

RSC Biomolecular Sciences No. 23
Innovations in Biomolecular Modeling and Simulations: Volume 1
Edited by Tamar Schlick

Published by the Royal Society of Chemistry, www.rsc.org

freedom and research initiatives, and wondering how they would become enforcers of a government security policy.

The Reagan administration not only wanted to restrict Soviet scientists from running simulations on supercomputers, it also did not want Soviet scientists close to supercomputers for fear they might learn how to build their own. But in 1987, two young physics students in Munich embarked upon just such a mission to build their own parallel supercomputer, although neither had formal training in this field. Not only did they figure it out, but their project cost came in around $60 000, much less than the $17.6 million retail price tag of the Cray-2, a popular supercomputer in the late 1980s. Their advisor, Klaus Schulten, chose to risk all the grant money he had on the project, even though he had no guarantees it would succeed and he was not an expert in parallel computing. While the parallel computer would make possible very large simulations, Schulten sums up why he agreed to the precarious plan: "I believe in people, and when I believe in people I let them go."

This article charts the history behind the software program NAMD. It starts in the late 1980s, when Klaus Schulten and two students attempted to adopt a fundamentally new technology, parallel computing, for use in their own research on proteins. A student revolt precipitated a software code designed specifically for parallel computers, NAMD, and its continual improvement over time in describing the behavior of biological macromolecules led Schulten to regard it as a "computational microscope." The story behind the computational microscope looks at the hurdles some key scientists faced and specifically the risk these scientists assumed.

2.2 Early Influences of Molecular Dynamics

In the mid-1980s Klaus Schulten was a professor at the Technical University of Munich and he had a wish list. He wanted to simulate on a supercomputer the behavior of a membrane protein, the photosynthetic reaction center, using molecular dynamics. Earlier in the previous decade, Martin Karplus's group at Harvard had done the first molecular dynamics simulation of a biomolecule.[2] Up to that time, molecular dynamics had been used to model hard spheres by Alder and Wainwright in the 1950s,[3] and other liquids in the 1960s and early 1970s.[4] But after much painstaking work, Andrew McCammon and Bruce Gelin, graduate students of Karplus's, prepared a computer program to model a small protein, the bovine pancreatic trypsin inhibitor.[5] After locating the necessary computing power in Europe during a workshop at CECAM, (Center Européen Calcul Atomique et Moléculaire), the trio published their landmark paper in 1977.

Schulten was aware of this work of McCammon, Gelin, and Karplus because he was finishing his graduate studies at Harvard, where Karplus was one of his advisors, and McCammon and Gelin were fellow students who sat right down the hall. The 1977 paper had a definite impact on Schulten, who was trained to use the theoretical and mathematical methods employed in chemistry and physics. "I realized that this computational approach," he says of molecular dynamics, "opens new doors to describe problems that you couldn't do with a purely theoretical approach that I had taken until that time."

But embracing a computational method instead of staying exclusively with pure theory had its costs, and he soon became known as "Computing Schulten." "I paid a big price for it, because basically during much of my career people thought I was stupid," reminisces Schulten. "Although I continued publishing mathematically-oriented papers, for every computing-oriented paper I had to redeem myself with ten theoretical papers, and I couldn't possibly do that."

In fact, selling the usefulness of the computational microscope and the molecular dynamics approach in its very early stages was also a battle Schulten sometimes had to wage. In 1985, while still a professor in Munich, Schulten went to a supercomputing center in Illinois to run some calculations, and returned to Germany with a movie illustrating a protein in motion, based on molecular dynamics simulations. When Schulten showed the movie, one of his colleagues became quite enraged. "He got so upset when he saw it, he almost wanted to physically attack me," Schulten recounts. "He told everybody this is the greatest rubbish he'd ever seen in his life. He was a crystallographer who thought basically of proteins as some kind of Gothic cathedral that were cast in stone."

2.3 Building a Parallel Computer

Despite the many struggles Schulten had to face, he was intent on using molecular dynamics for his work; he was sure it would lead him to new discoveries that would be valuable for science. "My love of scientific discovery," Schulten confirms, "made me do the dirty business of computing."

This was exactly what motivated Schulten in 1987 to make an audacious judgment call. He wanted to simulate the photosynthetic reaction center, which is a large protein that sits within a membrane. Simulating the protein by itself and neglecting the membrane and the liquid that surrounds it was not highly desirable because such isolation is not a natural environment for the protein. This protein is about 12 000 atoms and the membrane and water that surrounds it in its natural environment adds another 100 000 atoms to the tally. In the late 1980s no supercomputer was even close to capable of handling that task. Schulten thus decided to focus on understanding and simulating just part of a membrane in water and managed to run a calculation on a Cray-XMP, but it only covered a few picoseconds of time and it taught him one thing: he needed a supercomputer all to himself, perhaps for a year or more, to really understand the mechanism.

While Schulten, teaching physics at the Technical University of Munich, was wondering how he could commandeer a supercomputer all for himself, a young physics student at the same university was wondering how he could build faster computers. Actually, Helmut Grubmüller knew in theory how to make a computer faster, but was really wondering if he could get someone else to pay for it. Enrolling at the university in Munich in 1985, barely twenty years old,

Grubmüller was fascinated with how computers could elucidate nature. This fascination prompted him to begin soldering together a multi-processor computer in his second or third semester, one that would be much faster than what was available to regular students in those days. Although it was technically a parallel computer, Grubmüller just called it a multi-processor at that point. It was the mid-1980s and there was no World Wide Web to consult for details about how to construct his own. Instead, he read any books about microprocessors he could find, and talked to company representatives on the phone about technical details of the parts they sold. He cites as most important for making headway the data sheets of the chips he used as the processor, which were Motorola's 68 000 line. But he could only get so far on his own. "That very quickly blew my budget," recounts Grubmüller; "My private budget was a thousand dollars or so."

His persistent desire to build a faster computer led him to concoct an admittedly naïve idea. "I approached some of my professors and simply asked them if they would be willing to conduct a deal," Grubmüller says. "I would build a larger machine, larger computer, and they would pay for the hardware." Most of his professors told him that was crazy and rejected the idea right away.

Klaus Schulten was teaching Grubmüller a physics course in mechanics at the time, and he remembers the day Grubmüller visited Schulten's office, homemade multi-processor computer in hand. "He opened his bag and got this out and showed it to me," recalls Schulten. "He showed me the results and I fell almost off my chair." With the demonstration of some of the capabilities of the multi-processor, Schulten was impressed. "So I immediately integrated him in my group," says Schulten. Schulten told him there were other people in the group that also knew about computers.

One of those people was Helmut Heller who, like Helmut Grubmüller, was also self-taught about computers. Heller had caught Schulten's attention when he made a complicated fractal movie for a class Schulten was teaching. "He showed us how he modified the computer to do this calculation," Schulten relays. "Basically he went into this computer with wires and so, even today I couldn't tell you what he really did." Watching Heller toy so smoothly with the computer really left an impression on his professor. Schulten was pleased to invite Heller to join his group.

"Sometimes you take your chances," says Schulten on his decision to let Grubmüller and Heller build a parallel supercomputer. "And so that's what I did. And to this day I don't know if I would do it again." The risks were enormous for Schulten. Although he would have a supercomputer all to himself for a molecular dynamics simulation, if it somehow didn't work out, he would have nothing to show. He was not an expert in parallel computing at that time and had little basis for assessing the project's feasibility. Moreover, he would have to explain to the funding agency how he had sunk tens of thousands of Deutsche Marks into a failed computer when the money was earmarked for equipment that usually came with warranties.

Heller and Grubmüller, however, made a great team, even though what they were trying to do was no small feat. Not only did they have to build a supercomputer, but they also had to write software to work in parallel on it. Schulten was amazed by their ingenuity. "They thought in terms of solutions rather than difficulties," he recalls. For example, when the pair told Schulten they needed an oscilloscope to test that the computer was working properly, Schulten knew he didn't have the additional money for it. But the two students assured him that if they could find two or three oscilloscopes on the equipment trash heap, they could assemble one working oscilloscope out of parts, and that is what they did.

The pair decided to build their parallel computer, eventually named the T60, out of processors known as Transputers, because purportedly they were as easy to assemble as Lego blocks. The final computer had 600 processors. They had to hunt down each component of the supercomputer and solder everything together. This included obtaining and assembling the Transputers, RAM, power supply, fans, housing, voltage rails, circuit boards, and mounts. Other members of Schulten's group pitched in and soldered boards as well.

2.4 Transporting a Supercomputer

The project, to build and program the T60, began late in 1987. In the middle of its completion, Schulten took a new job in the United States, at the University of Illinois at Urbana-Champaign. Instead of shipping the T60 over to the United States, and waiting weeks for it to arrive and then go through customs, Schulten decided to just carry the computer in a backpack due to its portable size.

During the Cold War, the United States had long been at odds with its western European allies over the issue of restricting high technology items, like certain computers, to Soviet bloc countries. European nations did not want their commerce and profits with the Soviet bloc curbed by the United States, and even computer companies in the United States were worried that their overseas business would be curtailed by government restrictions. There existed a designated list of prohibited and restricted materials for export to communist countries, which was drawn up under the auspices of COCOM (Coordinating Committee on Export Controls), a committee formed after World War II to restrict arms to the Soviet Union. The list was agreed upon unanimously by the United States, NATO allies, and Japan. In 1984, after much rancorous exchange between the United States and western Europe, especially over computer exports, COCOM revised its list of goods and basically allowed personal computers to be freely exported, but not supercomputers, which it designated by their data processing rate.[6]

Not only did the Reagan administration want to restrict businesses, both domestic and foreign, from exporting supercomputing technology to the Soviets, it also wanted to keep Soviet scientists out of its newly formed supercomputing centers. In 1985, the National Science Foundation awarded $200 million in grant money to four universities to open supercomputer centers on their campuses. The move was deemed necessary, and authorized by Congress, to give academic researchers access to supercomputers that were usually restricted to defense work done by the Pentagon and National Laboratories.

But in 1986 the Reagan administration proposed prohibiting access to these supercomputer centers by Soviet scientists.[7] Such was the climate in the summer of 1988, when Schulten decided to risk taking his homemade supercomputer onto a transatlantic flight and carrying it through customs in Chicago.

Schulten was aware that, because of its data processing rate, the T60 computer had to be registered, especially when being exported or imported. He approached the customs officer and showed him the supercomputer. "What is this?" asked the customs officer.

"This is a computer," replied Schulten.
"Why are you showing it to me?"
"This is a very powerful computer and it needs to be registered with you."

Schulten then explained some technical details of the machine to the officer. Finally the customs officer looked at Schulten and said, "Please, just put it back and go."

2.5 Rejection

When Schulten arrived in Illinois, his students finished building the T60 and programming it. The program, named EGO, was written in OCCAM II, a language around which the Transputer had been designed. It was very important that EGO scale well because the membrane calculation that Schulten and his group wanted to run was a very big system. In fact, the calculation took twenty months of non-stop run time. The system they studied, a membrane made of lipids surrounded by water, consisted of 23 978 atoms.

With such incredible efforts applied to the membrane problem, Schulten was shocked when the article on it, with co-authors Helmut Heller and Michael Schaefer, was rejected within a week, even though the results agreed well with experiment. The paper contained theoretical biophysics results. Schulten acknowledges that to publish theory related to biology is very tricky, especially in high-impact journals. He says that even today he has a better chance of acceptance if he teams up with experimentalists.

The mystery of the rejection was never cleared up. "I still to this day don't know why it was rejected," says Schulten. "I was so furious that even though I published many papers in this journal, I told them I would never publish there anymore." Heller and Schaefer and Schulten eventually published in the *Journal of Physical Chemistry*, a paper that is now highly cited.[8]

The successful computation only whetted Schulten's appetite for studying larger and larger systems. But little did Schulten realize that his thirst for massive systems on parallel machines would lead to a student rebellion of sorts, and a software product called NAMD that would define his career.

2.6 Collaborations with Computer Scientists

In 1989 Schulten founded the Theoretical and Computational Biophysics Group at the Beckman Institute for Advanced Science and Technology, located at the University of Illinois. He received an initial two-year grant from the National Institutes of Health in 1990 for this center, a so-called presenter grant. By this time, Schulten knew a thing or two about parallel computing, compared to being a relative neophyte on the matter when he had sunk all his research money into a parallel machine back in Munich; he realized he could study massive biomolecules by taking advantage of parallel machines. There were already software codes, some even freely available since the mid-1980s, that did molecular dynamics, but Schulten realized that his needs far exceeded their capabilities.

In fact, Schulten realized his needs exceeded even his own capabilities as a computational biophysicist if he wished to carry forward his plan to study more massive biomolecules in their natural environments. Schulten decided to step outside his field. Motivated by three reasons, Schulten's judgment compelled him to cross disciplines and seek help from the computer scientists at his university.

First, up to 1991, the programming for molecular dynamics in Schulten's group had always been done by his physics students. After the two Helmuts wrote EGO for the home-built computer, another student wrote a molecular dynamics code for a parallel computer, the Connection Machine, housed nearby in Urbana at the NCSA (the National Center for Supercomputing Applications). Between the code for the T60 and the code for the Connection Machine, Schulten realized he needed software engineering skills in the group to make sure the software was written in a way that others could add to it and understand it in the future. Physics students simply weren't trained with such skills.

Second, Schulten cites that he realized programming techniques were changing, and especially that FORTRAN, developed in the 1950s, might not be his top choice as other languages started to gain prominence. "We needed better programming languages that were more systematic," declared Schulten.

Third, mastering code for parallel machines seemed daunting. The code for the Connection Machine didn't seem to be easily transferable to other parallel machines, and was difficult to write in general. The nuances of building a code that was portable to other parallel machines, at a time when so many different types of vendors were selling parallel computers, helped to convince Schulten to seek outside input. He wanted expertise on how to handle parallel machines, and he knew computer scientists could provide knowledge about a technology that was rapidly evolving and changing.

With the above threefold motivations in mind, and needing to write a strong renewal grant for his center – a big grant that would provide five years of funding instead of just two years – Schulten approached two computer scientists at the university in Illinois and asked them to be on the renewal proposal. One, Bob Skeel, was an expert in numerical algorithms, and the other, Sanjay Kalé, was an expert in parallel programming. In 1992 Schulten received a five-year grant from the National Institutes of Health and so began the genesis of the software code called NAMD.

2.7 Graduate Students Revolt

Actually, Schulten cites the early work on T60 as a seed of NAMD, coupled with a student revolt that happened in the early 1990s. At this time Schulten hoped to continue using the molecular dynamics codes that had been developed for the Connection Machine and the T60. Bob Skeel's graduate student, Mark Nelson, was studying the code for the Connection Machine, and Schulten's graduate students, Bill Humphrey and Andrew Dalke, were examining the code for T60. All three were vexed by the existing codes they were inspecting. "I banged my head against that code for a couple of months," recounts Nelson, "and was going nowhere. There were no comments, and all the variable names were abbreviations of German names, which really didn't help me."

The three graduate students, Nelson, Humphrey, and Dalke, together decided their task would be easier if they just wrote a new code from scratch. They surreptitiously worked for about a month until they had a viable code. Taking a big risk with their advisors, they presented at a group meeting their idea to create a whole new code and nix the existing ones, which were too complicated to follow. "They wanted to write it in C++," recalls Schulten. "They thought the professor will shoot them, for being so bold and telling him NO. They even put conditions on how they wanted to go forward." In fact, however, Schulten was thrilled at the initiative. "So I was actually very excited myself," recalls Schulten, "because I saw here a situation that was almost like the Helmuts before them. The students were going for it. They really wanted to realize it themselves."

2.8 Factors that Shaped NAMD

At this point, Schulten's group was well equipped to craft a software code from scratch, now that he had computer scientists on the team. Sanjay Kalé and his graduate students had developed a parallel programming language they named Charm++, which actually became very critical and valuable for the success of NAMD.

Many factors shaped the purposeful design of NAMD in 1994, not the least of which was that it was now well funded by the NIH grant. The graduate students insisted on using the language C++, a relative newcomer in the field of programming. "C++ offered a nice combination of having it be object-oriented," declares former student Mark Nelson, "but we could still write the key computational pieces in C and get the speed that we needed." Use of C++ in NAMD made it unique among molecular dynamics codes at the time.

Since NAMD was to be tailored for parallel machines, the developers had to take into account how the many processors would communicate with each other while running molecular dynamics calculations. Sanjay Kalé and Attila Gursoy, his graduate student and later his post-doc, were working on Charm++, a parallel programming environment based on C++, and using something called message-driven execution. It turned out that the structure of molecular dynamics as a parallel application was very well suited for what they were doing with message-driven execution. Charm++ utilized message-driven execution to effectively enable processors to talk to each other during molecular dynamics runs.

While C++ and Charm++ were unique elements to NAMD, the *from scratch* design was a hallmark. Sanjay Kalé cites the advantages the *from scratch* design afforded NAMD: the software scaled well and was able to handle whatever new machines were rapidly developing on the market. "That design has stood the test of time from then till now," Kalé reflects. "Pretty much the architecture of the parallel program has remained the same, even though at that time we were running on this HP cluster, roughly eight processors. And now we run on 200 000 plus processors."

And the vendors of the parallel computers were taking notice. From its inception, NAMD was offered to the community for free, and started accruing more and more users over the years. When it became popular IBM and other vendors started to pay attention to it. But that was not the only recognition NAMD received. Each year an international conference is held by computer scientists, called "Supercomputing" or more recently, simply SC. The Gordon Bell prize is given each year at the conference to recognize accomplishments in parallel computing. For NAMD and its developers, more affirmation came at SC2002 with the award of a Gordon Bell prize.

2.9 Continual Evolution of NAMD

By now the computational microscope had become sophisticated. In 1977, the first molecular dynamics simulation was done on a small protein in vacuum and lasted for 9.2 picoseconds. By 1996, NAMD was simulating a system with 36 000 atoms, the estrogen receptor with a segment of DNA in salt water, for 50 picoseconds.[9] The calculation ran for two or three days on an 8-processor cluster. In 2004, on an aquaporin-1 system of 81 065 atoms, the calculation ran for 22.4 hours on a 128 processor parallel machine.[10] By this time the molecular dynamics calculation on aquaporin-1 simulated 5 000 picoseconds of time.

With these kinds of advances the computational microscope took a foothold. "In the old days the computer simulation wasn't so accurate," recollects Schulten. "So it would have been hubris to just say, 'This is a computational microscope.' People would have laughed at us and said, 'This is a really bad microscope you have here.'"

When asked what forces drove the computational microscope to improve, Schulten is clear that there is a deeper cause than the obvious behind his motive for continually refining NAMD to do larger systems. "We didn't become larger because I wanted to beat my chest and say I can do more atoms than you, but rather for a clear intellectual reason. And the reason is, I think it is essential for the very nature of living systems, of living cells in particular, that you need to describe how biological macromolecules, in particular proteins, assemble and cooperate." Schulten ultimately wanted to look at not just a single protein, but at scores of proteins and the way they organize themselves and form societies in a living cell. "The key point," Schulten continues, "is simply that these macromolecules have to find each other in the right proportion, in the right geometry, at the right place in the cell, and they have to cooperate." And in order to describe this, NAMD had to improve to the point that it could describe tens or hundreds or thousands of proteins.

Oftentimes the computational microscope predicted correctly what experiment observed; sometimes it was ahead of experiment. One such time was an instance of using NAMD to run a steered molecular dynamics simulation. An ankyrin repeat is a certain sequence of amino acids and this sequence is found in more than 400 human proteins. In 2005, Marcos Sotomayor, David P. Corey, and Klaus Schulten studied the elastic properties of these ankyrin repeats using steered molecular dynamics, ultimately to better understand hearing and balance;[11] a year later, in 2006, experiment confirmed their simulation results. In a *Science* article in 2007, "Single-Molecule Experiments in Vitro and in Silico," Sotomayor and Schulten argue that *in silico* studies, like the one describing ankyrin repeats, are a very useful complement to *in vitro* experiments.[12] The term "*in silico*" refers to a study done by computer analysis. The justification of the computational microscope as a powerful tool only grew stronger as this *in silico* experiment preceded the *in vitro* experiment.

2.10 Conclusions

Over its sixteen-year history NAMD has elucidated the behavior of many large systems. A complete list is beyond the scope of this history; however, some of the systems covered are: the light harvesting complex II of photosynthetic purple bacteria,[13] steered molecular dynamics on the titin protein,[14] and examination of how aquaporins work to pass water or glycerol through a membrane but forbid protons from crossing the same membrane.[15] But to round out the history of NAMD, a focus on what the computational microscope has achieved in its study of viruses illuminates the legacy of NAMD in the last few years.

In 2009, the World Health Organization declared a pandemic of swine flu. Named H1N1pdm, it was quickly discovered that this strain of influenza had acquired resistance to Tamiflu, a drug that treats people recently infected with influenza. The avian flu virus, H5N1, was found to have resistance to Tamiflu as well. Using NAMD, and both molecular dynamics and steered molecular dynamics, workers at the University of Illinois and the University of Utah joined forces to ferret out the basis for this drug resistance.[16] Their 2010 finding helped pinpoint how mutations disrupt drug binding and provided a portal on how new drugs may evade the resistance mechanisms.

The satellite tobacco mosaic virus is one of the smallest known viruses; that fact, coupled with advances in computer power, enabled simulation of this entire life form using NAMD. Researchers at Illinois teamed up with those at the University of California, Irvine, to study the stability of the full virus particle, as well as its component parts, using molecular dynamics.[17] With results published in 2006, this was most likely the first all-atom simulation of an entire life form, encompassing over a million atoms. The findings established what factors are critical to the structural integrity of the complete virus and also what factors might guide assembly of the particle.

Viruses are very primitive particles and many only consist of a protein shell, or capsid, surrounding a strand of DNA or RNA. To reproduce, a virus enters a host cell and hijacks that cell's machinery to create more virus particles. While the capsid plays the role of protecting the virus, it also must somehow become unstable and release its internal components into the host cell for reproduction. Discerning the motion of the proteins that make up the capsid during such a release is highly desired and a study in 2006 examined the dynamics of various capsids.[18] A relatively new method, coarse-grained molecular dynamics, permitted simulations of entire capsids, which until that time had been out of reach. Members of Schulten's group refined the coarse-grained molecular dynamics technique and used NAMD to illustrate capsid stability for several viruses. With the new technique, it became possible to simulate capsids bigger than 10 nanometers, and look at transitions between stable and unstable structures, information useful for battling viral diseases.

The future looks bright for NAMD. Klaus Schulten's goal of understanding biological organization, the very essence of what makes a cell a living cell, is looming closely on the horizon, after forty years of dedication and creativity. "So now I can slowly do it," remarks Schulten on biological organization, "close to retirement. And other people behind me can use it now and can do it much better. And that makes me of course also happy." And the computational microscope continues to elucidate larger and larger systems. "Today we have a simulation running for real science work on 20 million atoms," Schulten says. "And we have a simulation just to be ready for the next generation of computers, with 100 million atoms." The combination of vision and risk taking has certainly paid off in terms of scientific discovery. From elucidating membrane proteins, to illuminating viruses to better combat disease, the computational projects made possible with parallel computers will continue to push boundaries. And as long as scientists take great personal risks to attempt the inconceivable, and continue to re-imagine new pathways for the computer to act as a scientific instrument, the future of the computational microscope seems certain to bring more enlightening discoveries.

Acknowledgements

This history could never have been written without interviews from the following scientists, who graciously offered up their time: Klaus Schulten, Helmut Grubmüller, Helmet Heller, Sanjay Kalé, Mark Nelson, and Jim Phillips.

References

1. D. E. Sanger, US to Restrict Supercomputer Use by Soviet Scholars, *N. Y. Times (N. Y. Ed.)*, 10 Feb 1986, A-13.
2. J. A. McCammon, B. R. Gelin and M. Karplus, *Nature*, 1977, **267**, 585.
3. B. J. Alder and T. E. Wainwright, *J. Chem. Phys.*, 1957, **27**, 1208.
4. F. H. Stillinger and A. Rahman, *J. Chem., Phys.*, 1974, **60**, 1545.
5. M. Karplus, *Annu. Rev. Biophys. Biomol. Struct.*, 2006, **35**, 1.
6. P. Mann, *Aviat. Week Space Technol.*, 21 Jan 1985, 108.
7. "Computer Rule Shift," *N. Y. Times (N. Y. Ed.)*, 1 Jan 1985, 1–38.
8. H. Heller, M. Schaefer and K. Schulten, *J. Phys. Chem.*, 1993, **97**, 8343.
9. M. Nelson, W. Humphrey, A. Gursoy, A. Dalke, L. Kalé, R. D. Skeel and K. Schulten, *International Journal of Supercomputer Applications and High Performance Computing*, 1996, **10**, 251.
10. F. Zhu, E. Tajkhorshid and K. Schulten, *Biophys. J.*, 2004, **86**, 50.
11. M. Sotomayor, D. P. Corey and K. Schulten, *Structure (Cambridge, MA, U. S.)*, 2005, **13**, 669.
12. M. Sotomayor and K. Schulten, *Science*, 2007, **316**, 1144.
13. J. Koepke, X. Hu, C. Muenke, K. Schulten and H. Michel, *Structure (Cambridge, MA, U. S.)*, 1996, **4**, 581.
14. P. E. Marszalek, H. Lu, H. Li, M. Carrion-Vazquez, A. F. Oberhauser, K. Schulten and J. M. Fernandez, *Nature*, 1999, **402**, 100.
15. E. Tajkhorshid, P. Nollert, M. Ø. Jensen, L. J. W. Miercke, J. O'Connell, R. M. Stroud and K. Schulten, *Science*, 2002, **296**, 525.
16. L. Le, E. H. Lee, D. J. Hardy, T. N. Truong and K. Schulten, *PLoS Comput. Biol.*, 2010, **6**, e1000939.
17. P. L. Freddolino, A. S. Arkhipov, S. B. Larson, A. McPherson and K. Schulten, *Structure (Cambridge, MA, U. S.)*, 2006, **14**, 437.
18. A. Arkhipov, P. L. Freddolino and K. Schulten, *Structure (Cambridge, MA, U. S.)*, 2006, **14**, 1767.

Force Fields and Electrostatics

CHAPTER 3

Towards Biomolecular Simulations with Explicit Inclusion of Polarizability: Development of a CHARMM Polarizable Force Field based on the Classical Drude Oscillator Model

C. M. BAKER, E. DARIAN AND A. D. MACKERELL Jr*

University of Maryland Baltimore, Department of Pharmaceutical Sciences, School of Pharmacy, 20 Penn St, Baltimore, MD 2101, USA
*Email: alex@outerbanks.umaryland.edu

3.1 Introduction

Over the past fifty years, molecular mechanics (MM) has become firmly established as an important tool in the study of biomolecular systems. It has revolutionized the understanding of these complexes systems by providing an atomic-resolution picture of molecular structure, dynamics, and their relation to function, something which is often difficult, or even impossible, to obtain by experimental methods alone. Since the first molecular dynamics (MD) simulation of a protein[1] confirmed that biomolecules are complex dynamic

RSC Biomolecular Sciences No. 23
Innovations in Biomolecular Modeling and Simulations: Volume 1
Edited by Tamar Schlick

Published by the Royal Society of Chemistry, www.rsc.org

systems,[2] almost every area of biochemistry has been touched by molecular simulation. From small molecule conformational preferences[3,4] to protein folding,[5] nucleic acid dynamics[6] and protein-ligand binding,[7] MD simulation, in combination with experiment, has shed new light on the dynamic processes by which biomolecules function.

Central to the MM approach is the concept of a force field, an empirically parametrized set of classical equations that is both mathematically simple and easily differentiable, allowing for the straightforward calculation of the forces acting on the atoms within a given system. The majority of force fields commonly used for the simulation of biomolecular systems, such as CHARMM,[8,9] AMBER,[10] GROMOS[11] and OPLS,[12] employ similar functional forms, an example of which is shown in Equation (3.1).

$$U(r) = \sum_{bonds} k_b(b-b_0)^2 + \sum_{angles} k_\theta(\theta-\theta_0)^2 + \sum_{dihedrals} k_\chi(1+\cos(n\chi-\delta)) + \sum_{\substack{nonbond\\ i\neq j}} \left(\varepsilon_{ij}\left[\left(\frac{R_{min,ij}}{r_{ij}}\right)^{12} - 2\left(\frac{R_{min,ij}}{r_{ij}}\right)^{6}\right] + \frac{q_i q_j}{4\pi\varepsilon_o r_{ij}}\right) \quad (3.1)$$

Equation (3.1) includes contributions arising from both bonded interactions, interactions that occur between covalently bound atoms, and nonbonded interactions, interactions that occur "through space". The bonded contributions are described by the bond, angle and dihedral terms, where b, θ and χ are the values of a given bond length, bond angle and dihedral angle at a given moment, and k_x, x_0, n and δ are empirical parameters to be fitted during the parameter optimization phase of the force field development. The remaining terms in Equation (3.1) represent the nonbonded interactions and consist of a Lennard-Jones (LJ) term to represent the van der Waals interactions between atoms, and a Coulomb term to represent the electrostatic interactions between atoms. In this case, r_{ij} is the distance between two atoms i and j, q_i is the charge on atom i, and $R_{min,ij}$ and ε_{ij} correspond to the value of r_{ij} at which the LJ interaction energy between atoms i and j is a minimum, and the depth of the corresponding LJ energy well, respectively. q_i, $R_{min,ij}$ and ε_{ij} are all empirical parameters that must also be fitted during the parameter optimization procedure. While Equation 3.1 includes all of the basic components of the commonly used force fields, additional terms may be included into a force field at the discretion of its developers. The CHARMM force field, for example, includes three additional terms.[13] The first represents improper torsions and the second is a Urey-Bradley term used to better describe angle bending via a 1,3 harmonic interaction.[14] The third term is a two-dimensional dihedral energy correction map termed CMAP, a grid-based term initially used to improve the treatment of the φ, ψ energy surface of the polypeptide backbone in proteins.[15–17]

Overall, it is the simplicity of these force fields that has ensured the success of MM methods. This simplicity greatly reduces the computational burden relative to quantum mechanical (QM) methods, rendering the study of large biomolecular systems computationally feasible. In combination with advances in computer hardware[18] and software,[19,20] the computational simplicity of MM force fields has facilitated once unimaginable achievements in biomolecular simulation,[21] such as the simulation of an entire virus[22] and the simulation of a protein on the millisecond timescale.[23] But while the simplicity of MM force fields has been their greatest blessing, it is also their greatest weakness. It is universally acknowledged that the results obtained from a computer simulation can only be as good as the accuracy of the model used to run the simulation. This is a problem that has been overcome, to a large extent, by the painstaking optimization of the empirical parameters present within the force fields to ensure accurate reproduction of experimental and QM target data.[24] However, there is a limit beyond which this approach cannot pass due to the inherent limitations in the form of the potential energy function. And with the increases in computer power indicated above, the need for simplicity is no longer as pressing as it once was.

One of the most striking simplifications adopted in the development of MM force fields lies in the treatment of electrostatic interactions. Within all of the force fields commonly used for the simulation of biomolecules, the electrostatic properties of a molecule are represented by effective point charges placed at the sites of the atomic nuclei. Energies of electrostatic interaction are then calculated using a Coulomb potential, as shown in Equation 3.1. In this approach, the electrostatic energy of the system is calculated as the sum of the individual atom–atom Coulomb interactions, leading to the use of the term "additive" to describe such force fields. This additive approach was originally employed due to computational considerations and has been remarkably successful,[25] but it neglects the distortion of the electron density around an atom or molecule under the influence of an external field; it neglects the polarizability of the molecule. This is important because the charge distribution, and hence electrostatic properties, of a real molecule can change significantly in response to a change in the molecule's surroundings. The dipole moment of a single water molecule, for example, varies from 1.85 D in the gas phase[26] to 2.1 D or higher in water clusters[27] and up to 2.9 D in the liquid phase.[28] For this reason, an additive force field including no explicit representation of polarizability is unable to provide an adequate representation of water in both the gas and liquid phases. Instead, in the development of non-polarizable force fields, polarizability is included implicitly by overestimating the gas phase dipole moment to give a better representation of liquid phase properties. This is an entirely reasonable approach for biomolecular systems, where gas phase properties are likely to be of little interest, but it does not solve the problem. A protein residue, for example, will experience a different dielectric environment within the hydrophobic core of a protein than it will on the protein surface, exposed to solvent. It is certain that its electrostatic properties will be affected as a result, and that this is an effect that will be overlooked by an additive force field. While this particular example is hypothetical, there are interactions of biological significance where polarizability is known to be important. The cation-π interaction[29] is one of the strongest known non-covalent interactions, energetically on a par with a hydrogen bond,[30] and has been shown to be important in a range of biological situations, including the stabilization of protein structure[31,32] and the stabilization of protein-DNA complexes.[33] Accurate modeling of the cation-π interaction has also been shown to require polarizability.[34] That polarizability should be important in interactions involving ions is unsurprising given their high charge density, and it has also been demonstrated that the inclusion of polarizability is important for the accurate modeling of the gramicidin ion channel.[35] It is expected that polarizability will prove to be important for the study of protein-ligand binding, enzyme catalysis and proton transport through aquaporins.[36] In general, it can be said with confidence that, without the explicit inclusion of polarizability within MM force fields, it will not be possible to obtain an adequate representation of a wide range of biomolecular systems.

3.2 Polarizable Force Fields

The need to include an explicit representation of polarizability within MM force fields has long been recognized, and multiple efforts towards this goal are currently underway.[37,38] There are also different schemes by which polarizability has been incorporated into molecular mechanics potentials.[38] Three of these schemes stand out for being particularly well studied by groups ultimately targeting the condensed phase simulation of large biomolecular systems. Those three schemes are based on induced dipoles, fluctuating charges and Drude oscillators.

The induced dipole model incorporates an explicit representation of polarizability by retaining the framework of fixed atomic charges and adding to this a set of inducible dipoles. These inducible dipoles are most commonly added to the atomic sites,[39] but can also be placed at the site of bonds.[40] The dipole moment, $\boldsymbol{\mu}_i$, induced at each point is then given by:

$$\boldsymbol{\mu}_i = \boldsymbol{\alpha}_i \left(\mathbf{E}_i^0 + \mathbf{E}_i^{\mathrm{p}} \right) \tag{3.2}$$

where $\boldsymbol{\alpha}_i$ is the polarizability tensor, $\mathbf{E}_i^0$ represents the field due to the permanent charges and $\mathbf{E}_i^{\mathrm{p}}$ is a perturbation to the field arising from the induced dipoles. In practice, the total field at any given time can be determined via a self-consistent procedure to minimize the polarization energy, though extended Lagrangian methods have been implemented to facilitate the application of induced dipoles in MD simulations, as discussed below. The induced dipole approach has the advantage of being intuitively logical, and relatively straightforward to parametrize, but suffers due to the vectorial nature of the point dipole, which complicates its implementation into existing MM packages.[41] That said, implementations of induced dipole methods into both AMBER[42] and CHARMM[43] have been reported, as have implementations due to the groups of Kaminski,[44,45] and Berne and Friesner.[46]

Fluctuating charge methods are those in which the polarizability of a molecule is represented by allowing the atomic partial charges to vary. This is achieved by assigning fictitious masses to the charges and making them additional degrees of freedom in the equations of motion, which can be treated *via* either an SCF or extended Lagrangian method, as described below. Charge is then allowed to "flow" between atoms until the instantaneous electronegativities are equalized. Such an approach is attractive because the resulting force field requires no additional interaction terms compared to a non-polarizable force field, but it does possess several drawbacks. Perhaps the most serious of these drawbacks is that the ability of the model to represent the different components of the polarizability tensor is limited. For example, a linear molecule can only be polarized along its axis, and a planar molecule cannot be polarized in a direction perpendicular to the plane. These models also considerably overestimate the polarizability of large molecules by allowing charge to flow between distant atoms at a low energy cost;[47] modifications are therefore necessary to make the scheme practical for the simulation of biomolecules.[47,48] Development of a CHARMM fluctuating charge force field for proteins and lipids is currently underway.[49,50]

Drude models incorporate polarizability using an approach based on the classical Drude oscillator model.[51] In this approach, polarizability is included by the addition of a set of massless charged particle, one attached to the atomic center of each of the polarizable atoms by a harmonic spring. The positions of the auxiliary, Drude, particles are then adjusted self-consistently to their local energy minima for any given configuration of atoms. This approach benefits from being both physically and chemically intuitive, with atoms reduced to pairs of point charges that can be considered to represent the nucleus and electron density, respectively. It is also easy to implement within existing force fields, but does suffer from a number of potentially significant drawbacks. These will be discussed in detail below.

3.3 The CHARMM Drude Polarizable Force Field

The CHARMM Drude polarizable force field is one example of a force field in which polarizability is included explicitly *via* the addition of Drude particles. At present, work is well advanced towards the development of a robust and reliable force field capable of modeling a range of different biomolecules across a wide variety of situations. The remainder of this chapter will focus on this model, as this is where efforts in our laboratory, in collaboration with the Roux group, have been concentrated.

3.3.1 The Model

Within the CHARMM Drude polarizable force field, all heavy (non-hydrogen) atoms are treated as being polarizable, while all H atoms have a fixed charge. The polarizability, α, of each heavy atom having charge q, is then represented by the addition of a Drude particle having charge q_D. The total charge on the atom remains q, but is partitioned over two sites: the "nucleus" has charge

$q_N = q - q_D$, and the Drude particle has charge q_D. Formally, it makes no difference whether q_D is positive[52] or negative,[53] but "chemical logic" prefers a scenario in which q_D, which represents the electronic degrees of freedom, is negative, leaving a net positive charge at the nucleus. The Drude particle is attached to the nucleus by a harmonic spring with force constant k_D. During an MD simulation in the absence of an external field, the Drude particle oscillates around the position of the nucleus, **r**, and, on average, the atom appears as a point charge with magnitude q. In the presence of a uniform electric field, **E**, the Drude particle oscillates around a position **r**+**d**, where **d** is a displacement from the nucleus that is calculated as:

$$\mathbf{d} = \frac{q_D \mathbf{E}}{k_D} \tag{3.3}$$

The average induced atomic dipole is then given by:

$$\boldsymbol{\mu} = \frac{q_D^2 \mathbf{E}}{k_D} \tag{3.4}$$

From this it follows that the isotropic atomic polarizability can be represented by Equation (3.5).

$$\alpha = \frac{q_D^2}{k_D} \tag{3.5}$$

From Equation (3.3) it is apparent that, for any value of α, the value of k_D can be chosen to ensure that the displacement, **d**, is much smaller that the interatomic distance, meaning that the resulting induced dipole approximates a point dipole. In practice, within the CHARMM Drude polarizable force field, k_D is chosen to be identical for all polarizable atoms, and takes a value of 500 kcal/mol/Å^2, which ensures that the point dipole approximation is valid at all times. In simple terms, this means that for each polarizable atom, it is necessary to optimize only one new parameter, α, when developing the electrostatic portion of the force field.

With the addition of the Drude particles, the potential energy function of the CHARMM Drude polarizable force field can be generalized as:[41]

$$U(\{\mathbf{r}\},\{\mathbf{d}\}) = U_{self}(\{\mathbf{d}\}) + U_{bond}(\{\mathbf{r}\}) + U_{elec}(\{\mathbf{r}\},\{\mathbf{d}\}) + U_{LJ}(\{\mathbf{r}\}) \tag{3.6}$$

where $U_{self}(\{\mathbf{d}\})$ represents the potential due to the harmonic springs connecting the Drude particles to their parent nuclei, $U_{bond}(\{\mathbf{r}\})$ represents the potential due to the bond, angle and dihedral terms, $U_{elec}(\{\mathbf{r}\},\{\mathbf{d}\})$ represents the potential due to all electrostatic interactions (nucleus–nucleus, nucleus–Drude, Drude–Drude) and $U_{LJ}(\{\mathbf{r}\})$ represents the LJ interactions between atoms. Equation (3.6) immediately reveals one of the key benefits of the Drude scheme for the explicit inclusion of polarizability: that the existing MM potential function requires very little modification.

Although the Drude potential function is relatively straightforward, performing MD simulations still presents some challenges. As Lamoureux and Roux have noted, "the main drawback of any treatment of induced polarization comes from the computationally expensive convergence of the self-consistent field (SCF) condition, which must be satisfied at each molecular dynamics time step to determine the magnitude of the induction and the atomic forces."[41] For a Drude model, this means that, for every new set of nuclear coordinates, the positions of the Drude particles must be minimized before Equation (3.6) can be evaluated (*i.e.* the Born-Oppenheimer approximation is satisfied *via* an SCF calculation).[54] Such a procedure adds significantly to the computational resources required for any given simulation, to an extent that would be prohibitive for large biomolecular systems. To overcome this problem, methods have been developed, for a range of polarizable force fields, in which the MD integrators include the polarizable degrees of freedom as dynamic variables.[55] Such approaches are termed "extended Lagrangian" methods. For the CHARMM Drude polarizable force field, Lamoureux and Roux[41] have developed an extended Lagrangian method in which a small mass, m_D, is taken from the atomic masses $\{m_i\}$ and assigned to each Drude particle. The motions of the atoms and the Drude particles, with positions $\{\mathbf{r}_i\}$ and $\{\mathbf{r}_{D,i} = \mathbf{r}_i + \mathbf{d}_i\}$, can then both be simulated dynamically:

$$(m_i - m_D)\ddot{\mathbf{r}}_i = \frac{\partial U}{\partial \mathbf{r}_i} \tag{3.7}$$

$$m_D\ddot{\mathbf{r}}_{D,i} = -\frac{\partial U}{\partial \mathbf{r}_{D,i}} \tag{3.8}$$

With Equations (3.7) and (3.8) in place, one would expect the motion of the Drude particles to be decoupled from the remaining motions if m_D is sufficiently small. While this is true, using a small value of m_D results in high frequency atom–Drude vibrational motions, which necessitates the use of a small integration timestep and negates any gains achieved by not using the full SCF procedure. This problem can be solved by controlling the amplitude of the oscillations of the Drude particles using a separate thermostat at temperature T*, where T* $<<$ T. As Lamoureux and Roux note, "the temperature T* should be small enough to leave almost no kinetic energy to the atom–Drude vibrations, yet large enough to allow the Drude particles to readjust to the room-temperature motion of the atoms."[41] This two-temperature approach has been shown to provide a good approximation to the SCF regime, and to allow simulations to proceed more than an order of magnitude faster than the equivalent simulations with full SCF calculations.[41] In practice, within the CHARMM Drude polarizable force field, the standard approach is to transfer a mass of 0.4 amu from real atoms to corresponding Drude particles and to use a T* value of 1 K.

Within the additive CHARMM force fields, nonbonded interactions between 1,2 and 1,3 pairs are not calculated. The reason for this is that these interactions are already implicitly included in the bond, angle and Urey-Bradley terms of the potential energy function. In the polarizable force field, however, the situation is not so straightforward. The positions of the Drude particles, and hence the induced dipole moments of the atoms, will continually change throughout the course of a simulation, and this effect cannot be captured by the non-varying bond and angle terms within the force field. Moreover, it is necessary to treat the cooperativity among induced dipoles in the molecule in order to obtain the correct anisotropic molecular dipoles from the collection of atomic dipoles. This means that 1,2 and 1,3 electrostatic interactions must be included within any polarizable model. For the CHARMM Drude polarizable force field, this is achieved *via* a modified electrostatic term that allows 1,2 and 1,3 screened dipole-dipole interactions in a fashion originally suggested by Thole.[56] The screening is implemented by effectively smearing the charges associated with the dipole moments such that, if two interacting dipoles i and j are composed of pairs of charges $q_i = \{q_{N,i}, q_{D,i}\}$ and $q_j = \{q_{N,j}, q_{D,j}\}$, the electrostatic interaction is represented by four terms of the type shown in Equation (3.9).[57]

$$U_{elec} = \frac{q_i q_j}{r_{ij}} \left[1 - \left(\frac{\bar{r}_{ij}}{2} \right) \exp - \bar{r}_{ij} \right] \tag{3.9}$$

where

$$\bar{r}_{ij} = a \frac{r_{ij}}{\sqrt[6]{\alpha_1 \alpha_2}} \tag{3.10}$$

Equation 3.10 contains a dimensionless constant, a, which is known as the "Thole factor". The value of a used in Equation 3.10 includes contributions from each of the atoms i and j, and is obtained *via* an additive combining rule. In early work on the CHARMM Drude polarizable force field, all atomic Thole factors were set to 1.3, meaning that all interacting pairs had a value of 2.6, which was based on the reproduction of the polarizability anisotropy in benzene.[56,58] More recently, it was shown that the use of atom-specific Thole factors facilitated a better reproduction of the polarizability tensor of N-methylacetamide (NMA), allowing for the dielectric of pure NMA to be accurately reproduced.[59] The same result was observed for N-containing heteroaromatics,[60] and in all subsequent parametrizations atomic Thole factors have been explicitly fitted as part of the electrostatic parameter optimization.

Anisotropy is incorporated into the electrostatic model *via* both the atomic charges and polarizabilities. Anisotropy in the charge density around hydrogen bond accepting atoms is accounted for by the addition of "lone pairs", which have been shown to facilitate an improved reproduction of H-bonding interactions with water molecules as a function of orientation.[61,62] These lone pairs consist of additional point charges, with the number of lone pairs on any one atom equivalent to the number of lone pairs on the equivalent real-world atom in the functional group of interest. For example, a neutral oxygen atom would have two lone pairs, a S atom two lone pairs, and a N one lone pair. The total charge of the atom, q, is typically transferred from the nuclear site to the lone pair sites, leaving a charge of 0 at the nucleus, though exceptions to this rule exist. When a Drude particle having charge q_D is added to the atom, the charge at the nucleus becomes $0-q_D=-q_D$.

The presence of the lone pair charge density away from the site of the atomic nucleus automatically breaks the spherical symmetry assumed by atom-centered models, and tests have shown that the addition of anisotropic atomic polarizabilities further improves the ability of the model to reproduce the anisotropy of interactions with the environment.[61] As a result, polarizabilities of all H-bond accepting atoms are treated as being anisotropic. This anisotropic polarizability is implemented by first noting that the Drude particle is bound to the nucleus of its parent atom by a harmonic spring having a potential:

$$U_{self} = \frac{1}{2} k_D \mathbf{d}^2 \tag{3.11}$$

Taking k_D to be a vector rather than a scalar, we can then write this as:

$$U_{self} = \frac{1}{2}\left([k_{D,11}]d_1{}^2 + [k_{D,22}]d_2{}^2 + [k_{D,33}]d_3{}^3\right) \tag{3.12}$$

where d_1, d_2 and d_3 are the projections of the Drude displacement vector **d** onto orthogonal axes defined using a local intramolecular reference frame. In effect, by treating k_D as a vector, the harmonic spring joining the Drude particle to its parent atom can be made more or less stiff in three orthogonal directions. The result is that the Drude particle can be displaced less easily in some directions than in others, and an explicit representation of the anisotropy of the polarizability has been incorporated. The three orthogonal components of the spring constant then become parameters to be fitted during the parameter optimization procedure. In practice, the stiffness of the bond along the reference frame is assigned a value of 1, and the other two components are optimized, subject to the constraint that the magnitude of the force constant must still equal 500 kcal/mol/Å^2.

One limitation of the Drude oscillator approach is the potential for over-polarization which, in its most extreme case, can result in the so-called "polarization catastrophe". In this scenario, the displacement of a Drude particle from its parent atom becomes large, on a par with inter-atomic separations, which inevitably results in unphysically large interaction energies and catastrophic failure of the simulation. This type of problem is particularly likely in a Drude polarizable force field because the Drude particle does not sit at the center of the LJ repulsive sphere, and is therefore not as well shielded from interactions with other charges as it would otherwise be. While this situation is always a risk, careful parametrization resulting in reasonable charge and polarizability values means that it should not, in most cases, be a problem. However, there are situations in which it becomes more likely, such as when studying atoms with unusually large polarizabilities, or with high atomic partial charges. Ions, for example, have been observed to be problematic. While there are a number of possible empirical solutions to this problem, in the CHARMM Drude polarizable force field the use of "an additional anharmonic restoring force" has been included "to prevent excessively large excursions of the Drude particle away from the atom".[63] This anharmonic term has the form:

$$U_{hyp} = K_{hyp}(\Delta R - \Delta R_{cut})^n \tag{3.13}$$

where the exponent n represents the order of the correction. To date, the correction has been applied with $n = 4$, which, as a second order correction to the polarizability, can be considered to represent the hyperpolarizability of the atom. The terms K_{hyp} and ΔR_{cut} in Equation (3.13) represent a force constant and the Drude displacement at which the anharmonic potential is switched on, respectively. Until now, all work that has applied this term has used values of K_{hyp} and ΔR_{cut} determined to be appropriate for the study of ions, specifically $K_{hyp} = 40\,000$ kcal/mol/Å^4 and $\Delta R_{cut} = 0.2$ Å.[63] Although there are many situations in which this anharmonic restoring force has not been required, the currently accepted "best practice" is that it should be included in all simulations performed with the CHARMM Drude polarizable force field.

The model described above performs well in a variety of situations, accurately reproducing a range of experimental and QM reference data after careful optimization of the relevant parameters. It has been observed, however, that the resulting force field yields a small but systematic error in the hydration free energies, predicting them to be too favorable by about 1 kcal/mol relative to experimental values.[64] This is potentially a serious problem. Water is the primary biological solvent, and if interactions with bulk water cannot be modeled accurately, there will be many biochemical processes than cannot be treated with suitable accuracy. As one example, a major component of the protein-ligand binding free energy is the desolvation penalty associated with removing the ligand from water.[65] This problem is not unique to the CHARMM Drude polarizable force field, having also been observed in a fluctuating charge polarizable force field,[66–68] and a number of reasons for its occurrence have been suggested.[64,67] One such suggestion is that the problem lies with the combining rules used to generate LJ parameters for unlike atom pairs from contributions due to the individual atoms. As LJ parameter combining rules, the CHARMM force fields use Equations (3.14) and (3.15), referred to as the Lorentz-Berthelot rules.[69]

$$R_{min,ij} = \frac{R_{min}}{2},i + \frac{R_{min}}{2},j \tag{3.14}$$

$$\varepsilon_{ij} = \sqrt{\varepsilon_i \times \varepsilon_j} \tag{3.15}$$

Since these CHARMM combining rules have previously been identified as sub-optimal,[70–72] work is currently underway in this laboratory to investigate whether an improved combining rule can be identified and applied, allowing for the accurate reproduction of hydration free energies without sacrificing the existing good agreement with other experimental data. While this work is in progress, an alternative scheme has been identified to allow for the accurate reproduction of hydration free energies. This scheme requires the inclusion of "pair-specific LJ parameters" between the water oxygen and solute non-hydrogen atoms. These pair-specific LJ parameters are used to override the standard LJ combining rules, meaning that values of $R_{min,ij}$ and ε_{ij} for a given atom pair are not calculated from individual contributions arising from each atom, but instead are specified directly. This approach allows for the inclusion of pair-specific LJ parameters for any atom pairs of choice, while nonbond interactions involving all other atom pairs are calculated using $R_{min,ij}$ and ε_{ij} values obtained via the standard combining rules. By introducing pair-specific LJ parameters for interactions between solute heavy atoms and water O atoms it is possible to correct the systematic error in the hydration free energies, without affecting any calculated properties that do not involve interactions between the solute molecule and water.

3.3.2 Parametrization Strategy

The parametrization strategy employed in the development of the CHARMM Drude polarizable force field is illustrated in Figure 3.1. The general procedure is that the biomolecule of interest is split into pieces, creating a set of small molecule analogues of the functional groups present within the biomolecule. This is illustrated in Figure 3.2 for DNA, where the oliogonucleotide is broken down until it is represented by three different classes of molecules: the bases; the sugar, and the phosphate. In the next step of the parametrization procedure, each of the small-molecule analogues is treated separately, with all bonding and nonbonding parameters optimized until a satisfactory agreement with reference QM and experimental data is obtained.

As Figure 3.1 suggests, the optimization of the parameters associated with the small molecule analogues of the functional groups present within the biomolecules represents the most complicated part of the parametrization procedure. The majority of the parameters that will be used in the final biomolecular simulations are developed at this stage, and close attention must be paid to ensure that they are as robust and reliable as possible. The first stage of the parametrization procedure involves the optimization of the electrostatic parameters. Beginning with a QM optimized geometry of the small molecule of interest, a series of QM electrostatic potentials (ESPs) are calculated, with each ESP perturbed by the presence of a single charge, providing a direct measure of the polarizability of the molecule. Initial guesses for the electrostatic parameters are based on charges from the additive CHARMM force fields,[8,9] polarizabilities from the work of Miller[73] and Thole factors set to 1.3. These are then simultaneously optimized, in an automated fashion,[74] using the FITCHARGE module within the CHARMM program.[75] If a conformationally flexible model is being considered, this procedure can be applied to several different

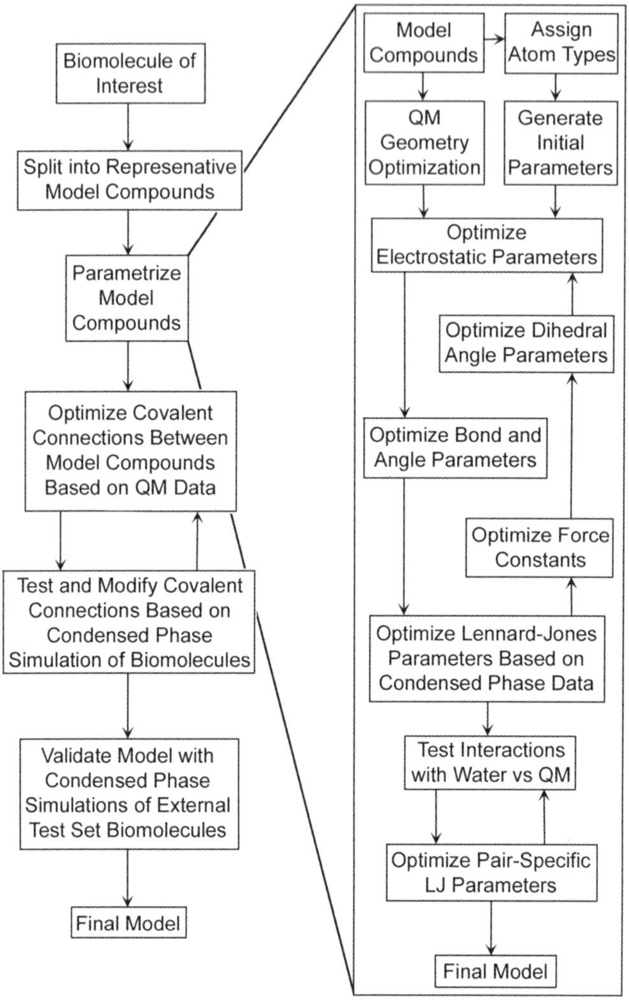

Figure 3.1 Schematic description of the parameter optimization procedure employed in the development of the CHARMM Drude polarizable force field for biomolecules.

conformations of the molecule, with the final electrostatic parameter values obtained by averaging over all of the conformations considered.

During the charge fitting procedure, the polarizability parameters are scaled by a factor varying from 0.6 to 1.0 to enable the accurate calculation of the liquid phase dielectric constant.[76] There is no way of knowing at this stage exactly what an appropriate scaling factor will be and it is necessary to estimate the polarizability scaling factor based on the similarity of the molecule(s) in question to previously parametrized molecules (Table 3.1). While the exact reason for the need for polarizability scaling is debated, it is believed to arise from real physical phenomena. Suggested explanations include the Pauli

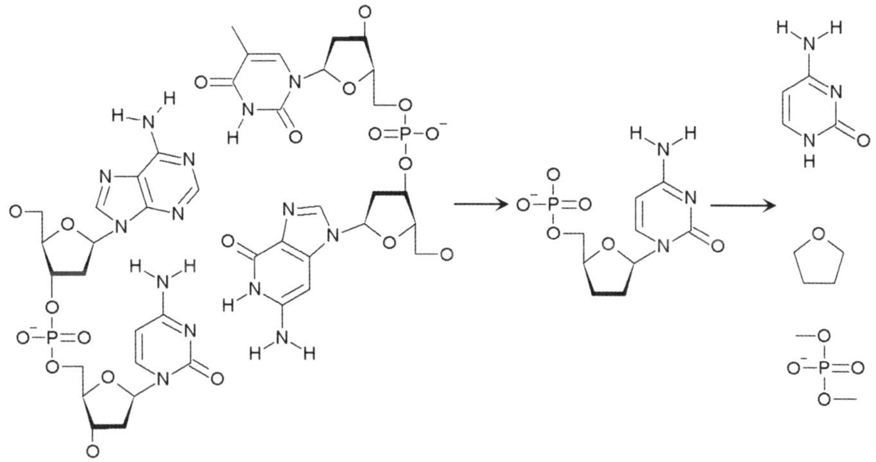

Figure 3.2 Breaking a biomolecule down to identify small molecule analogues of the functional groups present within the biomolecule. In the first step, the smallest repeating unit of DNA is identified, comprising a base, a sugar and a phosphate. In the second step, this repeating unit is further broken down into its constituent functional units.

Table 3.1 Polarizability scaling factors used in the optimization of small molecules within the CHARMM Drude polarizable force field.

Molecule	*Final Polarizability Scaling Factor*
Water	0.724
Alkanes	0.700
Alcohols	0.724
Ethers	0.850
Aromatics	0.724
N containing aromatics	0.850
S containing compounds	0.600/0.700/0.850
Amides	1.000

exclusion principle[77] and inhomogeneities in the electric field within the exclusion volume of the molecule.[78]

With the charges, polarizabilities and Thole factors in place, the final stage in the optimization of the electrostatic parameters is the fitting of the polarizability anisotropy parameters. Initial polarizability anisotropy parameters are assigned to all H-bond accepting atoms within the molecule, based on the previously optimized polarizability anisotropies of similar atom types. The polarizability anisotropy parameters are then optimized by considering the polarization response around H-bond accepting atoms. Perpendicular arcs are constructed in-plane and out-of-plane around all H-bond acceptor atoms and the total energy is calculated in the presence and absence of a perturbing ion of charge +0.5e placed at points along these arcs.[79] The polarization response is then defined as the difference between the total energies calculated with and without the perturbation ions. The components of the anisotropy tensor are varied until the best possible agreement is obtained with the results from equivalent QM calculations. As a validation of the resulting anisotropic polarizabilities, interactions with sodium cations on arcs placed around the H-bond accepting atoms are also calculated, and compared to results from equivalent QM calculations.

With electrostatic parameters in place, equilibrium bond length and angle parameters are targeted to reproduce crystallographic geometries taken from surveys of relevant crystal structures extracted from the Cambridge Structural Database.[80] Force constants for the bond stretching, angle bending and dihedral angle motions are optimized on the basis of vibrational spectra. Target data, consisting of both vibrational frequencies and assignments, is obtained from QM calculated infrared spectra.

To optimize dihedral parameters, QM scans of the relevant dihedral angles are calculated, with the corresponding parameters adjusted until the best possible reproduction of the QM scans is obtained. This process can be performed either manually or using an automated procedure such as Monte Carlo simulated annealing (MCSA),[81] with the latter approach particularly useful when molecules with multiple, correlated dihedral angles are considered.

For the optimization of LJ parameters, the primary target data are condensed phase thermodynamic properties. Ideally, these condensed phase data will consist of liquid phase enthalpies of vaporization and molecular volumes. However, such data is not always available and must, in these cases, be replaced by the best alternative data that can be identified. One such case is the nucleic acid bases.[82] To overcome the lack of liquid phase thermodynamic data, LJ parameters for the NA bases were initially taken from the structurally analogous aromatic molecules purine and pyridine, where they had been optimized based on liquid phase thermodynamic data.[60] LJ parameters were then adjusted to optimize the reproduction of H-bonding interactions in base-base dimers. While the H-bonding interactions of the bases are important, stacking interactions also make a significant contribution to the intermolecular interactions of the bases in DNA and RNA. These stacking interactions, however, are notoriously difficult to evaluate, even using QM methods,[83] and could not be explicitly considered as target data. Instead, crystal heats of sublimation, molecular volumes and lattice parameters are considered, the logic being that a set of parameters able to accurately reproduce both gas phase H-bonding and crystal phase thermodynamic properties must have achieved the correct balance between the H-bonding and stacking interactions. The actual optimization of the LJ parameters involves the manual iterative adjustment of the individual parameters until the best possible agreement with the target data is achieved, with the criterion that all experimental data should be reproduced to within a 2% error. In addition to condensed phase data, interactions with rare gas atoms (typically He and/or Ne) can be used in the optimization of LJ parameters.[84] While condensed phase data provide a measure of the overall LJ properties of a molecule, they provide little information about the properties of individual atoms. By considering interactions between rare gas and solute atoms, the LJ properties of individual atoms can be probed to ensure that an appropriate balance between the LJ parameters of the individual atoms has been achieved.

While the above discussion treats each class of parameter separately, the reality is that the parameters are highly interdependent. Changing the LJ parameters, for example, will affect the internal geometries that resulted from the optimization of the equilibrium bond length and angle parameters. For this reason, it is necessary to perform the parameter optimization in an iterative fashion, repeating the optimization of all bonded parameters whenever a new set of LJ parameters is adopted. This approach ensures an appropriate balance between the different parameters, resulting in an internally consistent force field. Once an acceptable set of parameters has been generated, extended liquid phase calculations are performed to assess the dielectric constant. As indicated above, the dielectric constant provides a good measure of whether the atomic polarizabilities obtained in the electrostatic parameter optimization have been appropriately scaled. If a systematic error is identified in the final, calculated dielectric constants for a given class of molecules, it is necessary to rescale the original atomic polarizabilities, and reoptimize the remaining internal and LJ parameters as necessary. This cycle should be repeated as often as is necessary to obtain the best possible agreement with all target data.

Within the additive CHARMM force field, atomic partial charges are derived based on the reproduction of energies and geometries of QM interactions with water.[85] This approach represents a computationally efficient method that, to a first approximation, accounts for the presence of environmental effects, and thereby explicitly incorporates such environmental effects into the parametrization scheme. Although this procedure has been replaced by the ESP fitting in the CHARMM Drude polarizable force field, the calculation of interactions with individual water molecules provides a useful validation of the newly optimized parameters for the polarizable force field. It is particularly important for assessing whether the empirical model can accurately reproduce the anisotropy of H-bond interactions. To test this, individual water molecules are placed at locations where they interact with various atoms of the solute molecule, and interaction energies and geometries are calculated. The resulting values are compared to the results obtained from equivalent QM calculations.

In addition to interactions with individual water molecules, it is important to consider the bulk hydration properties of the small molecules being parametrized. As was noted above, systematic errors in the reproduction of hydration free energies for small molecules have required the introduction of pair-specific LJ parameters.[64] The optimization of these parameters marks the final stage in the parametrization of a new class of small molecules. Target data for the optimization of pair-specific LJ parameters are experimental hydration free energies and, because the pair-specific LJ parameters are effectively free parameters, the chance of overfitting is high. To minimize this risk, a robust parametrization scheme has been developed.[62] And to reduce the chance of overfitting even further, the final set of pair-specific LJ parameters should be applied to an external test set of molecules. If the optimized parameters can treat the test set molecules as well as the training set molecules, then it is reasonable to assume that overfitting has not occurred. The introduction of these pair-specific LJ parameters will also have an impact on the interactions with individual water molecules and these values should be recalculated to check for any errors that have been introduced. If necessary, these steps should be repeated until a satisfactory level of agreement with both the hydration free energies and the QM interactions with individual water molecules is obtained.

Once parameters have been obtained for all of the individual small-molecule analogues of the biomolecule of interest, these small molecules are recombined, recreating the biomolecule. The parameters associated with the covalent connections between the different small-molecule analogues are then optimized, with the dihedral angle parameters requiring particularly careful attention. While this optimization is relatively straightforward, there is no guarantee that dihedral angle parameters able to reproduce gas phase QM properties of model compounds will also be suitable for condensed phase simulations of biomolecules. The new parameters must, therefore, be tested in condensed phase biomolecular simulations. Typically, a biomolecule will be simulated and properties such as dihedral angle distributions will be calculated and compared to known experimental data.[9] Where inadequacies are identified, the parameters will be modified accordingly, and re-tested on the gas phase model compounds. These steps will be repeated as many times as needed to gain a good agreement with both the gas phase QM data and the condensed phase experimental data.

3.3.3 Biomolecular Building Blocks

The majority of work performed on the CHARMM Drude polarizable force field to date has been focused in three specific areas. Firstly, the development of the Drude model and its associated theory, as well as its implementation within the CHARMM program[75] (the model has also been implemented into the program NAMD[86]). Secondly, the development of a robust and reliable parametrization procedure, as described in section 3.3.2. And thirdly, the optimization of the parameters associated with the small molecule analogues of the functional groups present within the proteins and the nucleic acids.

For the development of a comprehensive protein force field, parameters are required for all of the amino acid side chains, as well as the backbone and disulfide bridge. The chemical diversity of the amino acid side chains means that a large number of different functionalities must be parametrized and, to date, parameters have been published for CHARMM Drude polarizable models of alkanes,[87] alcohols,[88] aromatics,[89] N-containing aromatics,[60] S containing compounds[90] and amides.[59] Taken together, these compounds cover small molecule analogues of the backbone, the disulfide bridge and the majority of amino acid chains. Figure 3.3 provides a more specific description of the correspondence between the functional groups present within proteins and the small molecule model compounds used to develop parameters for those groups.

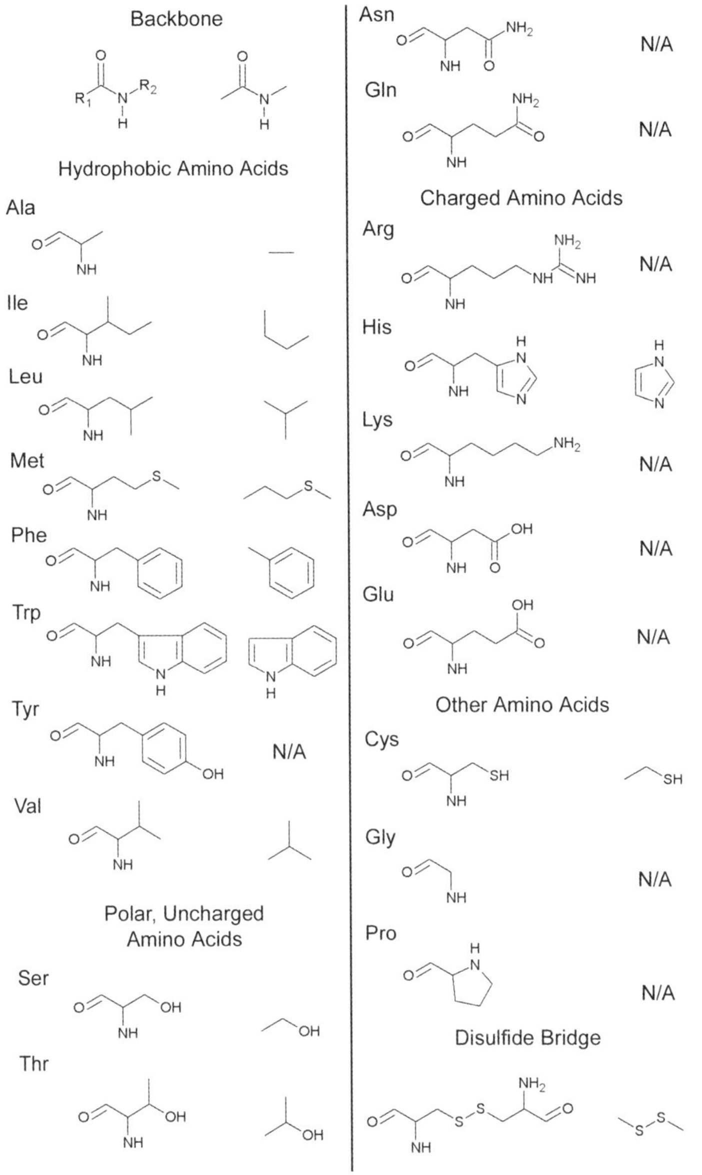

Figure 3.3 Components of proteins, and the small molecule analogues of these components for which CHARMM Drude polarizable force field parameters have been published to date. For protein components marked with N/A, no parameters have been published for small molecule analogues.

The protein fragments marked with N/A in Figure 3.3 are those for which parameters of a specific small molecule analogue have not been published: all are currently under development. Some of the remaining small molecules will be relatively easy to parametrize. Phenol, for example, has not been explicitly parameterized as an analogue of the tyrosine side chain but parameters for both benzene and toluene have been published.[89] Combining these parameters with published alcohol parameters[88] will be straightforward. Small molecule analogues of other side chains, such as proline or arginine for example, will present more of a challenge. While the remaining small molecule analogues are parametrized, work is also underway to optimize backbone parameters based on condensed phase simulation of alanine oligomers, and to optimize dihedral angle parameters for all amino acid χ dihedrals.

The optimization of parameters for the small molecule analogues of the functional groups present within the nucleic acids is simplified relative to that of the proteins by virtue of the fact that there are only five distinct nucleic acid bases, as opposed to twenty two proteinogenic amino acids. For this reason, the optimization of parameters for the small molecules associated with DNA and RNA is already complete, having required the development of parameters associated with five distinct classes of molecule: ethers;[76,91] phosphates; N-containing aromatic heterocycles;[60] nucleic acid bases,[82] and alcohols.[88]

The backbone sugar of the nucleic acids was represented by the small molecule tetrahydrofuran (THF), which was parameterized as part of a series of ether molecules that included a number of acyclic linear ethers, as well as the cyclic ethers THF and tetrahydropyran (THP).[76,91] While only THF appears in the nucleic acids, all of these ether molecules are constituent parts of biomolecules: the linear ether functionality features heavily in lipid molecules, and THP (along with THF) is an important component of the carbohydrates. Indeed, efforts towards the development of CHARMM Drude polarizable force fields for lipids[92] and carbohydrates are now underway, and have taken these ether molecules as some of their starting points.

The phosphate component of the nucleic acid backbone was represented by a single small molecule analogue, dimethylphosphate (DMP). Uniquely amongst small molecules parametrized to date, the inclusion of lone pairs within the DMP molecule was found to yield no significant improvement over an equivalent model containing no lone pairs. For this reason, lone pairs were omitted from the final set of DMP parameters. Polarizability anisotropy, however, was found to be extremely important in accurately reproducing interactions with sodium ions and individual water molecules.

As noted above, the development of parameters for the nucleic acid bases[82] was performed in conjunction with the development of parameters for a series of N-containing aromatic heterocycles that included purine and pyrimidine,[60] significantly reducing the number of LJ parameters to optimize based entirely on nucleic acid base properties. The final nucleic acid base parameter set provided a good illustration of the improvements in electrostatic properties that can be obtained by using a polarizable force field, an example of which is shown in Table 3.2. While the additive CHARMM27 model fails to predict correctly the direction of the change in dipole moment upon substitution of a methyl group for a H atom at the N9 position, the Drude model predicts this trend correctly, as well as providing a better reproduction of the components of the dipole moments in general.

Table 3.2 Gas phase dipole moments of methylated and unsubstituted purine nucleic acid bases, in Debye. The methylated bases were used for optimization of the CHARMM Drude polarizable force field, with the unsubstituted bases used for validation. QM values were calculated at the B3LYP/aug-cc-pVDZ//MP2/6-31G* level of theory.

	QM	*CH27*	*Drude*	*QM*	*CH27*	*Drude*
		Me-Ade			H-Ade	
X	−2.43	−1.42	−2.43	−1.91	−2.29	−1.80
Y	1.30	2.38	1.39	1.60	1.84	1.78
Z	0.63	−0.12	0.57	0.71	0.00	0.62
Total	2.83	2.77	2.86	2.59	2.94	2.61
		Me-Gua			H-Gua	
X	−0.45	2.13	−0.82	2.88	4.59	2.33
Y	−7.02	−6.31	−6.80	−5.90	−5.94	−5.89
Z	0.95	0.99	0.90	0.84	0.95	0.89
Total	7.10	6.73	6.91	6.62	7.56	6.40

As well as the optimization of parameters associated with functional groups present within biomolecules, it is also necessary to optimize parameters for other molecules important for biomolecular simulation. Most important is water, and the four site SWM4-NDP model[53] is currently the standard for simulations with the CHARMM Drude polarizable force field. This model, however, was developed before many of the present features of the Drude model had been adopted and it therefore lacks lone pairs and anisotropic polarizabilities. For this reason, work is currently underway to develop a new, five-site water model. Parameters have also been developed for a series of mono- and divalent ions.[63]

3.3.4 Case Study: The Nucleic Acids

As noted above, parameters have already been developed for all small molecule analogues of functional groups present within the nucleic acids. As Figure 3.1 suggests, work on the nucleic acids is now focused on the optimization, testing and refinement of parameters associated with the covalent connections between the various small molecules. This procedure is being performed in two distinct phases. Initially, parameters are being developed based on QM studies of a number of larger model compounds, with particular attention being paid to the optimization of dihedral angle parameters. For optimization of the dihedral parameters, QM potential energy scans (PES) are calculated around the dihedral angles of interest, and the dihedral parameters are then fitted to these data by the MM model *via*, for example, an automated procedure based on MCSA.[81] At this stage, calculations are also performed to test the electrostatic properties of these larger model compounds. This testing of electrostatic properties, such as dipole moment as a function of conformation, is necessary due to the potential for cooperative interactions between the induced dipole moments on the different small molecules. In the second step of the process, the resulting parameters are being tested and refined based on simulations of condensed phase nucleic acids, and their ability to reproduce experimental properties.

Before the optimization of these covalent connections is complete, it would be foolhardy to attempt to obtain meaningful results from MD simulations of any oligonucleotide using the CHARMM Drude polarizable force field. Nevertheless, with all nonbonded parameters now in place, it is possible to begin to apply the model to the study of biomolecules, and to begin to obtain some insight into the impact that the CHARMM Drude polarizable force field will have on the study of biomolecular systems. As a case study, the base-flipping process[93] will be considered.

As part of an ongoing project to study nucleotide excision repair (NER)[94] (E. Darian and A.D. MacKerell Jr., work in progress), the CHARMM27 additive force field has been used to study base flipping in a variety of DNA double helices having sequence [d(5′-CCATCG*C*N*ACC-3′).d(5′-GGT*N*′GCGATGG-3′)]. In this sequence, *N* and *N*′ can be either A or T, and G* represents the base that will flip out of the double helix, which is either G or an unnatural analogue of G. While a range of different sequences are being studied with the additive force field, this case-study will focus on the case where G* = G and *N* = A (*N*′ = T).

The base flipping process can be modeled using a reaction coordinate described by a pseudo-dihedral angle, D, defined in terms of the centers of mass of four groups of atoms, as shown in Figure 3.4.[95,96] With this pseudo-dihedral reaction coordinate defined, potentials of mean force (PMFs) have been calculated for base flipping. These calculations are based on a series of MD simulations performed in the presence of a biasing potential running from $D = 0°$ to $D = 360°$ in increments of 5°, with the weighted histogram analysis method (WHAM)[97] subsequently used to unbias the umbrella sampling simulations and extract the full PMF, as previously described.[95] At each window in the PMF calculations, the hydrated DNA structure is initially equilibrated, after which 1 ns of production MD simulation is performed, with atomic coordinates recorded every 10 ps. This results in a set of 72 CHARMM trajectories, each having D restrained to a different value.

For each of these 72 trajectories, we have subsequently extracted every set of coordinates and replaced the additive CHARMM27 nonbond parameters with the nonbond parameters from the CHARMM Drude polarizable force field. The Drude nonbond model, including both the electrostatic and LJ terms is then applied in a way that may be considered analogous to a single-point calculation used in QM calculations (*i.e.* energies calculated at a different level of theory than that used for the structure optimization). Once applied, the positions of all Drude particles are then minimized to achieve the SCF condition while keeping all other atoms fixed in their original positions. This procedure effectively generates a set of 72 pseudo-trajectories based on the CHARMM Drude polarizable force field, which can then be used to calculate a range of properties for comparison to results obtained using the additive force field. The results presented below represent the average value of the property of interest calculated over each coordinate file in the 72 individual trajectories.

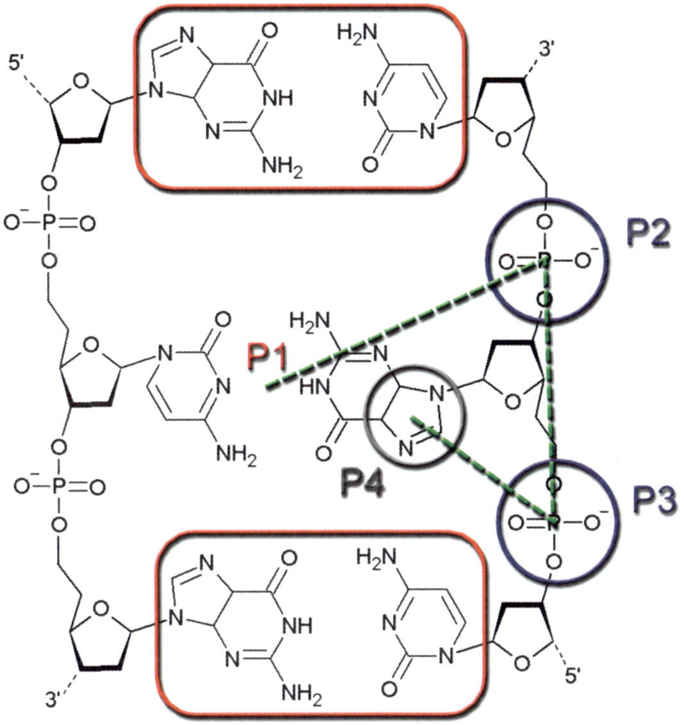

Figure 3.4 The reaction coordinate for base flipping is described in terms of a pseudo dihedral angle defined in terms of the centers of mass of four groups of atoms. Group 1 contains heavy atoms of the base pairs above and below the base that flips and has center of mass P1. Groups 2 and 3 contain the atoms in the phosphate groups before and after the base that flips, and have centers of mass P2 and P3. Group 4 contains the atoms in the 5-membered ring of the guanine base that flips and has center of mass P4.

It should be noted that the pseudo-trajectories, and hence any calculated properties, will not be identical to those that would be obtained from full MD simulations performed using the CHARMM Drude polarizable force field. They are sufficient, however, to allow us to begin to obtain insights into the impact that the polarizable force field will have on the study of biomolecular systems.

Considering the dipole moment of the guanine base as it flips out of the double helix (Figure 3.5) gives some immediate insights into the differences between the additive and polarizable force fields. With the additive CHARMM27 force field, the dipole moment of the base is approximately constant regardless of the value of D. This is unsurprising: without an explicit representation of polarizability, the molecule is unable to respond to any changes in the electric environment (the small variations in the dipole moment that are observed are due to small changes in the geometry of the base). With the Drude polarizable force field, the situation is very different. The dipole

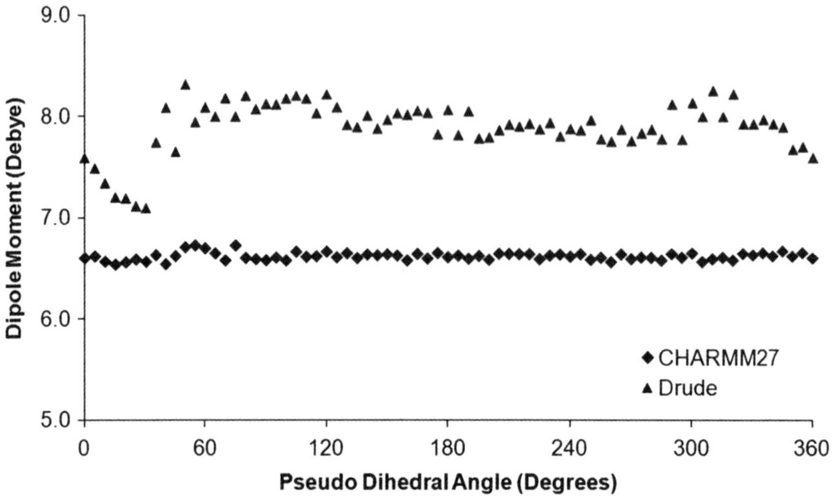

Figure 3.5 Dipole moment of the guanine base as it flips out of the DNA double helix. Values are shown as the mean over each trajectory.

moment of the guanine base is noticeably lower when it is inside the double helix than when it is exposed to the solvent. The dipole moments of bases not directly involved in the base flipping process show comparatively little change over the course of the base flipping process. However, considering the average dipole moment of these bases over all 72 trajectories is instructive. For the cytosine bases at positions 5 and 7 of chain A, directly before and after the base that flips, the average dipole moment over all trajectories is 10.22 D in both cases. For the cytosine at position 6 of chain A, which is the Watson-Crick partner of the base that flips, the average dipole moment over all trajectories is 10.09 D. This difference in average dipole moments may seem small, but performing a t test[98] reveals it to be statistically significant with a p value of 0.0001. We can conclude that the loss of the hydrogen bonds associated with the Watson-Crick pair results in a significant decrease in the dipole moment, compared to cytosine bases that maintain their Watson-Crick structure throughout the base flipping process.

The CHARMM Drude polarizable force field also predicts greater variability in the dipole moments of the bases when the DNA molecule is not undergoing conformational change. Figure 3.6 shows the average dipole moment of each of the individual bases, calculated using the CHARMM Drude polarizable force field, over the base flipping trajectory in which $D = 10^{\circ}$. This corresponds to the trajectory in which the flipping G base adopts the canonical Watson-Crick conformation. The same bases at different positions in the sequence have noticeably different dipole moments. Of the seven C bases in the sequence, for example, five are non-terminal and four of these have an average dipole moment of 10.2 D. The average dipole moments of the terminal C bases are higher, with values of 10.3 D and 10.4 D for the two terminal Cs. Interestingly,

	Chain A	Chain B	
10.4 +/- 0.4	C_1	G_{11}	*8.1 +/- 0.7*
10.2 +/- 0.5	C_2	G_{10}	*7.6 +/- 0.5*
6.1 +/- 0.5	A_3	T_9	*8.1 +/- 0.4*
8.1 +/- 0.4	T_4	A_8	*5.9 +/- 0.5*
10.4 +/- 0.4	C_5	G_7	*7.6 +/- 0.5*
7.7 +/- 0.5	G_6	C_6	*10.2 +/- 0.4*
10.2 +/- 0.5	C_7	G_5	*7.6 +/- 0.5*
6.1 +/- 0.5	A_8	T_4	*8.1 +/- 0.4*
6.1 +/- 0.4	A_9	T_3	*8.1 +/- 0.4*
10.2 +/- 0.5	C_{10}	G_2	*7.7 +/- 0.6*
10.3 +/- 0.5	C_{11}	G_1	*7.6 +/- 0.6*
	3'-TER	5'-TER	

Figure 3.6 Average dipole moments, in Debye, for each of the individual bases over all snapshots of the base flipping trajectory with $D = 10°$, in which all bases are in H-bonded Watson-Crick pairs. Calculated using the CHARMM Drude polarizable force field.

the G base with the largest dipole moment is also located at one of the termini, though the second terminal G base has an average dipole moment at the lower end of the observed range. The calculations performed here do not allow us to explain fully these observations, but they do offer an insight into the differences that will be obtained by using a polarizable force field. In general, the CHARMM27 force field predicts the bases to have smaller dipole moments than does the Drude polarizable force field, despite the polarizable force field predicting very similar (or smaller) gas phase polarizabilities. As again expected, the CHARMM27 dipole moments are the same for all bases regardless of their position in the DNA chain. With the additive force field, there is no way to assess whether a base at the end of the chain, for example, has different electrostatic properties than a base in the middle of the chain.

Taken together, these are exciting observations, confirming that the polarizable force field is able to provide insights, some obvious, some subtle, that are simply not available with an additive force field. Clearly, with the polarizable force field it is possible to obtain a different atomic-level picture of biomolecules.

3.4 Conclusions

Progress towards a polarizable force field for the simulation of biomolecules is well advanced. The theory of the Drude model has been developed to a point where, with careful parametrization, it allows for the accurate reproduction of a wide range of QM and experimental target data. It has also been implemented into CHARMM and NAMD, two of the most widely-used programs for molecular dynamics simulations. A second, equally important part of the development of the force field has been the optimization of empirical parameters targeted at the simulation of biomolecules. The first step in this process has been the optimization of parameters associated with small-molecule analogues of the functional groups present within biomolecules. This has been a painstaking but informative process that has resulted in high-quality parameters covering a range of different functionalities. For proteins, nucleic acids and lipids the majority of small molecule work is now complete. With this, the current focus is increasingly moving towards optimizing parameters for the covalent connections between the small molecules, which will ultimately allow full biomolecular simulations. While it is not yet meaningful to perform full MD simulations of these molecules, preliminary calculations offer a tantalizing glimpse of what the future offers. The CHARMM Drude polarizable force field already provides a different atomic-level picture to that obtained with a well-optimized additive force, hinting at the new insights that will be obtained once the final model is in place.

Acknowledgements

The authors thank the NIH for financial support through grants GM051501, GM070855 and GM0772558, DoD High Performance Computing, the Pittsburgh Supercomputing Center and the NSF/Teragrid for access to computational resources, and Prof. Benoit Roux and group members for a longstanding and productive collaboration.

References

1. J. A. McCammon, B. R. Gelin and M. Karplus, *Nature*, 1977, **267**, 585.
2. M. Karplus and J. A. McCammon, *Nat. Struct. Biol.*, 2002, **9**, 646.
3. N. A. Macleod, P. Butz, J. P. Simons, G. H. Grant, C. M. Baker and G. E. Tranter, *Isr. J. Chem.*, 2004, **44**, 27.
4. N. A. Macleod, P. Butz, J. P. Simons, G. H. Grant, C. M. Baker and G. E. Tranter, *Phys. Chem. Chem. Phys.*, 2005, **7**, 1432.
5. C. D. Snow, N. Nguyen, V. S. Pande and M. Grubele, *Nature*, 2002, **420**, 102.
6. N. K. Banavali, N. Huang and A. D. MacKerell Jr., *J. Phys. Chem. B*, 2006, **110**, 10997.
7. S. T. Wlodek, T. W. Clark, L. R. Scott and J. A. McCammon, *J. Am. Chem. Soc.*, 1997, **119**, 9513.
8. A. D. MacKerell Jr., D. Bashford, M. Bellott, R. L. Dunbrack Jr., J. D. Evanseck, M. J. Field, S. Fischer, J. Gao, H. Guo, S. Ha, D. Joseph-McCarthy, L. Kuchnir, K. Kuczera, F. T. K. Lau, C. Mattos, S. Michnick, T. Ngo, D. T. Nguyen, B. Prodhom, W. E. Reiher III, B. Roux, M. Schlenkrich, J. C. Smith, R. Stote, J. Straub, M. Watanabe, J. Wiórkiewicz-Kuczera, D. Yin and M. Karplus, *J. Phys. Chem. B*, 1998, **102**, 3586.
9. N. Foloppe and A. D. MacKerell Jr., *J. Comput. Chem.*, 2000, **21**, 86.

10. W. D. Cornell, P. Cieplak, C. I. Bayly, I. R. Gould, K. M. Merz Jr., D. M. Ferguson, D. C. Spellmeyer, T. Fox, J. W. Caldwell and P. A. Kollman, *J. Am. Chem. Soc.*, 1995, **117**, 5179.
11. W. R. P. Scott, P. H. Hünenberger, I. G. Tironi, A. E. Mark, S. R. Billeter, J. Fennen, A. E. Torda, T. Huber, P. Krüger and W. F. van Gunsteren, *J. Phys. Chem. A*, 1999, **103**, 3596.
12. W. L. Jorgensen, D. S. Maxwell and J. Tirado-Rives, *J. Am. Chem. Soc.*, 1996, **118**, 11225.
13. X. Zhu, P. E. M. Lopes and A. D. MacKerell Jr., Recent developments and applications of the CHARMM force fields, *WIREs Comput. Mol. Sci.*, 2012, **2**, 167–185. DOI: 10.1002/wcms.74.
14. H. C. Urey and C. A. Bradley Jr., *Phys. Rev.*, 1931, **38**, 1969.
15. A. D. MacKerell Jr., M. Feig and C. L. Brooks III, *J. Am. Chem. Soc.*, 2004, **126**, 698.
16. A. D. MacKerell Jr., M. Feig and C. L. Brooks III, *J. Comput. Chem.*, 2004, **25**, 1400.
17. M. Buck, S. Bouguet-Bonnet, R. W. Pastor and A. D. MacKerell Jr., *Biophys. J.*, 2006, **90**, L36.
18. D. E. Shaw, M. M. Deneroff, R. O. Dror, J. S. Kuskin, R. H. Larson, J. K. Salmon, C. Young, B. Batson, K. J. Bowers, J. C. Chao, M. P. Eastwood, J. Gagliardo, J. P. Grossman, C. R. Ho, D. J. Larardi, I. Kolossvary, J. L. Klepeis, T. Layman, C. Mcleavey, M. A. Moreas, R. Mueller, E. C. Priest, Y. B. Shan, J. Spengler, M. Theobald, B. Towles and S. C. Wang, *Commun. ACM*, 2008, **51**, 91.
19. D. E. Shaw, *J. Comput. Chem.*, 2005, **26**, 1318.
20. J. C. Phillips, R. Braun, W. Wang, J. Gumbart, E. Tajkhorshid, E. Villa, C. Chipot, R. D. Skeel, L. Kalé and K. Schulten, *J. Comput. Chem.*, 2005, **26**, 1781.
21. M. Vendruscolo and C. M. Dobson, *Curr. Biol.*, 2010, **21**, R68.
22. P. L. Freddolino, A. S. Arkhipov, S. B. Larson, A. McPherson and K. Schulten, *Structure*, 2006, **14**, 437.
23. D. E. Shaw, P. Maragakis, K. Lindorff-Larsen, S. Piana, R. O. Dror, M. P. Eastwood, J. A. Bank, J. M. Jumper, J. K. Salmon, Y. Shan and W. Wriggers, *Science*, 2010, **330**, 341.
24. A. D. MacKerell Jr., *J. Comput. Chem.*, 2004, **25**, 1584.
25. M. Karplus, *Acc. Chem. Res.*, 2002, **35**, 321.
26. D. R. Lide (ed), *CRC Handbook of Chemistry and Physics, 84th edn.*, CRC Press, Boca Raton, Fl, 2003.
27. J. K. Gregory, D. C. Clary, K. Liu, M. G. Brown and R. J. Saykally, *Science*, 1997, **275**, 814.
28. Y. S. Badyal, M.-L. Saboungi, D. L. Price, S. D. Shastri, D. R. Haeffner and A. K. Soper, *J. Chem. Phys.*, 2000, **112**, 9206.
29. D. A. Dougherty, *Science*, 1996, **271**, 163.
30. A. S. Reddy and G. N. Sastry, *J. Phys. Chem. A*, 2005, **109**, 8893.
31. J. P. Gallivan and D. A. Dougherty, *Proc. Natl. Acad. Sci. U.S.A.*, 1999, **96**, 9459.

32. L. K. Tsou, C. D. Tatko and M. L. Waters, *J. Am. Chem. Soc.*, 2002, **124**, 14917.
33. R. Wintjens, J. Liévin, M. Rooman and E. Buisine, *J. Mol. Biol.*, 2000, **302**, 395.
34. M. Aschi, F. Mazza and A. Di Nola, *J. Mol. Struct. (Theochem)*, 2002, **23**, 281.
35. T. W. Allen, O. S. Andersen and B. Roux, *Proc. Natl. Acad. Sci. U.S.A.*, 2004, **101**, 117.
36. A. Warshel, M. Kato and A. V. Pisliakov, *J. Chem. Theory Comput.*, 2007, **3**, 2034.
37. T. A. Halgren and W. Damm, *Curr. Opin. Struct. Biol.*, 2001, **11**, 236.
38. P. E. M. Lopes, B. Roux and A. D. MacKerell Jr., *Theor. Chem. Acc.*, 2009, **124**, 11.
39. J. R. Maple, Y. Cao, W. Damm, T. A. Halgren, G. A. Kaminski, L. Y. Zhang and R. A. Friesner, *J. Chem. Theory Comput.*, 2005, **1**, 694.
40. B. Ma, J.-H. Lii and N. L. Allinger, *J. Comput. Chem.*, 2000, **21**, 813.
41. G. Lamoureux and B. Roux, *J. Chem. Phys.*, 2003, **119**, 3025.
42. P. Cieplak, J. Caldwell and P. Kollman, *J. Comput. Chem.*, 2001, **22**, 1048.
43. W. Xie, J. Pu, A. D. MacKerell Jr. and J. Gao, *J. Chem. Theory Comput.*, 2007, **3**, 1878.
44. G. A. Kaminski, *J. Phys. Chem. B*, 2005, **109**, 5884.
45. C. M. Macdermaid and G. A. Kaminski, *J. Phys. Chem. B*, 2007, **111**, 9036.
46. H. A. Stern, G. A. Kaminski, J. L. Banks, R. Zhou, B. J. Berne and R. A. Friesner, *J. Phys. Chem. B*, 1999, **103**, 4730.
47. R. Chelli, P. Procacci, R. Righini and S. Califano, *J. Chem. Phys.*, 1999, **111**, 8569.
48. J. L. Banks, G. A. Kaminski, R. Zhou, D. T. Mainz, B. J. Berne and R. A. Friesner, *J. Chem. Phys.*, 1999, **110**, 741.
49. S. Patel and C. L. Brooks, *J. Comput. Chem.*, 2004, **25**, 1.
50. S. Patel, A. D. MacKerell Jr. and C. L. Brooks, *J. Comput. Chem.*, 2004, **25**, 1504.
51. P. Drude, *The Theory of Optics*, Green, New York, NY, 1902.
52. G. Lamoureux, A. D. MacKerell and B. Roux, *J. Chem. Phys.*, 2003, **119**, 5198.
53. G. Lamoureux, E. Harder, I. V. Vorbyov, B. Roux and A. D. MacKerell Jr., *Chem. Phys. Lett.*, 2006, **418**, 245.
54. P. J. van Maaren and D. van der Spoel, *J. Phys. Chem. B*, 2001, **105**, 2618.
55. M. Sprik and M. L. Klein, *J. Chem. Phys.*, 1988, **89**, 7556.
56. B. T. Thole, *Chem. Phys.*, 1981, **59**, 341.
57. S. Y. Noskov, G. Lamoureux and B. Roux, *J. Phys. Chem. B*, 2005, **109**, 6705.
58. P. T. van Duijnen and M. Swart, *J. Phys. Chem. A*, 1998, **102**, 2399.
59. E. Harder, V. M. Anisimov, T. Whitfield, A. D. MacKerell Jr. and B. Roux, *J. Phys. Chem. B*, 2008, **112**, 3509.
60. P. E. M. Lopes, G. Lamoureux and A. D. MacKerell Jr., *J. Comput. Chem.*, 2009, **30**, 1821.

61. E. Harder, V. M. Anisimov, I. V. Vorbyov, P. E. M. Lopes, S. Y. Noskov, A. D. MacKerell Jr. and B. Roux, *J. Chem. Theory Comput.*, 2006, **2**, 1587.
62. R. W. Dixon and P. A. Kollman, *J. Comput. Chem.*, 1997, **18**, 1632.
63. H. Yu, T. W. Whitfield, E. Harder, G. Lamoureux, I. Vorobyov, V. M. Anisimov, A. D. MacKerell Jr. and B. Roux, *J. Chem. Theory Comput.*, 2010, **6**, 774.
64. C. M. Baker, P. E. M. Lopes, X. Zhu, B. Roux and A. D. MacKerell Jr., *J. Chem. Theory Comput.*, 2010, **6**, 1181.
65. P. A. Sims, C. F. Wong, D. Vuga, J. A. McCammon and B. M. Sefton, *J. Comput. Chem.*, 2005, **26**, 668.
66. J. E. Davis, G. L. Warren and S. Patel, *J. Phys. Chem. B*, 2008, **112**, 8298.
67. J. E. Davis and S. Patel, *Chem. Phys. Lett.*, 2010, **484**, 173.
68. Y. Zhong and S. Patel, *J. Phys. Chem. B*, 2010, **114**, 11076.
69. M. P. Allen and D. J. Tildesley, *Computer Simulation of Liquids*, Oxford University Press, New York, NY, 1st edn, 1987, 21.
70. T. A. Halgren, *J. Am. Chem. Soc.*, 1992, **114**, 7827.
71. J. Delhommelle and P. Millie, *Mol. Phys.*, 2001, **99**, 619.
72. W. Song, P. J. Rossky and M. Maroncelli, *J. Chem. Phys.*, 2003, **119**, 9145.
73. K. J. Miller, *J. Am. Chem. Soc.*, 1990, **112**, 8533.
74. V. M. Anisimov, G. Lamoureux, I. V. Vorobyov, N. Huang, B. Roux and A. D. MacKerell Jr., *J. Chem. Theory Comput.*, 2005, **1**, 153.
75. B. R. Brooks, C. L. Brooks III, A. D. MacKerell Jr., L. Nilsson, R. J. Petrella, B. Roux, Y. Won, G. Archontis, C. Bartels, S. Boresch, A. Caflisch, L. Caves, Q. Cui, A. R. Dinner, M. Feig, S. Fischer, J. Gao, M. Hodoscek, W. Im, K. Kuczera, T. Lazaridis, J. Ma, V. Ovchinnikov, E. Paci, R. W. Pastor, C. B. Post, J. Z. Pu, M. Schaefer, B. Tidor, R. M. Venable, H. L. Woodcock, X. Wu, W. Yang, D. M. York and M. Karplus, *J. Comput. Chem.*, 2009, **30**, 1545.
76. C. M. Baker and A. D. MacKerell Jr., *J. Mol. Model.*, 2010, **16**, 567.
77. G. A. Kaminski, H. A. Stern, B. J. Berne and R. A. Friesner, *J. Phys. Chem. A*, 2004, **108**, 621.
78. B. Schropp and P. Tavan, *J. Phys. Chem. B*, 2008, **112**, 6233.
79. E. Harder, V. M. Anisimov, I. V. Vorobyov, P. E. M. Lopes, S. Y. Noskov, A. D. MacKerell Jr. and B. Roux, *J. Chem. Theory Comput.*, 2006, **2**, 1587.
80. F. H. Allen, *Acta. Crystallogr. B*, 2002, **58**, 380.
81. O. Guvench and A. D. MacKerell Jr., *J. Mol. Model.*, 2008, **14**, 667.
82. C. M. Baker, V. M. Anisimov and A. D. MacKerell Jr., *J. Phys. Chem. B*, 2011, **115**, 580.
83. M. O. Sinnokrot and C. D. Sherrill, *J. Phys. Chem. A*, 2004, **108**, 10200.
84. D. Yin and A. D. MacKerell Jr., *J. Comput. Chem.*, 1998, **19**, 334.
85. K. Vannommeslaeghe, E. Hatcher, C. Acharya, S. Kundu, S. Zhong, J. Shim, E. Darian, O. Guvench, P. Lopes, I. Vorobyov and A. D. MacKerell Jr., *J. Comput. Chem.*, 2010, **31**, 671.
86. W. Jiang, D. J. Hardy, J. C. Phillips, A. D. MacKerell Jr., K. Schulten and B. Roux, *J. Phys. Chem. Lett.*, 2011, **2**, 87.

87. I. V. Vorobyov, V. M. Anisimov and A. D. MacKerell Jr., *J. Phys. Chem. B*, 2005, **109**, 18988.
88. V. M. Anisimov, I. V. Vorobyov, B. Roux and A. D. MacKerell Jr., *J. Chem. Theory Comput.*, 2007, **3**, 1927.
89. P. E. M. Lopes, G. Lamoureux, B. Roux and A. D. MacKerell Jr., *J. Phys. Chem. B*, 2007, **111**, 2873.
90. X. Zhu and A. D. MacKerell Jr., *J. Comput. Chem.*, 2010, **31**, 2330.
91. I. Vorobyov, V. M. Anisimov, S. Greene, R. M. Venable, A. Moser, R. W. Pastor and A. D. MacKerell Jr., *J. Chem. Theory Comput.*, 2007, **3**, 1120.
92. E. Harder, A. D. MacKerell Jr. and B. Roux, *J. Am. Chem. Soc.*, 2009, **131**, 2760.
93. R. J. Roberts and X. Cheng, *Annu. Rev. Biochem.*, 1998, **67**, 181.
94. W. L. De Laat, N. G. J. Jaspers and J. H. J. Hoeijmakers, *Gene. Dev.*, 1999, **13**, 768.
95. N. K. Banavali and A. D. MacKerell Jr., *J. Mol. Biol.*, 2002, **319**, 141.
96. N. Huang, N. K. Banavali and A. D. MacKerell Jr., *Proc. Natl. Acad. Sci. U.S.A.*, 2023, **100**, 68.
97. S. Kumar, D. Bouzida, R. H. Swendsen, P. A. Kollman and J. M. Rosenberg, *J. Comput. Chem.*, 1992, **13**, 1011.
98. Student, *Biometrika*, 1908, **6**, 1.

CHAPTER 4

Integral Equation Theory of Biomolecules and Electrolytes

TYLER LUCHKO, IN SUK JOUNG AND DAVID A. CASE*

Department of Chemistry and Chemical Biology and BioMaPS Institute Rutgers University Piscataway, NJ 08854, USA
*Email: case@biomaps.rutgers.edu

4.1 Introduction

Computer simulations of biomolecules often use a simplified, implicit description of solvation effects, rather than explicitly representing individual water and ion molecules.[1] The most widely-adopted model treats water as a dielectric continuum (describing the average polarization of solvent dipoles arising from the charges in the biomolecule); the effects of mobile counter-ions are represented by a continuous charge density determined by the local electrostatic field. The resulting "Poisson-Boltzmann" equations can be solved numerically on a three-dimensional grid, or approximated in a variety of ways.[2–4] Such models are often very useful, but have a variety of limitations: they drastically average the response of water dipoles and ions to the fields created by solutes, miss most effects of atomic and molecular sizes, collapse all ion effects into a single ionic strength parameter, and fail to account for non-electrostatic aspects of solvation.

In many situations these approximations may be too severe, in which case a natural remedy is to treat the biomolecule, water and ions on the same, explicit level. Monte Carlo or molecular dynamics (MD) simulations of electrolytes, using empirically-calibrated force fields, (with or without biomolecular solutes)

RSC Biomolecular Sciences No. 23
Innovations in Biomolecular Modeling and Simulations: Volume 1
Edited by Tamar Schlick

Published by the Royal Society of Chemistry, www.rsc.org

are the most detailed and accurate methods for the theoretical study. Equilibrium simulations of bulk liquids, in principle, can be used to calculate all of the properties of the liquid at that thermodynamic state, but such simulations may require hundreds of ns of simulation time to converge.[5] Depending on the concentration and the precision desired, weeks or months of CPU time are necessary. Enhanced sampling techniques can ameliorate but not fully solve this problem. For example, sophisticated sampling techniques can be used to calculate the potential of mean force (PMF) between any two atomic sites in a finite concentration NaCl solution[6] but there are 10 site-site PMF combinations, so 10 different simulations are required.

There is a long-studied alternative approach to understanding the equilibrium properties of water and ions, based on the integral equation approach of Ornstein and Zernike.[7–9] These ideas were originally applied to atomic liquids, but have been extended to molecular solvents such as water by a variety of methods, most notably *via* the reference interaction site model (RISM, described below). The so-called three-dimensional version (3D-RISM) can be used to describe the interactions of solvent components (here we treat water and ions) with a chemical or biomolecular solute of arbitrary size and shape. Here we give an overview of the current status of such models, describing some aspects of "pure" electrolytes (water plus simple ions),[10,11] and of ionophores, proteins and nucleic acids in the presence of water and salts.[1] The applications of integral equation theories to study biomolecules in pure water have been reviewed elsewhere.[12–14] Here we focus primarily on interactions with water and dissolved salts; as a practical matter, the discussion is mostly limited to monovalent ions, since studies of divalent ions present many difficult problems that have not yet been addressed. This is not a comprehensive review, but covers a few recent examples that illustrate current issues.

4.1.1 Some History

Theories at the "primitive" level offer a very simplified picture in which an ion is defined as hard-sphere potential with a central charge embedded in a uniform dielectric medium.[10,15] In the limit of low concentration (<0.1 M) and an effective hard-sphere diameter $\sigma \ll \rho^{-1/3}$, where ρ is the number density, it is assumed that the ionic core has negligible contribution to the long-range radial distribution function (RDF). There is no analytic solution to even this simple model, and the most rigorous numerical solutions come from Monte Carlo simulations.[16–18] The linear and non-linear Poisson–Boltzmann (PB) are probably the most common approximations.[2–4,19,20]

The atomic Ornstein–Zernike (OZ) equation,[7,9] in which there is no orientational dependence, is sufficient for the primitive model but requires approximation of the so-called closure relation. The mean spherical approximation (MSA), which assumes a hard-sphere interaction and the correct long-range asymptotic behavior, is one such closure approximation and was solved for electrolyte solutions by Waisman and Lebowitz[21,22] and later extended to other potentials.[23] MSA is the simplest theory that accounts for hard-sphere ion-ion correlations, giving a structured RDF at sufficiently high concentrations. Observable thermodynamic quantities, such as osmotic coefficients, calculated from the theory compare well to Monte Carlo simulation though are not as good as the hyper-netted chain equation (HNC) closure.

HNC[24] is, arguably, the most successful closure for the restricted primitive model (where all ions have the same size) and is able to calculate many observables to a few percent of simulation.[16,25–30] The major shortcoming of this approximation is overstating the effect of attractive interactions while neglecting some repulsive interactions. As a result, there fails to be solutions in some regions of the phase diagram where they should exist, such as where compressibility changes quickly,[31,32] at very low concentrations or when interactions are very strong.[19,33,34] Even if a solution is found it may be incorrect; for divalent ions poor results are seen for concentrations less than 0.06 M. Nevertheless, the HNC description of the primitive model has been used to provide insights to a variety of problems, including ionic atmospheres around simple models of globular proteins (spheres)[35–38] or DNA (cylinders)[39] and suggests this may also be a good approach for more realistic models.

To go beyond the primitive model to the level of detail present in modern MD simulations requires including a molecular representation of water; a task that cannot be done in practice without some approximation to the OZ equation itself. Several statistical mechanical approaches have been developed that can, in principle, treat both water and ions at the same level, such as the diagrammatically proper interaction site model of Chandler–Silbey–Ladanyi[40,41] with extensions,[42,43] molecular-OZ[44,45] and density functional approaches.[46]

The extended reference interaction site model (XRISM),[47] in which orientational degrees of freedom are averaged over, was the first OZ-like theory to be applied to liquid water[48] and the study of ions at infinite dilution.[49,50] However, several problems were encountered with the dielectric constant predicted by XRISM, which was too small and inconsistent when ions at finite concentration were introduced. Still, finite concentration calculations have been done at very high temperature and pressure.[51–53]

These issues were addressed by dielectrically consistent RISM (DRISM),[54,55] which enforces a user input dielectric constant. As with XRISM, DRISM theory provides a detailed microscopic description of electrolyte solutions from which a rich array of thermodynamic observables can be computed, including osmotic coefficients,[50] partial molar volumes,[13,56–58] solvation entropies,[57,59–61] phase separation[62] and solute salting out.[63] Extensions to XRISM and DRISM have also enabled the study of time-dependent phenomena,[57,64–66] complex solvents[67,68] and been coupled with quantum mechanical methods.[69–71] As both XRISM and DRISM orientationally average all interactions such that only a separation dependence remains, they are collectively known as one-dimensional RISM (1D-RISM).

For many biological and industrial applications, the behavior of macromolecular solutes in electrolyte solvents is important and a number of 1D-RISM methods have been developed to this end. Examples include treating the solvent and entire macromolecule with 1D-RISM,[72–74] neglecting or coarse-graining parts of the solute,[75–77] applying the RISM equations to subunits in isolation[78,79] or by using RDFs from MD simulations in RISM thermodynamic expressions.[80] Such methods are generally not readily extensible to complex solutes without loss of detail. They also suffer from limitations of 1D-RISM in the accuracy of various properties, such as partial molar volumes, due to the loss of orientational information.

3D-RISM overcomes many of these limitations by only performing orientational averaging on the solvent,[81–84] making it an attractive choice for studying complex macromolecules at infinite dilution. In this approach the bulk solution is described at the DRISM level and a full 3D, atomistic description of the solute using conventional force fields is treated at the 3D-RISM level. Such an approach yields more accurate results, even for simple ions,[85–87] though it is generally applied to more complex solutes. As well as biomolecules, 3D-RISM has been used for a wide variety of different applications. For electrolyte solutions it has been used for molecular ions,[88] ion distribution along an infinite wall[89] and the study of carbon nanotube capacitors.[90] More generally, it has been used for crude oil,[91] room-temperature ionic liquids[92,93] and atomic force microscopy simulations.[94]

While complete equilibrium sampling of the solvent is performed by 3D-RISM, the solute is completely frozen and sampling of the solute degrees of freedom has to be undertaken in a separate step. 3D-RISM has been coupled using various schemes to quantum chemical calculations[84,95–98] and molecular dynamics (MD).[99,100] The MD implementations have also been extended to incorporate thermodynamic integration[101,102] and free energy perturbation (FEP).[101] 3D-RISM has also been used within the so-called MM/PBSA framework, MM/3D-RISM.[103,104]

This chapter gives an overview of the application of DRISM and 3D-RISM to simple, monovalent electrolyte solvents and their interaction with macromolecular solutes, using a few simple examples that qualitatively replicate various works in the literature. In Section 2 the relevant parts of the theory are outlined. Thermodynamic and structural properties of a pure electrolyte solvent, as described by DRISM, are discussed in Section 3. This is followed with Section 4, where we review several examples of complex macromolecules in ionic solutions.

4.2 Integral Equation Theory for Simple Liquids

4.2.1 Ornstein–Zernike Equation

In homogenous fluids, the spatial number density distribution, $\rho(\mathbf{r})$, is uniform and provides little information. In contrast, the number density of a particle, 2, relative to a fixed particle, 1, contains a wealth of information and, in the grand canonical ensemble, is given by

$$\rho^{(2)}(\mathbf{1},\mathbf{2}) = \frac{1}{\Xi}\sum_{N=2}^{\infty}\frac{z^N}{(N-2)!}\int \exp[-\beta V_N]d\mathbf{r}^{N-2} \tag{4.1}$$

where the position and orientation of molecular species are denoted by numbers (*e.g.* $\mathbf{1} = (\mathbf{r}_1, \Omega_1)$), $\beta = 1/k_bT$, k_b is Boltzmann's constant, T is the temperature, Ξ is the grand partition function, z^N and V_N are the N-particle partition function and potential. For additional information. the reader is referred to J.-P. Hansen and I. R. McDonald (1990).[105] For homogeneous fluids, the 2-particle density distribution is related to the pair distribution function (PDF), $g(\mathbf{1},\mathbf{2})$, by

$$\rho^{(2)}(\mathbf{1},\mathbf{2}) = \rho_1\rho_2 g(\mathbf{1},\mathbf{2}) \tag{4.2}$$

where ρ_1 and ρ_2 are the bulk number densities of particles 1 and 2. When orientational dependence is averaged out, $g(r)$ is known as the radial distribution function (RDF). Alternately, the PDF is also related to the potential of mean force, $w(\mathbf{1},\mathbf{2})$,

$$h(\mathbf{1},\mathbf{2}) + 1 = g(\mathbf{1},\mathbf{2}) = \exp(-\beta w(\mathbf{1},\mathbf{2})) \tag{4.3}$$

where we have also introduced the total correlation function (TCF), $h(\mathbf{1},\mathbf{2})$. Ornstein and Zernike had the insight to partition the TCF into direct interactions between the two particles and interactions mediated by the surrounding environment.[7–9] The result for homogenous, multicomponent, molecular liquids is the Ornstein–Zernike (OZ) integral equation

$$h_{ij}(\mathbf{1},\mathbf{2}) = c_{ij}(\mathbf{1},\mathbf{2}) + \sum_k \rho_k \int c_{ik}(\mathbf{1},\mathbf{3}) h_{kj}(\mathbf{3},\mathbf{2})\, d\mathbf{3} \tag{4.4}$$

where we denote molecular species by Latin letters (i, j, and k), $c(\mathbf{1},\mathbf{2})$ is the direct correlation function (DCF) and the integration is performed over all space. Eq. (4.4) is, in fact, the definition of the DCF. Physically, we can interpret the TCF as the sum of contributions from the direct interaction of the two particles (DCF) plus the interactions mediated by the surrounding particles (the right-hand-side summation). As both h and c are unknown functionals, a second, closure equation is required to find a solution.

4.2.2 Closures

The most general case for the closure equation is

$$g_{\alpha\gamma} = \exp\{-\beta u_{\alpha\gamma} + h_{\alpha\gamma} - c_{\alpha\gamma} + b_{\alpha\gamma}\} \tag{4.5}$$

where b is the so-called bridge function and we have dropped the functional arguments for brevity and generality. In practice, b is not known in closed form but only as an infinite series of functionals and is always subject to some approximation.[9] The number of closure approximations introduced in the literature is far to numerous to list here. Instead, we will restrict our attention to the hypernetted-chain equation (HNC)[24] and the related Kovalenko–Hirata (KH)[84] and partial series expansion of order-n (PSE-n)[106] closures, which have been the most widely used in 3D-RISM calculations.

In the HNC approximation, the bridge function is set to zero

$$g_{\alpha\gamma} = \exp\{-\beta u_{\alpha\gamma} + h_{\alpha\gamma} - c_{\alpha\gamma}\} \tag{4.6}$$

and may be obtained from a variety of mathematically and physically based approaches.[9] This has been found to produce very good results for ionic[16,55,107,108] and polar systems.[47,48,109] It also has an exact, closed form expression for the chemical potential when coupled with RISM theory.[110] However, it does have drawbacks, including thermodynamic inconsistencies,[111,112] poor results for neutral systems,[9,111] difficulties with particle size asymmetries[36] and difficulties converging solutions. To address the issue of convergence, Kovalenko and Hirata[84] developed a partially linearized closure. Regions of enhanced density were linearized, avoiding the exponential density response for strong potential interactions. This linearization was later generalized to a Taylor series[106]

$$g_{\alpha\gamma} = \begin{cases} \exp\left\{t^*_{\alpha\gamma}\right\} & \text{for } t^*_{\alpha\gamma} < 0 \\ \sum_{i=0}^{n} \frac{\left(t^*_{\alpha\gamma}\right)^i}{i!} & \text{for } t^*_{\alpha\gamma} \geq 0 \end{cases} \tag{4.7}$$

$$t^*_{\alpha\gamma} = -\beta u_{\alpha\gamma} + h_{\alpha\gamma} - c_{\alpha\gamma}$$

For $n = 1$ the KH closure is recovered and HNC is the limiting case as $n \to \infty$.

4.2.3 1D-RISM

Most modern biomolecular forcefields use interaction site models in which a molecule is composed of a number of sites, typically atoms, that interact in a pair-wise fashion. Such models offer a very effective way to deal with non-spherical molecules but require a practical method to apply Eq. (4.4) to molecular species with multiple sites. One approach (which, in practice, is restricted to molecules with a small number of sites) is to treat the molecules as rigid and site–site orientationally average the correlation functions for each site, reducing the equations to one dimension, *i.e.* orientational averaging is done about each site rather than, for example, averaging about the molecular center-of-mass. In the RISM approach this is achieved by first treating the DCF as decomposable into the sum of site–site direct correlation functions.[113,114]

$$c(\mathbf{1},\mathbf{2}) = \sum_{\alpha_1\gamma_2} c_{\alpha_1\gamma_2}\left(\left|\mathbf{r}_{\alpha_1} - \mathbf{r}_{\gamma_2}\right|\right) \tag{4.8}$$

where $c(\mathbf{1},\mathbf{2})$ is the DCF between molecules 1 and 2, and Greek letters (α and γ) are interaction sites on molecules 1 and 2, respectively. Molecules are assumed to be rigid, and their shape enters the theory through the intramolecular correlation matrix, represented in reciprocal space,

$$\hat{\omega}_{\alpha\gamma}(k) = \delta_{\alpha\gamma} + \left(1 - \delta_{\alpha\gamma}\right)\frac{\sin\left(kl_{\alpha\gamma}\right)}{kl_{\alpha\gamma}} \tag{4.9}$$

where δ is the Kronecker-δ function and $l_{\alpha\gamma}$ is the distance between sites in the same molecular species. For the same site, $\alpha=\gamma$, $l_{\alpha\alpha}=0$ and $\hat{\omega}_{\alpha\alpha}(k)=1$, while for sites belonging to different molecular species $\hat{\omega}_{\alpha\gamma}(k)=0$.

With the definition of the intramolecular correlation function, we can now express Eq. (4.4) in terms of interaction sites rather than molecules. The multicomponent 1D-RISM equation can be written explicitly for molecules 1 and 2 as

$$\begin{aligned} \rho_\alpha h_{\alpha\gamma}(r)\rho_\gamma = &\sum_{\lambda}^{N_{\text{site}}}\sum_{\beta}^{N_{\text{site}}} \omega_{\alpha\lambda}(r) * c_{\lambda\beta}(r) * \omega_{\beta\gamma}(r) \\ &+ \sum_{\lambda}^{N_{\text{site}}}\sum_{\beta}^{N_{\text{site}}} \omega_{\alpha\lambda}(r) * c_{\lambda\beta}(r) * \rho_\beta h_{\beta\gamma}(r)\rho_\gamma \end{aligned} \tag{4.10}$$

where * is the convolution operator and N_{site} is the total number of sites from all molecular species. Eq. (4.10) is generally written in the more compact matrix form

$$\begin{aligned} \boldsymbol{\rho}\mathbf{h}\boldsymbol{\rho} &= \boldsymbol{\omega} * \mathbf{c} * \boldsymbol{\omega} + \boldsymbol{\omega} * \mathbf{c} * \boldsymbol{\rho}\mathbf{h}\boldsymbol{\rho} \\ &= \left[1 - \boldsymbol{\omega} * \mathbf{c}\right]^{-1} \boldsymbol{\omega} * \mathbf{c} * \boldsymbol{\omega} \end{aligned} \tag{4.11}$$

with ρ being a diagonal matrix of scalar values, ω and $\mathbf{c}$ are matrices of radially dependent functions and all matrices are of size $N_{\text{site}} \times N_{\text{site}}$.

Eq. (4.11) works well for molecules limited to short-range interactions. Long-range (*i.e.* Coulombic) interactions cannot be directly Fourier transformed due to divergence at small k. To deal with such systems, the extended RISM (XRISM) method was developed utilizing a renormalization procedure.[47,109] The renormalization process still has numerical difficulties so an alternate approach, which analytically handles the long range asymptotics of the correlation functions, is now commonly used, and is discussed below (§2.5).

Even with correct handling of long-range interactions, XRISM gives the same trivial expression for the static dielectric constant for any local site-site closure with the asymptotics of Eq. (4.12),[115–117]

$$\epsilon = 1 + 3y = 1 + \frac{4}{3}\pi\beta\sum_i \rho_i\mu_i \tag{4.12}$$

where y is the dipole density and μ_i is the dipole moment of species i. Direct application of this equation to SPC/E models gives $\varepsilon < 20$ rather 71, given by simulation,[118] or 78.5, from experiment. This poor behavior is due to the long range asymptotics of HNC-like closures. To remedy this, Perkyns and Pettitt constructed a bridge-like correction to the hypernetted chain equation (HNC) that also provides consistency between solvent and solute routes to the dielectric constant, so-called dielectrically consistent RISM (DRISM).[54,55] While constructed as a bridge function, the correction is implemented in the XRISM integral equation such that Eq. (4.11) takes the form

$$\boldsymbol{\rho h \rho} = [1 - (\boldsymbol{\omega} + \boldsymbol{\chi}) * \mathbf{c}]^{-1}(\boldsymbol{\omega} + \boldsymbol{\chi}) * \mathbf{c} * (\boldsymbol{\omega} + \boldsymbol{\chi}) + \boldsymbol{\chi}. \tag{4.13}$$

$\boldsymbol{\chi}$ is bridge-like correction that corrects the long range behavior of the distribution functions such that they are consistent with the dipole moment density of the fluid and the desired dielectric constant input by the user. Note that the input dielectric constant should be that of the pure dipolar liquid without ions as the decrease in the dielectric arising from a finite concentration of mobile ions is implicitly included in the theory.[54,55,60]

DRISM also satisfies the Stillinger–Lovett moment conditions that must hold for any description of ionic solutions[119,120]

$$M_0 = -\sum_i \rho_i q_i^2 \tag{4.14}$$

$$M_2 = -\frac{3\epsilon}{2\pi\beta} \tag{4.15}$$

$$M_n = \sum_{i,j} \rho_i \rho_j q_i q_j \int h_{ij} r^n d\mathbf{r}. \tag{4.16}$$

where i and j are charged species. The zero moment, M_0, is essentially the charge neutrality condition while the second moment, M_2, provides a route to the dielectric constant.

Several alternate remedies to deal with the dielectric behavior have been proposed. Earlier attempts, sometimes known as ARISM, involved a scaling constant, A, for electrostatic interactions.[51,52,60,61,121] While sufficient for infinitely dilute ions, this approach suffered dielectric inconsistencies and failed to achieve the desired results for finite salt concentrations. In more recent approaches, several bridge functions have been derived that give non-trivial predictions for the dielectric constant within the RISM framework but these have not been widely used.[117,122]

4.2.4 3D-RISM

For macromolecular ions, which are composed of more than a few sites, the approximation of spherically symmetric distribution functions begins to break down. One approach is to use a full 3D description of the macromolecular solute, U, while using orientationally averaged distributions for the solvent, V. If the solute is at infinite dilution,[113] Eq. (4.4) can be rewritten as

$$h_{ij}^{\mathrm{VV}}(\mathbf{i},\mathbf{j}) = c_{ij}^{\mathrm{VV}}(\mathbf{i},\mathbf{j}) + \sum_k \rho_k^{\mathrm{V}} \int_{\Omega} c_{ik}^{\mathrm{VV}}(\mathbf{i},\mathbf{k}) h_{kj}^{\mathrm{VV}}(\mathbf{k},\mathbf{j})\, d\mathbf{k} \tag{4.17}$$

$$h_i^{\mathrm{UV}}(\mathbf{1},\mathbf{i}) = c_i^{\mathrm{UV}}(\mathbf{1},\mathbf{i}) + \sum_k \rho_k^{\mathrm{V}} \int_{\Omega} c_k^{\mathrm{UV}}(\mathbf{1},\mathbf{k}) h_{ki}^{\mathrm{VV}}(\mathbf{k},\mathbf{j})\, d\mathbf{k} \tag{4.18}$$

Eq. (4.17) gives the TCF of the bulk solvent, which is then used in Eq. (4.18) to obtain the distribution of the solvent about the solute. Note that in Eq. (4.18) particle 1 is the solute and particles i and j are solvent sites. Whereas for Eq. (4.17) there is an TCF and DCF for each site-site pair in the solvent, in Eq. (4.18) there is one TCF and DCF for each solvent site about the solute.

For a solvent composed of simple molecules like water and monovalent ions, Eq. (4.17) can be approximated by Eq. (4.13). This orientationally averaged description of the bulk solvent, coupled with the fundamental RISM approximation (Eq. (4.8)), gives the 3D-RISM equation[83,123]

$$h_\gamma^{\mathrm{UV}}(\mathbf{r}) = \sum_\alpha \int c_\alpha^{\mathrm{UV}}\left(\mathbf{r} - \mathbf{r}'\right) \chi_{\alpha\gamma}^{\mathrm{VV}}\left(r'\right) d\mathbf{r}' \tag{4.19}$$

where

$$\chi_{\alpha\gamma}^{\mathrm{VV}}(r) = \omega_{\alpha\gamma}^{\mathrm{VV}}(r) + \rho_\alpha^{\mathrm{V}} h_{\alpha\gamma}^{\mathrm{VV}}(r). \tag{4.20}$$

All information about the bulk solvent, including the dielectric behavior, is contained in $\chi_{\alpha\gamma}^{\mathrm{VV}}(r)$.

4.2.5 Long Range Asymptotics

Long-range interactions must be handled carefully and a number of alternatives to the renormalization procedure used in XRISM have been developed.[84,85,87,123–125] The method used here of analytically handling the long range asymptotics was originally developed for the OZ equation.[126,127] Here we present the 3D expressions from which 1D versions are easily obtained.[104,128]

For large r, the DCF has the asymptotics of

$$c(r) \to -\beta u(r) \approx -\beta u_{\mathrm{es}}(r) \text{ as } r \to \infty \tag{4.21}$$

where u is the pair potential and u_{es} is the electrostatic contribution. The electrostatic part of the potential is then partitioned in such a manner that the Fourier transform of the long-range component is easily found. For example,

$$u_\gamma^{(\mathrm{lr})}(\mathbf{r}) = u_\gamma - \sum_a \frac{Q_a^{\mathrm{U}} q_\gamma}{|\mathbf{r}-\mathbf{R}_a|}\,\mathrm{erfc}\left(\frac{|\mathbf{r}-\mathbf{R}_a|}{\eta}\right) \tag{4.22}$$

where q_γ is the partial charge on solvent site γ; Q_a^{U} and $\mathbf{R}_a$ are the partial charge and position of solute site a; η is a charge smearing parameter and erfc() is the complimentary error function. We can then write

$$c_\gamma^{(\mathrm{lr})}(\mathbf{r}) = -\beta u_\gamma^{(\mathrm{lr})}(\mathbf{r}) = -\frac{1}{k_b T}\sum_a \frac{Q_a^{\mathrm{U}} q_\gamma}{|\mathbf{r}-\mathbf{R}_a|}\,\mathrm{erf}\left(\frac{|\mathbf{r}-\mathbf{R}_a|}{\eta}\right) \tag{4.23}$$

$$c_\gamma^{(\mathrm{lr})}(\mathbf{k}) = -\frac{4\pi}{k_b T}\sum_a \frac{Q_a^{\mathrm{U}} q_\gamma}{k^2}\exp\left(-\frac{k^2\eta^2}{4} + i\mathbf{k}\cdot\mathbf{R}_a\right). \tag{4.24}$$

Eqs (4.23) and (4.24), together with Eq. (4.11) define the asymptotic form of h. However, the form of this is complicated and not suitable to handle analytically.[128] Rather, a simple equation with the correct asymptotic form is sufficient

$$\begin{aligned} h_j^{(\mathrm{lr})}(\mathbf{r}) = -\frac{1}{2\epsilon k_b T}\sum_a \frac{Q_a q_j}{|\mathbf{r}-\mathbf{R}_a|}\exp\left(\frac{\kappa_{\mathrm{D}}^2\eta^2}{4}\right) & \\ \left[\exp(-\kappa_{\mathrm{D}}|\mathbf{r}-\mathbf{R}_a|)\,\mathrm{erfc}\left(\frac{\kappa_{\mathrm{D}}\eta}{2} - \frac{|\mathbf{r}-\mathbf{R}_a|}{\eta}\right)\right. & \\ \left. -\exp(\kappa_{\mathrm{D}}|\mathbf{r}-\mathbf{R}_a|)\,\mathrm{erfc}\left(\frac{\kappa_{\mathrm{D}}\eta}{2} + \frac{|\mathbf{r}-\mathbf{R}_a|}{\eta}\right)\right] & \end{aligned} \tag{4.25}$$

$$h_j^{(\mathrm{lr})}(\mathbf{k}) = -\frac{4\pi}{\epsilon k_b T}\sum_a \frac{Q_a q_j}{k^2 + \kappa_{\mathrm{D}}^2}\exp\left(-\frac{k^2\eta^2}{4} + i\mathbf{k}\cdot\mathbf{R}_a\right) \tag{4.26}$$

where $\kappa_{\mathrm{D}} = \sqrt{4\pi\beta\epsilon\rho_j q_j^2}$ is the contribution to the inverse Debye length of ionic species j. Note that the TCF long-range asymptotics are only applied for ionic solvent species.

These equations are applied before and after the Fourier transforms while iteratively solving the RISM equations. That is, before transforming $c_\gamma(\mathbf{r})$ to reciprocal space, the long-range component, Eq. (4.23), is subtracted, leaving only the well-behaved short-range component. After the transform, the long-range component, Eq. (4.24), is added back and $h_\gamma(\mathbf{k})$ is computed. Transforming $h_\gamma(\mathbf{k})$ back to real space is done in an analogous manner. In 1D-RISM the divergence for $k = 0$ is handled analytically by omitting $k = 0$

and $r = 0$ from the numerical solution and then extrapolating their values from the converged solution. This extrapolated value is then applied in 3D-RISM.

4.2.6 Solvation Free Energy

Among the most useful thermodynamic properties for a solute/solvent system is the solvation free energy, ΔG. In the case of infinite dilution the solvation free energy is the excess chemical potential, $\Delta\mu$. In general, this is calculated using the Kirkwood charging formula where a coupling parameter, λ, varies from 0 to 1 and controls the interaction between the solute and solvent, an approach first used by Morita and Hiroike.[129] For a solute site, α,

$$\Delta G_\alpha = \Delta\mu_\alpha = \sum_\gamma \rho_\gamma \int_0^1 d\lambda \int d\mathbf{r} \frac{\partial u_\gamma^{\mathrm{UV}}(\mathbf{r};\lambda)}{\partial\lambda} g_\gamma^{\mathrm{UV}}(\mathbf{r};\lambda). \tag{4.27}$$

Numerical integration of this formula requires solutions for several values of λ. For simplicity and computational efficiency, we would like to obtain $\Delta\mu$ from a single calculation with $\lambda = 1$. However, the explicit dependence on g implies that such an expression will be closure dependent and may not even exist for a given closure. That is, if there is no exact differential for Eq. (4.27) then $\Delta\mu$ is path dependent and not a state function. For XRISM coupled with HNC,[110] KH[84] and PSE-n[106] closures the integration can be done analytically for a solute at infinite dilution. For the PSE-n closure, from which HNC and KH expressions are easily obtained,

$$\Delta\mu_\alpha = k_\mathrm{b}T \sum_\lambda \rho_\lambda \int \frac{h_{\alpha\lambda}^2}{2} - c_{\alpha\lambda} - \frac{h_{\alpha\lambda}c_{\alpha\lambda}}{2} - \frac{\left(t_{\alpha\lambda}^*\right)^{n+1}}{(n+1)!}\Theta(h_{\alpha\lambda})d\mathbf{r} \tag{4.28}$$

where Θ is the Heaviside function. Eq. (4.28) is valid for 3D-RISM calculations, where the solute is at infinite dilution, and has been used in the past for solutes at finite concentration with 1D-RISM and gives values in good agreement with experiment and simulation, though the application is not strictly valid. At finite concentrations of molecular solute, i, the equation becomes[130]

$$\begin{aligned}\Delta\mu_i = k_b T\Bigg\{ & \sum_j \rho_j \delta_{ik} F_{ii} + \sum_j \sum_{k=j+1} \left(\rho_j\delta_{ik} + \rho_k\delta_{ji}\right) F_{jk} \\ & - \frac{1}{2(2\pi)^3}\int \left[\mathrm{Tr}\left(\hat{\boldsymbol{\omega}}^i(k)\hat{\mathbf{c}}^{ii}(k)\right) - \mathrm{Tr}\left\{\left(\mathbf{I} - \hat{\boldsymbol{\omega}}(k)\hat{\mathbf{c}}(k)\boldsymbol{\rho}\right)^{-1}\frac{\partial\hat{\boldsymbol{\omega}}(k)\hat{\mathbf{c}}(k)\boldsymbol{\rho}}{\partial\rho_i}\right\}\right]k^2dk\Bigg\}\end{aligned} \tag{4.29}$$

where **I** is the identity matrix and

$$F_{ij} = \sum_{\alpha \in i} \sum_{\gamma \in j} \int c_{\alpha\gamma}(\mathbf{r}) - \frac{h_{\alpha\gamma}^2(\mathbf{r})}{2} + \frac{\left(t_{\alpha\gamma}^*\right)^{n+1}}{(n+1)!} \Theta(h_{\alpha\gamma}) \mathrm{d}\mathbf{r}. \tag{4.30}$$

Though DRISM is an important improvement over XRISM, the Helmholtz free energy is not stationary w.r.t. the TCF or DCF for HNC-like closures, meaning that Eq. (4.29) is not valid for DRISM.[130] This means that the solvation free energy, for example, is not a state function and it is not strictly meaningful to calculate the solvation free energy either through numerical or analytic means. The implications of this recent result have not been fully explored.

4.2.7 Numerical Approaches

Both 1D- and 3D-RISM equations must be solved numerically, with their respective closures, on grids of the appropriate dimension. 1D-RISM calculations generally use a grid spacing on the order of 0.01–0.025 Å and a maximum separation of ~800 Å for electrolyte solutions. The large separations are necessary to capture the long range behavior the TCF function for use in 3D-RISM. In 3D-RISM, grid spacings of 0.25–0.5 Å with a solvation box 20–60 Å larger than the solute are typical but the 3D grids required mean that the memory demands are significantly higher than for 1D-RISM.

Most of the iterative schemes for solving the RISM equations can be used for both 1D- and 3D-RISM. The simplest and slowest of these is Picard iteration in which, for example, a solution for **h** is generated from Eq. (4.11), (4.13) or (4.19) as appropriate and feed into the selected closure equation to generate **c**. The process is repeated until the desired residual is reached,

$$R = \sqrt{\frac{1}{N_{\text{point}} N_{\text{site}}} \sum_{i}^{N_{\text{point}}} \sum_{\alpha}^{N_{\text{site}}} \left(g_\alpha[i] - g_\alpha^0[i]\right)^2} \tag{4.31}$$

where N_{point} and N_{site} are the number of grid points and solvent sites, $g[\,]$ is the grid containing the current closure solution, Eq. (4.7) for example, and $g^0[\,]$ is calculated from the relevant RISM equation, Eq. (4.11), (4.13) or (4.19). Other approaches include Labík–Malijevský–Voňka,[131] wavelet treatments,[132] generalized minimal residual (GMRes),[133] dynamic relaxation[134] and modified direct inversion of the iterative subspace (MDIIS).[134] Multi-grid methods have also been developed for 1D-RISM[135] and 3D-RISM.[133] It is typical to converge solutions to $R < 10^{-10}$ for 1D-RISM and $R < 10^{-5}$ for 3D-RISM.

While XRISM, DRISM and 3D-RISM have been implemented many times independently, at the time of writing we are aware of only a few publicly distributed computer codes. 3D-RISM has been integrated into the Amber and AmberTools molecular modeling suite (http://ambermd.org).[100,136] In this distribution, XRISM and DRISM are available through a stand-alone program, which can handle arbitrarily complex solvents, producing structural and thermodynamics output. Bulk solvent data from 1D-RISM are used for 3D-RISM calculations either through a stand-alone program or through the nucleic acid builder (NAB) programming language.[137] Coupling with the NAB MD engine allows 3D-RISM to use standard biomolecular force fields to analyze static solute snapshots or perform MD calculations. A related version of 1D- and 3D-RISM has been implemented in the Amsterdam Density Functional (ADF) Package.[96] As this implementation couples 3D-RISM with quantum chemical density functional theory, it allows the calculation of thermodynamics, electronic properties and modeling of chemical reactions in a molecular liquid. A stand-alone 3D-RISM distribution is also available from the Pettitt group (http://imd.uh.edu/).[124,133] A number of different closures and iterative solvers are implemented and various post-processing analysis programs are included. An implementation of XRISM and ARISM is available from Benoît Roux (http://thallium.bsd.uchicago.edu/RouxLab/rism.html) and pre-dates the development of DRISM. It is not coupled with any other simulation methods but does allow a number of different thermodynamic quantities to be calculated. A version of this code is also available in the CHARMM molecular modeling suite.[138]

4.3 Pure Solvent Structure and Thermodynamics

The study of microscopic electrolyte structure, especially in the biologically relevant 0.1 M range, has been challenging for both simulation and experiment. In both cases, the small number of ionic particles makes sampling difficult. Few experimental techniques can investigate the structure beyond the first solvation shell.[139,140] In fact, even in methods such as neutron diffraction, which can observe beyond the first solvation shell, simulation methods like empirical potential structure refinement must be used to meaningfully interpret experimental data.[141] DRISM does not suffer in either of these respects and it is useful to compare the structure observed here with what is known from experiment. Our discussion follows the work of several other studies[55–57,60] and the details of the calculations involved may be found in Section A. The reader may consult these earlier papers for a more comprehensive look at electrolytes treated with RISM theory.

Figure 4.1 shows the solvent structure around K^+ and Cl^- ions at 0.6688 m (1:83 ion to water ratio) with the first shell peak position and coordination numbers given in Table 4.1. The RDFs show the structure of each site-site pairing in great detail and may be directly compared to neutron diffraction studies such as Mancinelli *et al.*[139] Density distributions shown in Figures 4.1(c) and (d) put into perspective the relative amounts of each particle type. Direct integration up to the first minimum, $r_{\min}$, of these distributions yields the first shell coordination numbers of site γ about α

$$N^C_{\alpha\gamma} = 4\pi\rho_\gamma \int_0^{r_{\min}} g_{\alpha\gamma}(r) r^2 dr. \tag{4.32}$$

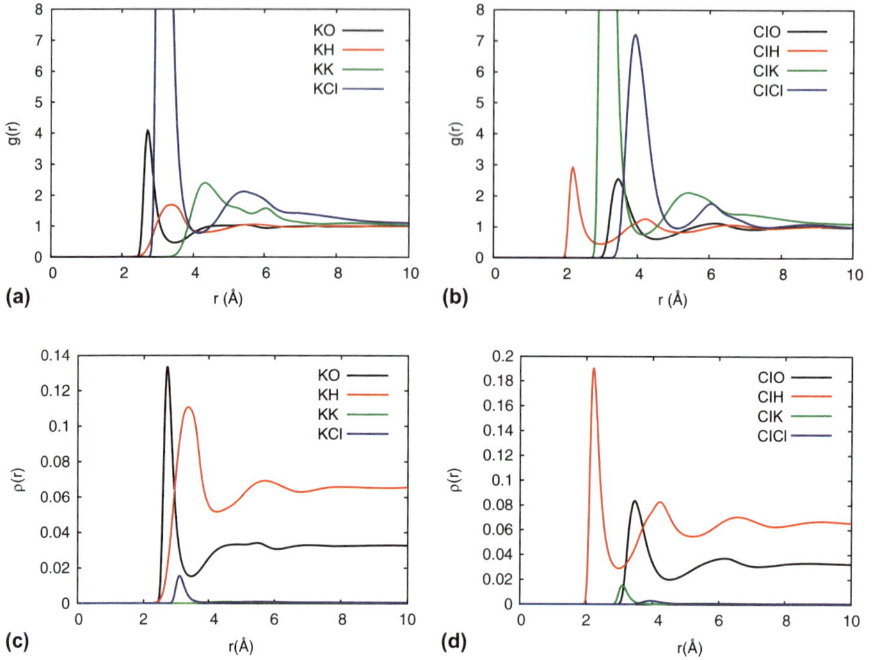

Figure 4.1 Complete solvation structure of 0.6688 m KCl in cSPC/E water around K^+ and Cl^- sites as solved by DRISM-HNC.

Table 4.1 Site-site first shell peak position and coordination numbers for K^+ and Cl^- at 0.6688 m from DRISM and experiment.[139] No experimental error was given for the K^+–Cl^- first peak position.

	1st peak position (Å)		*Coordination number*	
Site pair	*DRISM*	*Experiment*	*DRISM*	*Experiment*
K^+–O	2.700	2.65 ± 0.18	5.46	6.0 ± 1.2
K^+–H	3.350	3.25 ± 0.22	17.88	16.5 ± 2.0
K^+–K^+	4.325	—	0.37	—
K^+–Cl^-	3.100	3.0	0.71	0.5 ± 0.6
Cl^-–O	3.450	3.14 ± 0.17	11.07	7.0 ± 1.1
Cl^-–H	2.200	2.18 ± 0.04	5.62	6.1 ± 1.0
Cl^-–Cl^-	3.925	—	0.47	—

Moving radially outward, K^+ is surrounded by a ring of 5.5 coordinating water oxygens followed by 18 water hydrogens. This compares well with the 6.0 ± 1.2 oxygens and 16.5 ± 2.0 hydrogens seen in experiment at the same concentration. Due to the large radius of Cl^- relative to water, the first peak in $g_{K^+Cl^-}$, corresponding to a contact ion pair (CIP) between K^+ and Cl^-, is present just outside the first oxygen peak and contains 0.7 Cl^-. However, the anion-cation interactions are somewhat overemphasized: the first $g_{K^+Cl^-}(r)$ peak (off-scale in Figure 4.1) is 38.7 compared to ~22 in experiment. This is due to the HNC closure, which is known to over emphasize strong electrostatic attraction. The first shell of the $g_{K^+K^+}(r)$ contains 0.4 K^+; note that this is a solvent separated ion pair (SSIP) with either a water molecule or a Cl^- mediating the interactions. Likewise, Cl^- is surrounded with 5.6 water hydrogens, 11.1 water oxygens, 0.7 K^+ and 0.5 Cl^- compared to 6.1 ± 1.0 hydrogens, 7.0 ± 1.1 oxygens observed in experiment. The over-representation of oxygen in the first hydration peak of Cl^- is due to the distortion of the tetrahedral water structure in DRISM and XRISM calculations. For pure water, the first shell O–O coordination number from RISM theory is >9 rather than the observed 4.4, meaning that the tetrahedral structure has collapsed. This is due to orientational averaging and a poor representation of the molecular potential of water. Though no experimental value was given for the Cl^-–Cl^- coordination number (likely due to very noisy data) it should be noted that the experimental height of the first peak, ~3.5, was much smaller than the 7.2 given by DRISM. The origin of this discrepancy is less clear.

Another significant feature in Figure 4.1 is the asymmetry of cation-cation and anion-anion interactions. That is, $g_{++} \neq g_{--}$ as in experiment. This asymmetry is present even when the cation and anion Lennard–Jones parameters are the same, unlike the primitive model.[15,60] 1D-RISM obtains the correct result because molecular model of water is included and takes different orientations about the oppositely charged ions.

Even at 0.6688 m, far from saturation, we observe that first solvation shell of each ion contains a counter-ion more often than not. These are, in fact, CIP and displace a water molecule from the solvation shell. As the concentration of ions approached to saturation in experiment (4.3 M, or 1:13 ion to water ratio, for KCl), the number of CIP observed in experiment increases to 1.3 ± 1.1 and the coordination number of oxygens and hydrogens decreases to 4.8 ± 1.6 and 15.0 ± 2.9 for K^+ and 5.7 ± 1.5 and 4.9 ± 1.5 for Cl^-. DRISM calculations, depending on the Lennard–Jones parameters and closure used, typically cannot approach fully saturated concentrations[62] but a similar trend is observed.

While DRISM does not give any direct orientational information, we can make some inferences from the relative peak positions of water sites. The absolute difference in the positions of the first peak maxima for $g_{K^+O}(r)$ and $g_{K^+H}(r)$ is 0.65 Å and 1.25 Å for $g_{Cl^-O}(r)$ and $g_{Cl^-H}(r)$. Given that the O–H bond length for SPC/E water is 1 Å we see that the waters are oriented with oxygens in contact with K^+ and hydrogens pointed directly away. The first hydrogen shell is quite broad, suggesting a fair amount of variance in this orientation. Around Cl^- there appears to be an 180° Cl^-–H–O angle. The fact that the distance between the peaks is larger than the O H bond lcngth is duc to the collapsed structure of the water. Recall that the coordination number of the first oxygen shell is roughly double that of the first hydrogen shell so slightly less than half of these oxygens will be bonded to a first shell hydrogen. This, admittedly, crude analysis is consistent with experimental observation and is the cause of the anion-cation asymmetry previously noted.

The experimental picture for how ions change the structure of water, particularly beyond the first solvation shell, is still hotly debated (for a recent review see Marcus).[140] Kinoshita and coworkers have approached this problem with DRISM,[63] which we follow but with the Joung–Cheatham ion parameters and cSPC/E water and compare to a neutron diffraction study by Mancinelli *et al.*[142] Figure 4.2 summarizes the results for the change in water structure with increasing concentration of KCl. As the concentration increases, oxygens move to the first O–O shell from the location of the second O–O peak (~4.5 Å). The decrease in density is still observed with the present water model even though the second peak is not visible. This is also seen in the work by Kinoshita and Hirata,[63] where they use a different water model (PR-SPC/E) that does display a small second O–O peak (see Appendix A.1 for details of the models). The shifting of the second peak into the first appears qualitatively correct, but it is difficult to compare with experiment due to the presentation of the data and the different concentration ranges considered. The changes in Figure 4.2(b) are the same magnitude as those in (4.2a) so we would expect an observable decrease in hydrogen bonding near saturation. Indeed, a shortening of the hydrogen-bond distance by 0.1 Å in the experiment is reported along with a broadening and slight decrease in amplitude of the first O–H peak. Mancinelli *et al.*[142] also report little change in the H–H structure. The very small changes seen in Figure 4.2(c) would suggest that these would be very difficult to observe given the statistical noise in the experiment and subsequent simulations.

After determining the solvent structure, it is possible to calculate a wide variety of thermodynamic observables. Among the most useful, and what we will focus on here, are solvation free energies. Table 4.2 compares the solvation free energy of several monovalent ions at infinite dilution from MD, DRISM-HNC, and DRISM-KH calculated with Eq. (4.28). RISM is a statistical mechanical sampling method, as is MD, and should, ideally, reproduce MD results exactly if the same potential is used. Generally, the agreement is good though certain shortcomings do present themselves. First, the results using HNC are uniformly better than KH, which attenuates attractive interactions between water and the ion and broadens regions of density enhancement, increasing the total solvation free energy. However, this result does depend on

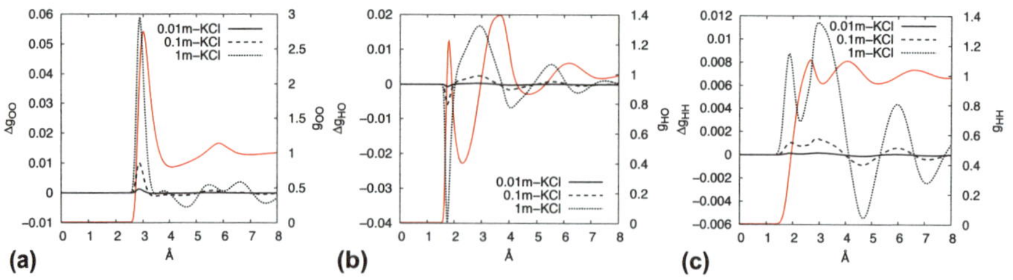

Figure 4.2 Electrolyte density dependence of water structure from DRISM-PSE-3: a) g_{OO}, b) g_{OH} and c) g_{HH}. Pure water RDFs are shown in red and concentration dependent differences are in black. $\Delta g_{\alpha\gamma}(r) = g_{\alpha\gamma}^{salt}(r) - g_{\alpha\gamma}^{pure}(r)$.

Table 4.2 Solvation free energies of ions in infinitely dilute solution decomposed into solvent polarization and non-polar contributions. DRISM-HNC and -KH results are compared with explicit MD simulations with the same parameters.

	Explicit MD Simulation			*DRISM-HNC*			*DRISM-KH*		
Ion	*Total*	*Polar*	*Non-polar*	*Total*	*Polar*	*Non-polar*	*Total*	*Polar*	*Non-polar*
Li	−113.3	−113.5	0.20	−111.5	−112.4	0.9	−108.4	−109.4	1.1
Na	−88.4	−88.9	0.48	−86.3	−88.6	2.3	−85.0	−87.5	2.5
K	−71.0	−71.8	0.81	−69.4	−73.3	3.9	−68.6	−73.0	4.4
Cl	−89.3	−96.0	6.6	−79.5	−93.6	14.2	−78.0	−93.5	15.4
Br	−82.7	−89.5	6.8	−73.7	−89.0	15.3	−72.1	−88.9	16.8
I	−74.4	−81.9	7.5	−64.4	−82.2	17.8	−62.6	−82.2	19.6

the parameters used and other groups have reported better results with the KH closure.[130,143] We also observe that, for all closures RISM predicts a solvation free energy that is too positive, *i.e.* with too weak an ion-solvent interaction. For HNC this varies from a difference of <2 kcal/mol for Li^+ to 10 kcal/mol for I^-. The source of this error can be readily found by decomposing the solvation free energy into solvent polarization and non-polar components. The non-polar component is simply the solvation free energy of the charge-free solute (the ionic charge is set to zero), ΔG_α^{np}. The polar component is then the difference between the full solvation free energy and the non-polar component

$$\Delta G_\alpha^{pol} = \Delta G_\alpha^{sol} - \Delta G_\alpha^{np}. \tag{4.33}$$

This is the same approach as using thermodynamic integration to "turn-on" the LJ potential of the ion and then the Coulomb potential. In RISM, TI is done analytically with Eq. (4.28) so the decomposition requires a single additional calculation. From this calculation we see that DRISM both over- and underestimates the solvent polarization free energy for the various ions. The maximum absolute error is 2.4 kcal/mol for HNC (Cl^-) and 4.1 for KH (Li^+). The dominant source of error is from the non-polar component, having a maximum error of 10.2 kcal/mol and 12.1 kcal/mol for HNC and KH respectively. This is a well-known problem of HNC-like closures is due to the lack of a repulsive bridge term.[144] Corrective bridge terms have been applied to electrolyte solutions[69,71,121] and pure water.[145,146] It is worth noting that the difference between KH and HNC is also primarily due to the non-polar component. In fact, the polar component of KH is not uniformly higher than HNC as one might expect, given that this component arises from attractive Coulomb interactions.

When dealing with finite concentrations it becomes useful to consider activity coefficients rather than absolute free energies, especially when comparing to experiment. DRISM can calculate activity coefficients for individual ion species but these cannot be directly measured.[147] To facilitate comparison, we calculate molal mean activity coefficients, γ^s, using the following equation[148]

$$k_b T \ln \frac{\rho_m}{\rho_0} + \Delta\mu_m - \Delta\mu_0 = k_b T \ln \gamma^s \quad (4.34)$$

where ρ_m and $\Delta\mu_m$ indicate the number density of solvent, and excess chemical, respectively. Sub-index m represents molal concentration of the salt and sub-index zero represents the status of pure water; s is the number of ions that a single salt molecule is separated into in water (*e.g.* $s = 1 + 1$ for NaCl, $s = 1 + 2$ for $MgCl_2$). $\Delta\mu_0$ is calculated with Eq. (4.28) while $\Delta\mu_m$ should use Eq. (4.29), though this is approximate for DRISM. For the discussion here, we follow[143] instead and use Eq. (4.28) for $\Delta\mu_m$. Figure 4.3 shows mean activity coefficients from experiment, DRISM-HNC, DRISM-KH, extended Debye–Hückel (EDH) and Debye–Hückel limiting law (DHLL).[10] EDH estimates activity coefficients of ions (γ) with the following equation.

$$\ln \gamma = -\frac{A|z_+ z_-|I^{1/2}}{1 + BI^{1/2}} \quad (4.35)$$

I is the ionic strength of the solution and z_+ and z_- are the charge numbers of the cation and anion. A and B are the parameters of the theory. While A is directly measurable from other thermodynamic variables, B is rather empirical. In DHLL, B is considered to be zero. For a further discussion of EDH and DHLL, see Fawcett.[10] As expected, DHLL is only valid at very low concentrations, typically <0.01 m. EDH, fitted with radii of $B_{NaCl} = 1.5270$ and $B_{KCl} = 1.2882$, reproduces the experimental data remarkably well. DRISM results are sensitive not only to the LJ parameters used but to the closure as well. For the Joung–Cheatham parameters,[149] HNC gives values close to experiment, approaching those of EDH. KH, on the other hand, shows less

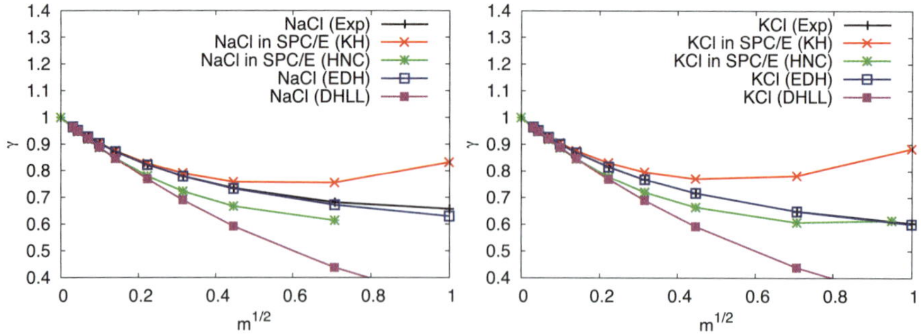

Figure 4.3 Mean activity coefficients of NaCl and KCl salts at various concentrations from experiment (Exp), DRISM-HNC (HNC), DRISM-KH (KH), Debye–Hückel limiting law (DHLL) and extended Debye–Hückel (EDH).

satisfactory results, particularly at high concentration. Recall that the ion parameters used here were optimized for MD simulations and activities were not used as part of the optimization procedure. Certainly, the HNC results are very encouraging but results are quite sensitive to LJ parameters. In the first detailed DRISM study of electrolytes, Perkyns and Pettitt[55] used DRISM-HNC with PR-TIP3P and found for their NaCl parameters a monotonically decreasing activity coefficient up to 4 M, contrary to experiment. Yoshida *et al.*[143] used a PR–SPC water model with OPLS ion parameters and DRISM-KH and achieved excellent results for mono- and divalent ions. However, the largest concentration tested was 0.1 m.

4.4 Macromolecules and Other Solutes

The 3D-RISM model has been used for many studies of solutes in electrolytes. It is not possible to give any comprehensive overview here, but several excellent reviews are available.[13,14,123] To give a flavor of some opportunities and challenges in this area, we discuss some results here for crown ethers and engineered proteins that bind monovalent cations, the local ion atmosphere around DNA and assess the quality of solvent distributions produced by 3D-RISM.

4.4.1 Ion Binding to Crown Ethers

12-crown-4-ether[150,151] is a simple molecule that displays selective binding of monovalent and divalent cations and has been a test system for many RISM methods. In the earliest example, XRISM-HNC was used simply to identify the relative stability of conformers in water, carbon tetrachloride and acetonitrile.[152] Due to the dielectric limitations of XRISM, ionic solutions were not considered. A hybrid Monte Carlo (MC)/DRISM-HNC approach was used to study the binding of Na^+, K^+, Rb^+, and Cs^+ to fully flexible 12-crown-4-ether and 18-crown-6-ether in methanol and acetonitrile.[153] The approach was able to calculate binding free energies, potentials of mean force, dissociative limits, free energy surfaces and reaction pathways. Compared to experiment, absolute binding free energies were off by several kcal/mol for methanol while only Rb^+ differed significantly from experiment in terms of relative free energies. For acetonitrile the results were much worse. Of course, it is difficult to know what errors are due to theory and what are due to the model employed.

A number of 3D-RISM approaches have been used for simple ion binding to crown-ethers. Miyata, Ikuta, and Hirata used the binding of K^+ to a fully flexible 18-crown-6-ether to demonstrate MD/3D-RISM[99,100] coupled with free energy perturbation (FEP) and thermodynamic integration (TI) of a coupling parameter[101] or spatial TI.[102] In all cases, both the crown-ether and K^+ were treated at infinite dilution in pure water. Separate coupling parameters were used for LJ and electrostatic interactions. Binding free energies were obtained by sequentially modifying coupling parameters for FEP (-5.2 ± 0.2 kcal/mol). For TI the LJ and electrostatic interactions were modified sequentially

(-5.1 ± 0.2 kcal/mol) and simultaneously (-4.9 ± 0.3 kcal/mol). For spatial TI calculations, the reaction coordinate was chosen as the distance between the geometric centers of the two solutes. While the separation was constrained, neither the orientation nor the conformation of the crown-ether was restricted in any way. As well as giving as binding free energy of -5.7 ± 0.7 kcal/mol, the potential of mean force was determined with its decomposition in entropic, energetic and solvation free energy terms. This decomposition shows that the crown-ether/K^+ Coulomb interaction is almost exactly balanced by the desolvation free energy, both being approximately 60 kcal/mol in magnitude. This cancellation is likely the largest source of numerical error in the spatial TI approach and very accurate solvation methods are needed to obtain an acceptable result. Furthermore, by visualizing solvent distributions at various points along the reaction coordinate, they were able to observe the desolvation process and calculate the presence of a single water molecule in the binding site when it was not filled by K^+. They also calculated the number of waters in the first solvation shell and attributed the barrier in the PMF to the partial desolvation of K^+ before it was able to sufficiently interact with the oxygens in the crown ether. It is encouraging that all the methods agree within error. For comparison, the experimentally determined value for K^+ binding is -2.8 ± 0.2 kcal/mol. As the authors point out, this may in part be attributed to the parameters used for the calculation. There are a few MD studies of the absolute binding free energy of this system[154–156] but none have used the exact same parameters.

Selective ion binding to 12-crown-4-ether was qualitatively investigated using 3D-RISM solvent densities.[157] We have repeated the calculations for Li^+ and K^+ using our solvent and ion parameters (see Appendix A.3 for details). As seen in Figure 4.4(a), the very high relative density Li^+ (>8 times the bulk density) in the center of the 12-crown-4-ether indicates that it is strongly associated, in agreement with experiment.[151] In fact, the left-most figure ($g(\mathbf{r})>2$) exhibits multilayering in the Li^+ distribution while the cross-sections indicate that the pore of the crown-ether is completely accessible. K^+, despite the large electrostatic attraction, is too large and cannot approach the center of the crown-ether ring (see Figure 4.4(b)). Though the profile perpendicular to the ring shows some high density regions ($g(\mathbf{r})>6$), these are much smaller and not within the ring structure. Indeed, the $g(\mathbf{r})>8$ isosurface is almost completely absent as is the multilayer density structure at $g(\mathbf{r})>2$. It is clear that K^+ is not "recognized" by 12-crown-4 ether. When Ikuta *et al.*[157] repeated the calculation with divalent ions, Mg^{2+} and Ca^{2+}, the multilayer distribution was greatly enhanced, but only Mg^{2+} was recognized.

The factors determining ion recognition are electrostatic stabilization, steric core repulsion and desolvation cost. For Li^+ and Mg^{2+} there is minimal core repulsion and the effects of desolvation are easily overcome. K^+ is simply too large and steric repulsion is enough to exclude it from the ring. Ca^{2+} can be accommodated by the ring structure, but the penalty for desolvating the ion and the ring interior (which is otherwise hydrated) is simply too strong. When 18-crown-6-ether is used, the balance of these factors changes such that all of

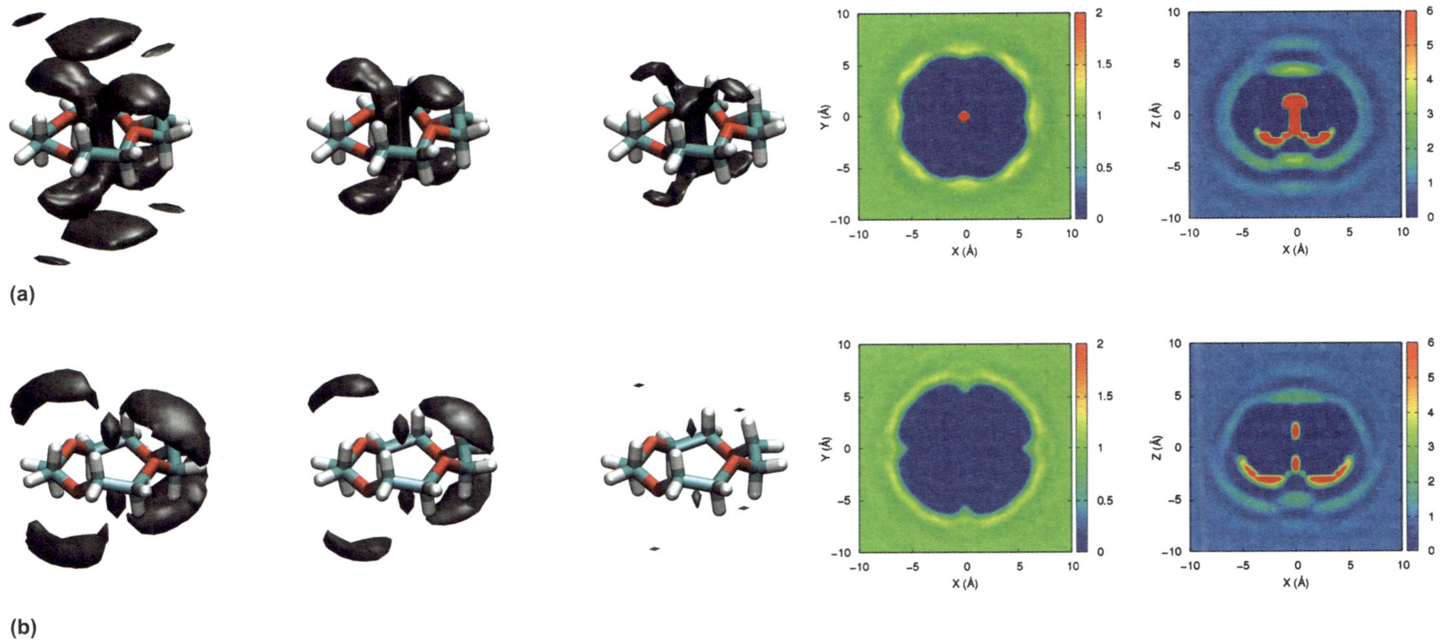

Figure 4.4 Binding of a) Li^+ and b) K^+ ions on 12-crown-4 ether. From left to right the first three images show the cation isosurfaces $g(\mathbf{r})$ equal to 2, 4 and 8 respectively. The two right-most images show cross sections through the center-of-mass of the crown-ether parallel and perpendicular to the ring. The value of $g(\mathbf{r})$ is indicated by the false coloring. Li^+ binds in the center of the crown ether while K^+ is populated slightly off along the z-axis by 1.6 Å from the center.

the ions are recognized.[158] How the ions bind is quite different though. Li^+, Mg^{2+} and, to a lesser extent, Ca^{2+} form a ring-like density in the crown-ether center, suggesting that some partial solvation may remain and that steric conflicts are absent. The K^+ density is strongly peaked in the center, much like that for Li^+:12-crown-4-ether.

While Ikuta *et al.*[157] did not make quantitative calculations with regards to binding affinities, their method was able to reproduce the binding selectivity of the receptor with very little computational effort while providing physical insights. Our own calculations repeating this work were performed on commodity desktop workstations (Quad–Core AMD Opteron™ processor 2354) and required less than a minute of CPU time for the two 3D-RISM-KH calculations. The prerequisite DRISM-KH calculations also required just a few minutes and can be reused for any subsequent 3D-RISM-KH calculations that use the same solvent conditions.

4.4.2 Ion Binding to Proteins

Though RISM methods have been used for a wide variety of protein systems, electrolyte solutions have only been used in a handful of studies. Perhaps the earliest such study compared XRISM-MSA to MC in a study of calcium binding to calbindin and several mutants.[77] Using a coarse-grained potential, the 722 all-atom representation of calbindin was reduced to 49 sites. Furthermore, rather than including water in the simulation, the primitive model was used so dielectric properties of XRISM could be neglected. Even with this level of simplification, pK shifts in good accord with experiment were found using both methods. Using DRISM with a simplified model of the *Streptomyces lividans* potassium channel (KcsA), Cui and Smith were able to show ion selectivity in the eight most favorable occupation states.[75,76] However, the numerical results were generally not in good accord with previous MD simulations. The potassium channel of *Paramecium bursaria* chlorella virus (Kcv) and several mutants were studied with MD and the resulting structures analyzed with 3D-RISM-KH.[159] Visualizing $g_{K^+}(\mathbf{r})$, it was possible to directly observe two very different physical mechanisms that changed the behavior of the mutants compared to the wild-type. In the case of the KB-KcvP13A mutant, the ion-gate was widened, resulting in a hyperactive channel. In contrast, the KB-KcvΔN8 mutant had a large free energy basin outside the ion gate, effectively allowing a potassium ion to bind and plug the channel.

Yoshida and coworkers studied selective ion binding of human lysozyme mutants using electrolyte density distributions from 3D-RISM-KH [160,143]. Previous experimental work had designed several small mutations to human lysozyme that imparted a Ca^{2+} binding site.[161–164] The experimental work showed that the wild-type and Q86D mutant forms exhibited no ion binding, the A92D mutant selectively bound Na^+ and the double mutant Q86D/A92D preferentially bound Ca^{2+} over Na^+. 3D-RISM-KH solvent distributions for 0.01 m NaCl, KCl and $CaCl_2$ were solved for the X-ray crystal structures of the wild type and all mutants, including apo and holo (Ca^{2+} bound) structures for

Q86D/A92D. Ions were qualitatively assessed to be bound if significant ion distributions with $g(\mathbf{r}) > 5$ were observed in the binding site. In accord with experiment, no significant K^+ densities were found for any system nor were any such distributions for any ions around the wild-type or Q86D mutant. For the A92D mutant, Na^+ binding was observed, as expected, but so too was Ca^{2+} binding, with a density distribution very similar to that of Na^+. A similar situation was observed for the double mutant Q86D/A92D in both apo and holo forms. However, for these two structures the maximum values of $g(\mathbf{r})$ for Ca^{2+} (apo: 19.1, holo: 63.3) were twice that of Na^+ (apo: 10.6, holo: 32.2), suggesting that Ca^{2+} is preferred though no mixed NaCl/$CaCl_2$ competition calculations were performed to test this. This is consistent with Na^+ weakly binding the apo structure but not being able to stabilize the holo form, which appears to be the picture from experiment. The ability of the method to qualitatively identify binding sites for specific ligands with no prior knowledge is encouraging, but the lack of experimental observables makes further assessment of the method difficult. For practical, predictive studies, the number of ions in the binding site, equilibrium constants and the relative stability of the mutants in different solutions are useful quantities and may be obtained from 3D-RISM calculations but were not included in the study.

4.4.3 Ion Atmosphere Around Duplex DNA

DNA presents additional modeling challenges due to its relatively high intrinsic charge, sensitivity to salt concentration and elongated shape. The aspect ratio of DNA is not a particular problem for 3D-RISM, but correctly predicting the interaction of simple ions and DNA is a non-trivial task and essential to studies of DNA behavior. Following Howard, Lynch and Pettitt,[108] we assess the quality of 3D-RISM-KH results compared to all atom MD simulations and experiment on DNA. Our calculations differ in some details (as described in Appendix A.4), but the main conclusions agree though distribution peaks should be broader and some local solvent structure will be modified due to the use of different DNA sequences.

Figure 4.5(a) shows the first solvation shell of 1 M KCl around the Dickerson docecamer at $g_\alpha(\mathbf{r}) = 4$. In this first shell, the solvent structure is dominated by the presence of oxygen and K^+ as these are the most highly peaked distributions. Note that despite the visual dominance of K^+, its number density never approaches that of oxygen or hydrogen; K^+ simply has the highest peaked distribution function relative to the bulk. Hydrogen has a very even distribution and is barely visible at $g_H(\mathbf{r}) = 4$. Cl^- is largely depleted close to the solute but there remain some prominent peaks in the major and minor grooves.

K^+ is primarily associated with exposed DNA oxygen atoms, giving a very high distribution around the backbone. The broad and highly peaked distribution represents a substantial population of K^+ ions that screen the highly charged backbone, making the duplex stable. Though it cannot be seen in Figure 4.5(a), there is also extensive hydrogen bonding between water and the backbone oxygens. Water hydrogens forms domes around the backbone

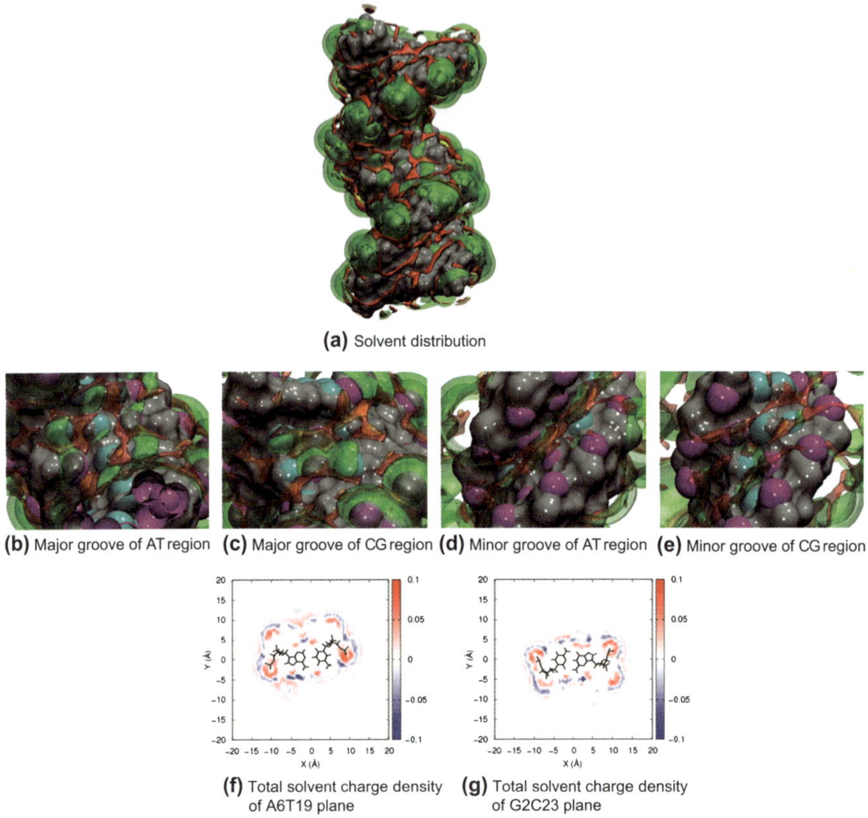

Figure 4.5 Water and 1 M KCl distribution around the Dickerson DNA dodecamer. (a) Isosurfaces of the 3D distribution of oxygen (red), hydrogen (yellow), K^+ (green) and Cl^- (orange) sites show the first solvation layer. Displayed isosurfaces are for $g(\mathbf{r}) = 4$. The (b), (c) major and (d), (e) minor grooves of (a) on both AT and CG base-pair regions reveals detailed distribution patterns of solvent. Solute oxygen and nitrogen, are colored magenta and cyan. Solvent hydrogen is not displayed and the backbone of the solute molecule for (d) and (e) is truncated to enhance visibility. (f) and (g) A cross-section of the multilayered charge density (e Å^{-3}) is visualized on the plane of A6T19 and G2C23 base pairs.

oxygens, similar to those of K^+, and bring water oxygens with them that form part of the first shell but are oriented away from the backbone. Due to the size of K^+, its center of mass is located farther away from the backbone oxygens than the water hydrogens but is closer than the water oxygens; a distribution similar to that of K^+ around Cl^- (see Figure 4.1). As in that case, K^+ is in contact with the negatively charged site and partially desolvated, as are all ions in the first solvation shell.

The major groove distribution is sequence dependent. As K^+ is found around exposed oxygens, it tends to form bands in the central AT region (Figure 4.5(b)) and clusters in the CG regions (Figure 4.5(c)). As in MD simulations, four hydration sites along the plane of the base pair are identified by local oxygen densities. Two sites sit on the edge of the groove by the phosphate backbone. The relative position of the two middle oxygens is determined by either the presence of thymine methyl or the positions of polar nitrogens and oxygens. Located at oxygen sites that do not overlap with K^+ sites are small, well-defined Cl^- peaks. There are typically 2–3 of these sites per base pair.

Sequence dependence is even more significant in the minor groove where the AT groove width is smaller than for CG. This difference in groove width allows only a single row of waters to populate the AT groove (Figure 4.5(d)) while the CG groove (Figure 4.5(e)) contains two distinct water channels. Further, in the AT groove there is a well-ordered pattern of K^+ sites associated with polar oxygens and nitrogens separated by small Cl^- peaks. The wider CG groove shows a distribution similar to the major grooves in that K^+ sites follow DNA oxygens while Cl^- sites are found around water sites that do not overlap with the K^+ peaks. Howard *et al.*[108] found an almost continuous K^+ distribution in the CG minor groove; however, this may be due to the sequence ordering that they used.

The complex picture of ion and water distribution given by 3D-RISM and MD also gives a complex solvent charge density as shown in Figure 4.5(f) and 4.5(g). This multilayer charge density is a result of both the distribution of ions and the orientation of waters and is in contrast to the simplified electric double layer of DNA solvation.[89]

4.4.4 Solvent Distributions

All observables in 3D-RISM are calculated from the 3D distributions that are explicitly solved for. It is important to assess the quality of these distributions and the physical picture they present. Ultimately, 3D-RISM should reproduce the solvation distributions observed around real macromolecules. While experimental methods like small angle X-ray scattering (SAXS) provide meaningful constraints on what the solvation distribution around a given solute should look like, direct comparison to the detailed distribution obtained from 3D-RISM is not possible.[165] In fact, methods like the Poisson–Boltzmann–Langevin and 3D-RISM equation may be used to refine and interpret SAXS data.[166] Furthermore, RISM is an approximate method for statistically sampling a particular model and should agree as closely as possible with the most accurate sampling method we have, MD, even in the cases where the model being studied gives incorrect results. To this end, a few studies have compared MD and 3D-RISM PDFs. As previously discussed, Howard *et al.*[108] have compared ion and water distributions around DNA. However, 3D-RISM-HNC was calculated on a single static structure while the benchmark MD simulations used fully flexible solutes. Averaged MD solvent distributions such as these blur important details and limit the utility of the comparison.

Yonetani *et al.*[167] addressed solute flexibility by performing MD simulations on rigid and flexible DNA in explicit water and KCl (0.15 M). 3D-RISM-KH calculations without ions in the solvent were performed on the rigid structure and on a subset of structures from the flexible MD simulation.

The 3D-RISM-KH distributions from the flexible structures were then averaged. Qualitatively, distributions from the two methods are very similar. Aside from overall visual similarity, superposition of the two distributions shows very good agreement, including important features such as spine hydration. However, the 3D-RISM-KH distributions were overly broad and did not show the range in density seen in the MD distributions. This is probably due to the KH closure as it is known to create overly broad distributions. In fact, $g(\mathbf{r})$ isosurfaces were set to 1.7 for 3D-RISM-KH and 2.25 for MD for comparisons between the two methods to compensate for the low peak height. Another shortcoming was the difference in fine solvation detail. For example, while the 3D-RISM-KH water oxygen distribution along the DNA spine qualitatively agreed with MD, the hydrogen distribution did not resemble MD results. It is difficult to determine if this was due to the overly broad distributions in 3D-RISM-KH or if orientational averaging of the solvent played a larger role.

In a detailed quantitative analysis of solvent distributions, Stumpe *et al.*[168] compared pure water 3D mass density distributions from MD and 3D-RISM-KH around a rigid GroEL/ES complex using a variety of water models, using the same conditions in both calculations. Ions were included in the MD equilibration phase at "physiological concentration" and were unrestrained (as were protein side-chains) but were fixed, along with the protein, for both production MD and 3D-RISM-KH calculations. MD calculations were also performed in the NVT ensemble, facilitating comparison to 3D-RISM. As with other studies, the 3D-RISM-KH 3D distributions were found to be broader and less highly peaked than those from MD. The range of densities found in the 3D distributions was also compared using density histograms (fraction of all grid points with a given density). Explicit TIP3P water had a low, broad Gaussian distribution with a width of $\sigma = 0.1593\,\mathrm{g/cm^3}$, whereas 3D-RISM-KH distributions were sharply peaked around the bulk water density and had too narrow a distribution to meaningfully fit a Gaussian. The broad distributions for 3D-RISM-KH systematically under solvated side chains but not free ions. This gave water densities that were typical 2–4% too low in a 3 Å shell around the side chains. The effect was reduced as the size of the shell was increased until it reversed at 10 Å and the 3D-RISM-KH densities became roughly 1% too high. The location of density peaks was also very similar, though some tolerance was required to get acceptable agreement. For the top 1% highest density grid sites, there was roughly a 50% agreement between MD and 3D-RISM-KH. When the top 0.1% highest density sites were considered, this dropped to approximately 25%. By allowing a tolerance of 2 Å, the agreement became >90% and >50% respectively. Overall, different models using the same method had better agreement than the same model with different methods.

4.5 Conclusions

We have focused our attention here on 1D- and 3D-RISM studies of electrolytes and their interaction with solutes. Even though such studies are rather new, some general conclusions can be reached.

The structure of KCl obtained from DRISM-HNC is almost completely consistent with experiment. All available positions for RDF first shell peaks and the respective coordination numbers are within error of experiment except for Cl^-–O where both values are over estimated. Thus, the relative positions and orientations of the sites are correct. However, ion–ion peak heights are over estimated, in some cases by as much as a factor of two. This may be due to the HNC closure but it is not clear. The response of water structure to the addition of salt is somewhat less successful though it is difficult to make a quantitative assessment due to the magnitude of the changes and the limited concentration range to which we are able to obtain solutions. In particular, it is not clear if the shift of waters from the location of the second peak to the first is in agreement between experiment and theory. This is in part model dependent as there are also discrepancies between DRISM and experiment for the pure water O–O RDF and coordination number.

The overall computed thermodynamic behavior of ions in solution compares well to explicit simulation and experiment. The absolute solvation free energies at infinite dilution had a range of errors from 2 to 12 kcal/mol depending on the ion and closure. This error is almost completely attributable to non-polar interactions with water, a known problem of HNC-like closures. HNC uniformly gives better results than KH at both infinite dilution and for activity coefficients at all concentrations, though this result depends on the water and ion parameters that were used.

Using the same ion parameters, though with the KH closure, the distribution of KCl and NaCl around 12-crown-4 ether was solved. The difference in these distributions clearly shows the ion selectivity of this simple host as Na^+ is found through the interior of the ring but K^+ is not. Ikuta *et al.*[157] found in their original study that 12-crown-4 ether was also able to discriminate between Mg^{2+} and Ca^{2+}, allowing only the former into the binding site. The same methodology was applied to mutants of human lysozyme engineered to bind Ca^{2+}.[143] It was shown that the binding behavior was consistent with experiment for all four isotypes. Selective ion binding is non-trivial as ion parameters are not selected to reproduce this behavior. Rather, they are taken from common molecular force fields and the subtle differences between association energetics and entropic and desolvation penalties must be correctly accounted for by 3D-RISM. Studies such as this are remarkably fast. In the case of 12-crown-4 ether, complete equilibrium distributions were obtained in less than 30 seconds.

The structure of DNA is highly sensitive to its ionic environment. Accurately reproducing this environment is necessary for modeling the molecule itself. The ionic atmosphere calculated with 3D-RISM-KH is detailed, complex and in broad agreement with MD simulations and experiment. Charge neutralization of the backbone is immediately apparent. The solvation of the major and minor grooves reflects that seen in simulation but readily displays the complex arrangement of counter-ions and co-ions. Together this gives a multilayered charge density, much more nuanced and detailed than the electric double layer often discussed.

Despite the many successes presented in this review there remain problems with the theory. In general, distribution peak predicted by 1D- and 3D-RISM are too broad, though much of this is attributable to the KH closure. All of the closures discussed here overestimate the non-polar solvation free energy and, though we have not discussed it, there are problems with thermodynamic consistency.[169,170]

In assessing these issues it is important to separate the theory (RISM) from the model (force field) as much as possible. This has been done to a limited extent by comparing "correct" distributions computed by MD as a gold standard for 3D-RISM to reproduce. The most thorough studies have used the KH closure and found broad peaks with underestimated heights in distributions that are, otherwise, qualitatively correct. The location of the most prominent peaks is correct but the variance in density is far too small.

Improvements to the theory have either considered the closure or a post-calculation re-weighting of the distributions. Though a number of closures have been proposed,[124,144,145] it is not yet clear how successful these will be. Furthermore, all appear to have path dependent free energies. Another approach has been to use a parameterized expression for the solvation free energy rather than the analytically correct one.[171,172] While this has shown success reproducing experimental results for solvation free energies, it is so far not useful for calculating other thermodynamic properties.

DRISM and 3D-RISM offer a qualitatively and, often, quantitatively correct, rich, detailed picture of electrolytes and their interaction with solutes. The method has already proven to be a valuable tool for investigating many physical problems. Its future utility depends on the further active development of the theory.

4.6 Methods

4.6.1 Parameters

Parameters for all solvent calculations use the coincident SPC/E (cSPC/E) water model[100] and the SPC/E ion models of Joung and Cheatham.[149] DNA calculations used the Amber ff99bsc0 force field,[173] while for crown-ether calculations the parameters of Ikuta *et al.*[157] were used.

Commonly used three-site water models, such as SPC/E[174] and TIP3P,[175] omit LJ parameters for the hydrogen sites. However, the nature of the site–site potential energy function in 1D-RISM requires non-zero parameters to prevent catastrophic attraction between oppositely charged sites. All the results in this chapter are reproduced using the coincident SPC/E (cSPC/E) water model,[100] which is the same as SPC/E water but hydrogen atoms have additional centers for van der Waals interactions such that

$$\sigma_{\mathrm{H}} = \sigma_{\mathrm{O}} - 2b_{\mathrm{OH}} = 1.1658\,\text{Å} \tag{4.36}$$

$$\epsilon_{\mathrm{H}} = 0.1\epsilon_{\mathrm{O}} = 0.01553\ \text{kcal/mol} \tag{4.37}$$

where O and H denote water oxygen and hydrogen, σ and are the usual Lennard–Jones variables, $b_{\rm OH}$ is the bond oxygen-hydrogen bond length. A commonly used alternate is Pettitt and Rossky's van der Waals parameters [48], $\sigma_{\rm H}^{\rm PR} = 0.4$ Å and $\epsilon_{\rm H}^{\rm PR} = 0.046$ kcal/mol. In this chapter, the parameter set is notated by prefix "PR-" For example, PR-SPC/E has the same parameters as SPC/E but with $\sigma_{\rm H}^{\rm PR}$ and $\epsilon_{\rm H}^{\rm PR}$ parameters for the hydrogen sites. Both van der Waals parameters for the hydrogens have their own advantages and disadvantages. cSPC/E water lacks a $g_{\rm OO}(r)$ peak at ~4.5 Å and has a first shell coordination number >9. The PR–SPC/E model has the second O–O peak at ~4.5 Å but still does not achieve the correct water structure with a coordination number >6 and generally produces less satisfactory $g_{\rm OH}(r)$, $g_{\rm HH}(r)$ and solvation free energies.

4.6.2 Bulk Electrolytes

For 1D-RISM calculations, correlation functions were numerically constructed using 32768 grid points with grid spacing of 0.025 Å. Therefore, the radius of the solvation sphere was 819.2 Å. Temperature was set to 298 K. As an input parameter, molar or number densities of particles are required for RISM calculations. As water density varies as a function of electrolyte concentration, to get the best input parameters available, we used interpolated partial densities of electrolytes at 293 K which were measured for simple solutions.[176] Though there is a 5 K difference from the input temperature, this is a marked improvement over a fixed water density, regardless electrolyte concentration, typically used. Dielectric constant for the electrolyte was adopted from that of pure water, 78.5 instead of using explicit MD simulation data.

4.6.3 Crown-ether

For 3D-RISM, solvent buffer size was 14 Å from the edge of a solute molecule and grid gap was 0.5 Å. Force field parameters of the crown-ether were employed from the original paper.[157] Geometry optimization was performed with the level of B3LYP/6-31G(d) keeping D_{2d} molecular symmetry as specified by Ikuta *et al.*[157] 3D-RISM was solved with the KH closure.

4.6.4 DNA

For 3D-RISM, solvent buffer size was 14 Å from the edge of a solute molecule and grid gap was 0.25 Å. We used Dickerson dodecamer[177] (d(CpGpCpGpApApTpTpCpGpCpG), PDB ID: 1BNA) as a solute molecule to visualize solvent structure around DNA. 3D-RISM was complemented by KH closure.

Acknowledgements

We would like to thank Andriy Kovalenko, B. Montgomery Pettitt and Georg Schmeer for useful discussions. This work was supported by NIH grant GM57513.

References

1. M. Feig, *Modeling Solvent Environments. Applications to Simulations of Biomolecules*, Wiley-VCH, Weinheim, 2010.
2. B. Honig and A. Nicholls, *Science*, 1995, **268**, 1144–1149.
3. K. Sharp and B. Honig, *J. Phys. Chem.*, 1990, **94**, 7684–7692.
4. N. Baker, D. Bashford and D. Case, in *New Algorithms for Macromolecular Simulation*, ed. B. Leimkuhler, C. Chipot, R. Elber, A. Laaksonen, A. Mark, T. Schlick, C. Schuette and R. Skeel, 2006, pp. 263–295.
5. A.-P. E. Kunz and W. F. V. Gunsteren, *J. Phys. Chem. B*, 2011, **115**, 2931–2936.
6. I. V. Khavrutskii, J. Dzubiella and J. A. Mccammon, *J. Chem. Phys.*, 2008, **128**, 4106.
7. L. S. Ornstein and F. Zernike, *Proc. Akad. Sci. (Amsterdam)*, 1914, **17**, 793.
8. L. S. Ornstein and F. Zernike, in *The equilibrium theory of classical fluids: a lecture note and reprint volume*, W. A. Benjamin, Inc., 1964, pp. III-2–16.
9. J.-P. Hansen and I. R. McDonald, *Theory of Simple Liquids*, Academic Press, London, Great Britain, 2nd edn., 1990, ch. 5, pp. 97–144.
10. W. R. Fawcett, *Liquids, Solutions and Interfaces*, Oxford University Press, New York, 2004.
11. M. R. Wright, *An introduction to aqueous electrolytes*, John Wiley and Sons, Chichester, England, 2007.
12. F. Hirata, *Molecular theory of solvation*, Kluwer Academic Publishers, 2003.
13. J. J. Howard and B. M. Pettitt, *J. Stat. Phys.*, 2011, **145**, 441.
14. N. Yoshida, Y. Kiyota, Y. Ikuta, T. Imai and F. Hirata, in *Modeling Solvent Environments. Applications to Simulations of Biomolecules*, ed. M. Feig, Wiley-VCH, Weinheim, 2010, pp. 31–54.
15. J.-L. Barrat and J.-P. Hansen, *Basic concepts for simple and complex liquids*, Cambridge University Press, Cambridge, UK, 2003, ch. 3, pp. 69–106.
16. J. C. Rasaiah, D. N. Card and J. P. Valleau, *J. Chem. Phys.*, 1972, **56**, 248.
17. G. Hummer and D. M. Soumpasis, *J. Chem. Phys.*, 1993, **98**, 581.
18. D. N. Card and J. P. Valleau, *J. Chem. Phys.*, 1970, **52**, 6232.
19. V. Vlachy, *Annu. Rev. Phys. Chem.*, 1999, **50**, 145.
20. J. Chen, C. L. Brooks and J. Khandogin, *Curr. Opin. Struc. Biol.*, 2008, **18**, 140–148.
21. E. Waisman and J. L. Lebowitz, *J. Chem. Phys.*, 1972, **56**, 3086.

22. E. Waisman and J. L. Lebowitz, *J. Chem. Phys.*, 1972, **56**, 3093.
23. J. S. Høye, J. L. Lebowitz and G. Stell, *J. Chem. Phys.*, 1974, **61**, 3253.
24. T. Morita, *Progress of Theoretical Physics*, 1958, **20**, 920.
25. B. H. Lee, V. Vlachy, L. B. Bhuiyan, C. W. Outhwaite and M. Molero, *Mol. Phys.*, 2003, **101**, 2969.
26. P. N. Vorontsov-Veliaminov, A. M. Eliashevich, J. C. Rasaiah and H. L. Friedman, *J. Chem. Phys.*, 1970, **52**, 1013.
27. B. Larsen, *J. Chem. Phys.*, 1978, **68**, 4511.
28. J. C. Rasaiah and H. L. Friedman, *J. Chem. Phys.*, 1968, **48**, 2742.
29. J. C. Rasaiah and H. L. Friedman, *J. Chem. Phys.*, 1969, **50**, 3965.
30. J.-P. Hansen and I. R. McDonald, *Theory of Simple Liquids*, Academic Press, London, Great Britain, 2nd edn., 1990, ch. 10, pp. 364–405.
31. A. T. Peplow, R. E. Beardmore and F. Bresme, *Phys. Rev. E*, 2006, **74**, 046705.
32. L. Belloni, *Phys. Rev. Lett.*, 1986, **57**, 2026.
33. J. C. Rasaiah, *J. Chem. Phys.*, 1972, **56**, 3071.
34. T.-H. Chung and L. L. Lee, *J. Chem. Phys.*, 2009, **130**, 4513.
35. G. I. Guerrero-García, E. González-Tovar, M. Lozada-Cassou and F. de J. Guevara-Rodríguez, *J. Chem. Phys.*, 2005, **123**, 4703.
36. Y. V. Kalyuzhnyi, V. Vlachy, M. F. Holovko and G. Stell, *J. Chem. Phys.*, 1995, **102**, 5770.
37. Y. V. Kalyuzhnyi, L. Blum, M. F. Holovko and I. A. Protsykevytch, *Physica A*, 1997, **236**, 85.
38. B. Hribar, Y. V. Kalyuzhnyi and V. Vlachy, *Mol. Phys.*, 1996, **87**, 1317.
39. E. Gonzales-Tovar, M. Lozada-Cassou and D. Henderson, *J. Chem. Phys.*, 1985, **83**, 361.
40. D. Chandler, R. Silbey and B. M. Ladanyi, *Mol. Phys.*, 1982, **46**, 1335.
41. P. J. Rossky and R. A. Chiles, *Mol. Phys.*, 1984, **51**, 661–674.
42. K. M. Dyer, J. S. Perkyns and B. M. Pettitt, *J. Chem. Phys.*, 2005, **123**, 204512.
43. K. M. Dyer, J. S. Perkyns, G. Stell and B. M. Pettitt, *Mol. Phys.*, 2009, **107**, 423–431.
44. L. Blum and A. J. Torruella, *J. Chem. Phys.*, 1972, **56**, 303.
45. L. Gendre, R. Ramirez and D. Borgis, *Chem. Phys. Lett.*, 2009, **474**, 366–370.
46. T. Sumi and H. Sekino, *J. Chem. Phys.*, 2006, **125**, 034509–034509.
47. F. Hirata and P. Rossky, *Chemical Physics Letters*, 1981, **83**, 329–334.
48. B. M. Pettitt and P. Rossky, *J. Chem. Phys.*, 1982, **77**, 1451–1457.
49. F. Hirata, P. J. Rossky and B. M. Pettitt, *J. Chem. Phys.*, 1983, **78**, 4133.
50. B. M. Pettitt and P. J. Rossky, *J. Chem. Phys.*, 1986, **84**, 5836.
51. G. Hummer, D. M. Soumpasis and M. Neumann, *Mol. Phys.*, 1992, **77**, 769.
52. G. Hummer and D. Soumpasis, *Mol. Phys.*, 1992, **75**, 633.
53. M. V. Fedotova, *J. Mol. Liq.*, 2010, **153**, 9–14.
54. J. S. Perkyns and B. M. Pettitt, *Chem. Phys. Lett.*, 1992, **190**, 626–630.
55. J. S. Perkyns and B. M. Pettitt, *J. Chem. Phys.*, 1992, **97**, 7656–7666.

56. T. Imai, H. Nomura, M. Kinoshita and F. Hirata, *J. Phys. Chem. B*, 2002, **106**, 7308–7314.
57. S. Chong and F. Hirata, *J. Phys. Chem. B*, 1997, **101**, 3209–3220.
58. Y. Harano, T. Imai, A. Kovalenko, M. Kinoshita and F. Hirata, *J. Chem. Phys.*, 2001, **114**, 9506–9511.
59. G. N. Chuev, M. V. Fedorov, S. Chiodo, N. Russo and E. Sicilia, *J. Comput. Chem.*, 2008, **29**, 2406–2415.
60. J. Perkyns and B. M. Pettitt, *Biophys. Chem.*, 1994, **51**, 129–146.
61. H. A. Yu, B. Roux and M. Karplus, *J. Chem. Phys.*, 1990, **92**, 5020.
62. J. Perkyns and B. M. Pettitt, *J. Phys. Chem.*, 1994, **98**, 5147–5151.
63. M. Kinoshita and F. Hirata, *J. Chem. Phys.*, 1997, **106**, 5202–5215.
64. F. O. Raineri, H. Resat, B. C. Perng, F. Hirata and H. L. Friedman, *J. Chem. Phys.*, 1994, **100**, 1477–1491.
65. S. Chong and F. Hirata, *J. Chem. Phys.*, 1998, **108**, 7339–7349.
66. S. Chong and F. Hirata, *J. Chem. Phys.*, 1999, **111**, 3654–3667.
67. T. Yamazaki, A. Kovalenko, V. V. Murashov and G. N. Patey, *J. Phys. Chem. B*, 2010, **114**, 613–619.
68. S. Bruzzone, M. Malvaldi and C. Chiappe, *Phys. Chem. Chem. Phys.*, 2007, **9**, 5576.
69. G. Chuev, S. Chiodo, S. Erofeeva, M. Fedorov, N. Russo and E. Sicilia, *Chem. Phys. Lett.*, 2006, **418**, 485.
70. A. Ikeda, D. Yokogawa, H. Sato and S. Sakaki, *International Journal of Quantum Chemistry*, 2007, **107**, 3132.
71. S. Chiodo, G. N. Chuev, S. E. Erofeeva, M. V. Fedorov, N. Russo and E. Sicilia, *International Journal of Quantum Chemistry*, 2007, **107**, 265.
72. M. Kinoshita, Y. Okamoto and F. Hirata, *J. Comput. Chem.*, 1998, **19**, 1724–1735.
73. S. Ten-No, F. Hirata and S. Kato, *Chem. Phys. Lett.*, 1993, **214**, 391.
74. A. Mitsutake, M. Kinoshita, Y. Okamoto and F. Hirata, *Chem. Phys. Lett.*, 2000, **329**, 295.
75. Q. Cui and V. H. Smith, Jr., *Chem. Phys. Lett.*, 2002, **365**, 110.
76. Q. Cui and V. H. Smith, Jr., *Mol. Phys.*, 2005, **103**, 191.
77. B. Svensson and C. E. Woodward, *J. Phys. Chem.*, 1995, **99**, 1614–1618.
78. R. Ishizuka, G. A. Huber and J. A. McCammon, *J. Phys. Chem. Lett.*, 2010, **1**, 2279–2283.
79. B. M. Pettitt and M. Karplus, *Chem. Phys. Lett.*, 1985, **121**, 194.
80. H. Freedman and T. N. Truong, *Chem. Phys. Lett.*, 2003, **381**, 362.
81. D. Beglov and B. Roux, *J. Chem. Phys.*, 1995, **103**, 360.
82. D. Beglov and B. Roux, *J. Phys. Chem. B*, 1997, **101**, 7821–7826.
83. A. Kovalenko and F. Hirata, *Chem. Phys. Lett.*, 1998, **290**, 237–244.
84. A. Kovalenko and F. Hirata, *J. Chem. Phys.*, 1999, **110**, 10095.
85. A. Kovalenko and F. Hirata, *J. Chem. Phys.*, 2000, **112**, 10391–10402.
86. A. Kovalenko and F. Hirata, *J. Chem. Phys.*, 2000, **112**, 10403.
87. T. Kloss and S. M. Kast, *J. Chem. Phys.*, 2008, **128**, 134505.
88. A. Kovalenko and F. Hirata, *J. Phys. Chem. B*, 1999, **103**, 7942–7957.

89. J. J. Howard, J. S. Perkyns and B. M. Pettitt, *J. Phys. Chem. B*, 2010, **114**, 6074–6083.
90. A. Tanimura, A. Kovalenko and F. Hirata, *Chem. Phys. Lett.*, 2003, **378**, 638–646.
91. S. R. Stoyanov, S. Gusarov and A. Kovalenko, *Mol. Simulation*, 2008, **34**, 953–960.
92. C. Chiappe, M. Malvaldi and C. S. Pomelli, *J. Chem. Theory Comput.*, 2010, **6**, 179–183.
93. S. Bruzzone, M. Malvaldi and C. Chiappe, *J. Chem. Phys.*, 2008, **129**, 4509.
94. M. Harada and M. Tsukada, *Phys. Rev. B*, 2010, **82**, 35414.
95. Q. Du, D. Beglov, D. Wei and B. Roux, *Solvation and polarization of the N-methyl amine molecule in aqueous solution: A combined study of quantum mechanics and integral equation theory in three dimensions*, 2003.
96. S. Gusarov, T. Ziegler and A. Kovalenko, *J. Phys. Chem. A*, 2006, **110**, 6083–6090.
97. N. Yoshida and F. Hirata, *J. Comput. Chem.*, 2006, **27**, 453–462.
98. T. Kloss, J. Heil and S. M. Kast, *J. Phys. Chem. B*, 2008, **112**, 4337–4343.
99. T. Miyata and F. Hirata, *J. Comput. Chem.*, 2008, **29**, 871–882.
100. T. Luchko, S. Gusarov, D. R. Roe, C. Simmerling, D. A. Case, J. Tuszynski and A. Kovalenko, *J. Chem. Theory Comput.*, 2010, **6**, 607–624.
101. T. Miyata, Y. Ikuta and F. Hirata, *J. Chem. Phys.*, 2010, **133**, 044114.
102. T. Miyata, Y. Ikuta and F. Hirata, *J. Chem. Phys.*, 2011, **134**, 044127.
103. N. Blinov, L. Dorosh, D. Wishart and A. Kovalenko, *Biophys. J.*, 2010, **98**, 282–296.
104. S. Genheden, T. Luchko, S. Gusarov, A. Kovalenko and U. Ryde, *J. Phys. Chem. B*, 2010, **114**, 8505–8516.
105. J.-P. Hansen and I. R. McDonald, *Theory of Simple Liquids*, Academic Press, London, Great Britain, 2nd edn., 1990, ch. 2, pp. 13–44.
106. S. M. Kast and T. Kloss, *J Chem Phys*, 2008, **129**, 236101.
107. J. P. Hansen and I. R. McDonald, *Phys. Rev. A*, 1975, **11**, 2111.
108. J. J. Howard, G. C. Lynch and B. M. Pettitt, *J. Phys. Chem. B*, 2011, **115**, 547–56.
109. F. Hirata, B. M. Pettitt and P. Rossky, *J. Chem. Phys.*, 1982, **77**, 509.
110. S. Singer and D. Chandler, *Molecular Physics*, 1985, **55**, 621–625.
111. L. Verlet and D. Levesque, *Physica*, 1962, **28**, 1124–1142.
112. G. S. Rushbrooke and P. Hutchinson, *Physica*, 1961, **27**, 647.
113. F. Hirata, in *Molecular theory of solvation*, ed. F. Hirata, Kluwer Academic Publishers, 2003, ch. 1, pp. 1–60.
114. D. Chandler and H. Andersen, *J. Chem. Phys.*, 1972, **57**, 1930–1937.
115. J. S. Høye and G. Stell, *J. Chem. Phys.*, 1976, **65**, 18.
116. D. Chandler, *J. Chem. Phys.*, 1977, **67**, 1113.
117. F. O. Raineri and G. Stell, *J. Phys. Chem. B*, 2001, **105**, 11880–11892.
118. O. Gereben and L. Pusztai, *Chem. Phys. Lett.*, 2011, **507**, 80–83.
119. F. H. Stillinger and R. Lovett, *J. Chem. Phys.*, 1968, **49**, 1991.

120. D. Y. C. Chan, D. J. Mitchell, B. W. Ninham and B. A. Pailthorpe, *J. Chem. Phys.*, 1978, **69**, 691.
121. M. V. Fedorov and A. A. Kornyshev, *Mol. Phys.*, 2007, **105**, 1.
122. P. González-Mozuelos, *J. Phys. Chem. B*, 2006, **110**, 22702–22711.
123. A. Kovalenko, in *Molecular theory of solvation*, ed. F. Hirata, Kluwer Academic Publishers, 2003, ch. 4, pp. 169–276.
124. J. S. Perkyns, G. C. Lynch, J. J. Howard and B. M. Pettitt, *J. Chem. Phys.*, 2010, **132**, 064106.
125. M. Kinoshita and F. Hirata, *Journal of Chemical Physics*, 1996, **104**, 8807–8815.
126. J. F. Springer, M. A. Pokrant and F. A. Stevens, *J. Chem. Phys.*, 1973, **58**, 4863.
127. G. M. Abernethy and M. J. Gillan, *Molecular Physics*, 1980, **39**, 839.
128. J. W. Kaminski, S. Gusarov, T. A. Wesolowski and A. Kovalenko, *J. Phys. Chem. A*, 2010, **114**, 6082–6096.
129. T. Morita and K. Hiroike, *Prog. Theor. Phys.*, 1960, **23**, 1003.
130. G. Schmeer and A. Maurer, *Phys. Chem. Chem. Phys.*, 2010, **12**, 2407–17.
131. S. Woelki, H.-H. Kohler, H. Krienke and G. Schmeer, *Phys. Chem. Chem. Phys.*, 2008, **10**, 898–910.
132. G. N. Chuev and M. V. Fedorov, *J. Chem. Phys.*, 2004, **120**, 1191–1196.
133. J. J. Howard, J. S. Perkyns, N. Choudhury and B. M. Pettitt, *J. Chem. Theory Comput.*, 2008, **4**, 1928–1939.
134. A. Kovalenko, S. Ten-no and F. Hirata, *J. Comput. Chem.*, 1999, **20**, 928–936.
135. V. P. Sergiievskyi, W. Hackbusch and M. V. Fedorov, *J. Comput. Chem.*, 2011, **32**, 1982–1992.
136. D. A. Case, T. E. Cheatham III., T. Darden, H. Gohlke, R. Luo, K. M. Merz, A. Onufriev, C. Simmerling, B. Wang and R. J. Woods, *J. Comput. Chem.*, 2005, **26**, 1668–1688.
137. T. Macke and D. Case, in *In Molecular Modeling of Nucleic Acids*, ed. N. Leontes and J. J. SantaLucia, American Chemical Society, 1998, pp. 379–393.
138. B. R. Brooks, C. L. Brooks, A. D. Mackerell, L. Nilsson, R. J. Petrella, B. Roux, Y. Won, G. Archontis, C. Bartels, S. Boresch, A. Caflisch, L. Caves, Q. Cui, A. R. Dinner, M. Feig, S. Fischer, J. Gao, M. Hodoscek, W. Im, K. Kuczera, T. Lazaridis, J. Ma, V. Ovchinnikov, E. Paci, R. W. Pastor, C. B. Post, J. Z. Pu, M. Schaefer, B. Tidor, R. M. Venable, H. L. Woodcock, X. Wu, W. Yang, D. M. York and M. Karplus, *J. Comput. Chem.*, 2009, **30**, 1545–614.
139. R. Mancinelli, A. Botti, F. Bruni, M. A. Ricci and A. K. Soper, *J. Phys. Chem. B*, 2007, **111**, 13570–13577.
140. Y. Marcus, *Chem. Rev.*, 2009, **109**, 1346–1370.
141. A. K. Soper, *Mol. Phys.*, 2001, **99**, 1503–1516.
142. R. Mancinelli, A. Botti, F. Bruni, M. A. Ricci and A. K. Soper, *Phys. Chem. Chem. Phys.*, 2007, **9**, 2959–2967.

143. N. Yoshida, S. Phongphanphanee and F. Hirata, *J. Phys. Chem. B*, 2007, **111**, 4588–95.
144. A. Kovalenko and F. Hirata, *J. Chem. Phys.*, 2000, **113**, 2793–2805.
145. Q. H. Du, D. Beglov and B. Roux, *J. Phys. Chem. B*, 2000, **104**, 796–805.
146. C. M. Cortis, P. J. Rossky and R. A. Friesner, *J. Chem. Phys.*, 1997, **107**, 6400–6414.
147. P. Hunenberger and M. Reif, *Single-Ion Solvation: Experimental and Theoretical Approaches to Elusive Thermodynamic Quantities*, Royal Society of Chemistry, London, 2011.
148. H. Kokubo, J. Rösgen, D. Bolen and B. Pettitt, *Biophys. J.*, 2007, **93**, 3392–3407.
149. I. Joung and T. Cheatham III, *J. Phys. Chem. B*, 2008, **112**, 9020–9041.
150. C. J. Pedersen, *J. Am. Chem. Soc.*, 1967, **89**, 2495–2496.
151. C. J. Pedersen, *J. Am. Chem. Soc.*, 1967, **89**, 7017–7036.
152. Y. L. Ha and A. K. Chakraborty, *J. Phys. Chem.*, 1994, **98**, 11193–11203.
153. K. F. Schmidt and S. M. Kast, *J. Phys. Chem. B*, 2002, **106**, 6289–6297.
154. L. X. Dang and P. A. Kollman, *J. Am. Chem. Soc.*, 1990, **112**, 5716–5720.
155. L. X. Dang and P. A. Kollman, *J. Phys. Chem.*, 1995, **99**, 55–58.
156. M. C. Zwier, J. W. Kaus and L. T. Chong, *J. Chem. Theory Comput.*, 2011, **7**, 1189–1197.
157. Y. Ikuta, Y. Maruyama, M. Matsugami and F. Hirata, *Chem. Phys. Lett.*, 2007, **433**, 403.
158. Y. Maruyama, M. Matsugami and Y. Ikuta, *Condens. Matter Phys.*, 2007, **10**, 315–322.
159. S. Tayefeh, T. Kloss, G. Thiel, B. Hertel, A. Moroni and S. M. Kast, *Biochem.*, 2007, **46**, 4826–4839.
160. N. Yoshida, S. Phongphanphanee, Y. Maruyama, T. Imai and F. Hirata, *J. Am. Chem. Soc.*, 2006, **128**, 12042–12043.
161. K. Inaka, R. Kuroki, M. Kikuchi and M. Matsushima, *J. Biol. Chem.*, 1991, **266**, 20666–71.
162. R. Kuroki, Y. Taniyama, C. Seko, H. Nakamura, M. Kikuchi and M. Ikehara, *P. Natl. Acad. Sci. USA*, 1989, **86**, 6903–7.
163. R. Kuroki, S. Kawakita, H. Nakamura and K. Yutani, *P. Natl. Acad. Sci. USA*, 1992, **89**, 6803–7.
164. R. Kuroki and K. Yutani, *J. Biol. Chem.*, 1998, **273**, 34310–5.
165. S. A. Pabit, K. D. Finkelstein and L. Pollack, *Method Enzymol.*, 2009, **469**, 391–410.
166. F. Poitevin, H. Orland, S. Doniach, P. Koehl and M. Delarue, *Nucleic Acids Research*, 2011.
167. Y. Yonetani, Y. Maruyama, F. Hirata and H. Kono, *J. Chem. Phys.*, 2008, **128**, 185102.
168. M. C. Stumpe, N. Blinov, D. Wishart, A. Kovalenko and V. S. Pande, *J. Phys. Chem. B*, 2011, **115**, 319–328.
169. G. Stell, *Mol. Phys.*, 1969, **16**, 209.

170. A. Santos, R. Fantoni and A. Giacometti, *J. Chem. Phys.*, 2009, **131**, 181105.
171. D. S. Palmer, A. I. Frolov, E. L. Ratkova and M. V. Fedorov, *J. Phys.: Condens. Matter*, 2010, **22**, 492101.
172. A. I. Frolov, E. L. Ratkova, D. S. Palmer and M. V. Fedorov, *J. Phys. Chem. B*, 2011, **115**, 6011–22.
173. A. Pérez, I. Marchán, D. Svozil, J. Sponer, T. Cheatham III, C. Laughton and M. Orozco, *Biophys. J.*, 2007, **92**, 3817–3829.
174. H. Berendsen, J. Grigera and T. Straatsma, *J. Phys. Chem.*, 1987, **91**, 6269–6271.
175. W. L. Jorgensen, J. Chandrasekhar, J. D. Madura, R. W. Impey and M. L. Klein, *J. Chem. Phys.*, 1983, **79**, 926–935.
176. D. Lide, *Boca Raton, FL: CRC*, 2010.
177. H. R. Drew, R. M. Wing, T. Takano, C. Broka, S. Tanaka, K. Itakura and R. E. Dickerson, *Proc. Natl. Acad. Sci. USA*, 1981, **78**, 2179–83.

CHAPTER 5

Molecular Simulation in the Energy Biosciences

XIAOLIN CHENG,[a,b] JERRY M. PARKS,[a,b] LOUKAS PETRIDIS,[a,b] BENJAMIN LINDNER,[a,b] ROLAND SCHULZ,[a,b] HAO-BO GUO,[a,b] GOUNDLA SRINIVAS[a,b] AND JEREMY C. SMITH*[a,b]

[a] UT/ORNL Center for Molecular Biophysics, Oak Ridge National Laboratory P. O. Box 2008 Oak Ridge, TN 37831, USA; [b] Department of Biochemistry and Cellular and Molecular Biology, University of Tennessee, M407 Walters Life Sciences, 1414 Cumberland Avenue, Knoxville TN 37996, USA
*Email: smithjc@ornl.gov

5.1 Introduction

Molecular simulation can be used to understand key physical processes in the energy biosciences. Both molecular mechanical (MM) and quantum mechanical (QM) simulation techniques provide atomic-detailed insight into mechanisms at the core of research in bioenergy and bioremediation.

This article describes molecular simulation in the energy biosciences in two sections: Methods and Applications. In the Methods section, we provide a synopsis of current progress in developing simulation techniques that make efficient use of large-scale supercomputers. This is done with two examples: scalable implicit solvation models and scaling molecular dynamics (MD) to O(100k) cores. In the Applications section, we discuss modeling and

RSC Biomolecular Sciences No. 23
Innovations in Biomolecular Modeling and Simulations: Volume 1
Edited by Tamar Schlick

Published by the Royal Society of Chemistry, www.rsc.org

simulation of cellulosic biomass, an effort aimed at shedding light on biomass recalcitrance to hydrolysis (a bottleneck in biofuel production) and simulations describing the fate of mercury in contaminated biogeochemical systems. We outline research aimed at understanding the dynamics and function of the proteins and enzymes that confer mercury resistance to bacteria.

5.2 Methodological Development: An Order N Poisson-Boltzmann Solver

In modeling biological systems, it is important to take into account long-range electrostatic contributions from the surrounding environment. However, explicit inclusion of solvent molecules in a simulation adds significant extra cost to the calculation, and also leads to scalability problems on massively parallel computers. An alternative approach is to model solvent effects implicitly using Poisson-Boltzmann (PB) theory. In PB, the electrostatic potential $\varphi(r)$ and charge density $\rho(r)$ are related *via* the following differential equation that is based on a combination of the Poisson and Debye-Hückel theories[1]:

$$\nabla \cdot \varepsilon(r)\nabla\phi(r) - \kappa^2 \sinh\phi(r) = -4\pi\rho(r) \tag{5.1}$$

where $\varepsilon(r)$ is the position-dependent dielectric, and $\kappa^2 \sim \beta I$, β is a constant and I is the ionic strength. Although it is impossible to find analytical solutions to the PB equation for biomolecular systems with complex molecular surfaces and charge distributions, various numerical techniques have been developed over the past two decades, making PB an extremely useful tool in studying the stability, dynamics, and association of biomolecules in solution.[1–4] However, with the ever-increasing need to address larger subcellular structures in dynamical simulations, presently available PB models still face significant challenges.

Here, we focus on a boundary integral equation (BIE) based numerical solver for the linearized PB equation. For numerical solution of the PB equation, the BIE method has several advantages over other widely used numerical methods, by using, for example, a 2D surface mesh instead of a 3D volume mesh, together with natural (physical) instead of artificial boundary conditions, and by making possible convenient force calculations. In particular, the dimension reduction from 3D to 2D offers tremendous advantages in terms of computational efficiency (*e.g.* in the storage, number of operations, and the condition number of the resulting algebraic systems) as well as accuracy as the system size increases. However, due to the prohibitive computational cost associated with surface integrations, the application of the BIE in PB has so far been limited. We have therefore developed an efficient BIE/PB algorithm that employs a new version of the fast multipole method (FMM) for matrix-vector multiplication within a Krylov subspace method, achieving order N complexity for both speed and memory performance.[4,5]

When Green's second identity is applied, traditional boundary integral equations for the linearized PB equation for a single domain (molecule) can be written as

$$\phi_p^{\text{int}} = \oint_S \left[G_{pt} \frac{\partial \phi_t^{\text{int}}}{\partial n} - \frac{\partial G_{pt}}{\partial n} \phi_t^{\text{int}} \right] dS_t + \frac{1}{\varepsilon_{\text{int}}} \sum_k q_k G_{pk}, \quad p, k \in \Omega, \tag{5.2}$$

$$\phi_p^{ext} = \oint_S \left[-u_{pt} \frac{\partial \phi_t^{ext}}{\partial n} + \frac{\partial u_{pt}}{\partial n} \phi_t^{ext} \right] dS_t \quad p \in \bar{\Omega} \tag{5.3}$$

where ϕ_p^{int} is the interior potential at position p of the molecular domain Ω, q_k is the kth source charge, $S = \partial\Omega$ is the molecular boundary. ϕ_p^{ext} is the exterior potential at position p, ε_{int} is the interior dielectric constant, t is an arbitrary point on the boundary, and n is the outward normal vector at t. G_{pt} and u_{pt} are the fundamental solutions of the corresponding Poisson and PB equations, respectively. When point p approaches the surface S, by satisfying the boundary conditions $\phi^{\text{int}} = \phi^{ext}$ and $\varepsilon_{\text{int}}(\nabla\phi^{\text{int}} \cdot n) = \varepsilon_{ext}(\nabla\phi^{ext} \cdot n)$, Eqs (5.2) and (5.3) become a set of self-consistent BIEs,

$$\alpha_p f_p = \oint_S^{PV} \left[\varepsilon G_{pt} h_t - \frac{\partial G_{pt}}{\partial n} f_t \right] dS_t + \frac{1}{\varepsilon_{\text{int}}} \sum_k q_k G_{pk}, \quad p \in S, \tag{5.4}$$

$$(1 - \alpha_p) f_p = \oint_S^{PV} \left[-u_{pt} h_t + \frac{\partial u_{pt}}{\partial n} f_t \right] dS_t, \quad p \in S, \tag{5.5}$$

where PV denotes the principal value integral to avoid the singularity when $t \to p$, $f = \phi^{ext}$, $h = \nabla\phi^{ext} \cdot n$, and $\varepsilon = \varepsilon_{ext}/\varepsilon_{\text{int}}$. α_p depends on the local surface geometry at node p, and is equal to $1/2$ for a smooth surface. For the vertex of a polyhedron, $\alpha_p = A_p/4\pi$, where A_p is the interior solid angle at p. The derivative BIEs (dBIEs) can be obtained by linearly combining Eqs (5.4) and (5.5), and their derivative forms (for the smooth surface case).

$$\left(\frac{1}{2\varepsilon} + \frac{1}{2} \right) f_p = \oint_S^{PV} \left[(G_{pt} - u_{pt}) h_t - \left(\frac{1}{\varepsilon} \frac{\partial G_{pt}}{\partial n} - \frac{\partial u_{pt}}{\partial n} \right) f_t \right] dS_t + \frac{1}{D_{ext}} \sum_k q_k G_{pk} \quad p \in S \tag{5.6}$$

$$\left(\frac{1}{2\varepsilon} + \frac{1}{2} \right) h_p = \oint_S^{PV} \left[\left(\frac{\partial G_{pt}}{\partial n_0} - \frac{1}{\varepsilon} \frac{\partial u_{pt}}{\partial n_0} \right) h_t - \frac{1}{\varepsilon} \left(\frac{\partial^2 G_{pt}}{\partial n_0 \partial n} - \frac{\partial^2 u_{pt}}{\partial n_0 \partial n} \right) f_t \right] dS_t + \frac{1}{D_{ext}} \sum_k q_k \frac{\partial G_{pk}}{\partial n_0} \quad p \in S \tag{5.7}$$

where n is the unit normal vector at point t and n_0 is the unit normal vector at point p. The dBIEs lead to a well-conditioned (Fredholm second kind) system of algebraic equations. When Krylov subspace methods are applied to such systems, the number of iterations is bounded, independent of the number of nodes in the discretization. Hence, the total number of operations required to solve discretized Eqs (5.6) and (5.7) is a constant (the number of iterations) multiplied by the operations required for a matrix vector multiplication. To accelerate the matrix vector multiplication, we have applied the new version of fast multipole method (FMM)[6,7] that requires $O(N)$ operations with an optimized prefactor, and the linear equation system can therefore be solved in asymptotically optimal $O(N)$ time. Unlike our previous implementation,[4] in which a uniform oct-tree structure is used, the new implementation uses an adaptive tree structure.[5] This is an important improvement as all the unknowns in BIE/PB are located on molecule surfaces.

The accuracy of the new algorithm was first examined by direct comparison with the analytical solution of a point charge located at the center of a spherical cavity. For a relatively uniform element distribution, the adaptive (AFMM) and non-adaptive (FMM) algorithms were found to exhibit almost the same level of accuracy as shown in Figure 5.1. Both energy and potential errors decrease monotonically with decreasing (finer) element size. Energy and potential calculations converge at a relative error $<0.2\%$ when the surface mesh resolution is below 0.25 Å^2 (the surface area of a single triangular panel).

When compared with the non-adaptive solver, the AFMM-based solver shows significantly more efficient memory usage. This is because, in a non-adaptive tree structure, when the elements are only distributed on the surface as in BEM, a large number of boxes beyond the molecular surface are empty,

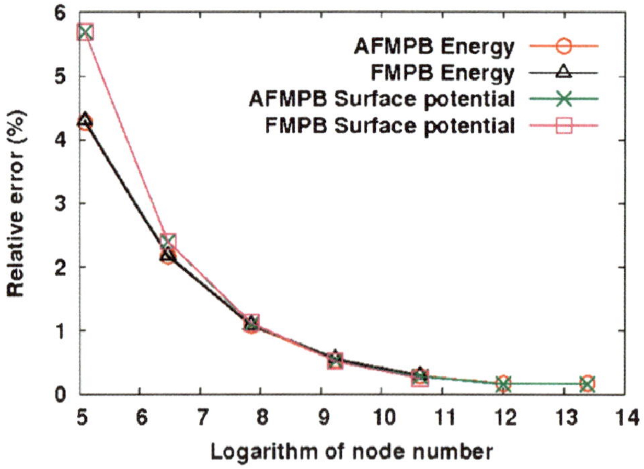

Figure 5.1 Accuracy of energy and potential calculations with the non-adaptive (FMM) and adaptive (AFMM) solvers. The relative errors of potentials are averaged over all surface node points. Reprinted with permission from B. Lu, X. Cheng, J. Huang and J. A. McCammon, An Adaptive Fast Multipole Boundary Element Method for Poisson-Boltzmann Electrostatics, *J. Chem. Theory Comput.*, 2009, **5**, 1692–1699.

Table 5.1 Performance Comparisons for a Spherical Cavity Case at Various Discretization Resolutions.

Number of elements	*CPU (s)*		*Memory (Megabytes)*		*Max level*	
	AFMM	*FMM*	*AFMM*	*FMM*	*AFMM*	*FMM*
162	0.05	0.13	2.7	2.7	3	2
642	0.21	0.62	7.0	7.9	4	3
2562	0.89	2.66	24.4	54.0	5	3
10 242	4.63	11.44	113.3	241.0	6	4
40 962	19.26	57.73	511.8	935.0	7	5
1 63 842	78.35	—	2152.1	—	8	—
6 55 362	1051.20	—	7900.7	—	9	—

leading to unnecessary memory usage for storing these empty boxes and their associated expansion coefficients. By only considering the memory usage by FMM for a uniform distribution of particles on a spherical surface, we estimate a memory reduction of $\sim 2^{n-1}$ fold for an n-level depth of tree structure. Although this simplified analysis is not strictly valid for PB calculations of biomolecules, the overall trend remains similar (Table 5.1).

Furthermore, the adaptive FMM algorithm allows an improved load-balance for partitioning the elements into local (calculated directly) and far field (calculated using expansion coefficients) contributions, whereas in the non-adaptive FMM the partition is greatly limited by the power growth of memory (*i.e.* $\sim 8^n$ in an *oct*-tree data structure) as the number of tree levels n increases. As shown in Table 5.1, with 6 55 362 surface elements, the adaptive solver can handle a maximum tree level of 9 without causing memory overflow, while the non-adaptive algorithm can only handle a maximum level of 6. When the calculation is performed at level 6, too many elements are assigned to the local list for direct calculation, which significantly compromises the overall efficiency. For most of the systems tested, an average speedup of 2–3 fold has been observed by using the adaptive FMM. For larger systems, even better performance is expected.

5.2.1 A Tree Code-Based Generalized Born (tGB) Method

The generalized Born (GB) model[8–11] has been widely used as an alternative to the numerical solution of the PB equation, especially in molecular dynamics simulations. In GB, the electrostatic solvation energy is given by the sum of pairwise interactions between the solute atoms using the so-called effective Born radii of individual atoms to take into account the polarization effects of the bulk solvent. The GB algorithm consists of two major time-consuming steps: the calculation of the effective Born radii and the summation of interactions over all atom pairs, both of which currently scale as order N^2, where N is the number of atoms in the system. Therefore, for biomolecules with >10 000 atoms, GB remains extremely computationally demanding. It is even less competitive than explicit solvent models, for which fast algorithms such as the FFT-based particle-mesh Ewald method[12] have been widely applied to achieve an overall $O(N\log N)$ complexity. For the Born radii calculations,

FFT-accelerated methods have been developed to reduce the arithmetic complexity of the algorithm.[13] However, to the best of our knowledge, no linear or super-linear algorithm has been proposed for calculating the summation of interactions over all atom pairs, probably due to the complex interaction kernel used in GB. Here, we focus on describing an O(NlogN) method for the summation of pairwise interactions based on a tree-code algorithm.

In the common form of the GB model, the electrostatic solvation free energy, ΔG_{elec}, is computed by the sum over all pairs of the atoms,[9]

$$\Delta G_{elec} = \frac{1}{2}\left(\frac{1}{\varepsilon_i} - \frac{1}{\varepsilon_0}\right)\sum_{i,j=1}^{N} \frac{q_i q_j}{\sqrt{r_{ij}^2 + \alpha_i\alpha_j \exp(-4r_{ij}^2/\alpha_i\alpha_j)}} \tag{5.8}$$

where q_i and q_j are the atomic charges, ε_i and ε_0 are the interior and exterior permittivity, r_{ij} is the distance between atoms i and j, N is the total number of atoms in the system, and α_i and α_j are the effective Born radii of atoms i and j. The effective Born radius α_i of atom i is usually defined as

$$\alpha_i = -\frac{1}{2}\left(\frac{1}{\varepsilon_i} - \frac{1}{\varepsilon_0}\right)\frac{q_i^2}{\Delta G_{elec}} \tag{5.9}$$

where ΔG_{elec} is the self electrostatic solvation energy of atom i in the same dielectric environment but in the absence of other charges. Ideally, the self-energy would be computed by solving the Poisson equation numerically, which leads to the so-called "perfect" effective Born radius.[14] This approach is, however, impractical for irregular geometries due to its high computational cost. Instead, most existing GB methods turn to a formulation of the electrostatic energy in terms of integration over the solute dielectric regions based on the Coulomb-field approximation, which assumes the electric displacement flux remains in the Coulombic form when the dielectric constant varies from ε_i and ε_0 upon solvation. The Coulomb-field approximation, however, has been found to heavily overestimate the Born radii for off-center atoms. To address this difficulty, an alternative R6 volume integral model has been proposed by Grycuk,[15,16]

$$\frac{1}{\alpha_i^3} = \frac{1}{a_i^3} - \frac{3}{4\pi}\int_{\Omega_{in}\backslash A_i} \frac{d^3r}{|r - r_i|^6} \tag{5.10}$$

where A_i is the vdW sphere of atom i and a_i is its radius.

For relatively large biomolecules, numerical integration over the solute dielectric regions in Eq. (5.10) is still too computationally demanding. Therefore, pairwise approximations have been proposed to compute the integral by a sum of contributions over individual atomic spheres.[17] The pairwise summation, however, overestimates the solute region because of the possible overlaps between atoms. To compensate for this error, Hawkins *et al.* proposed an empirical correction in which the Born radii were scaled using fitted parameters.[18] Rather than fitting the scaling parameters, Gallichio and Levy[19] later introduced a method to compute these parameters from the geometry of the molecule based on the Poincare formula for the union of N spheres. Using Gallichio and Levy's analytical pairwise descreening approximation, the integral in Eq. (5.10) then takes the following form, in which the molecular volume is decomposed into a set of vdW spheres.

$$\frac{1}{\alpha_i^3} = \frac{1}{a_i^3} - \frac{3}{4\pi}\sum_j s_{ji} H_{ji}(a_i, a_j, r_{ij}) \quad \text{with} \quad H_{ji}(a_i, a_j, r_{ij}) = \int_{A_j \backslash A_i} \frac{d^3 r}{|r - r_i|^6} \tag{5.11}$$

where the sum runs over all the spheres with inter-particle distance r_{ij} greater than a_i, and $A_j \backslash A_i$ denotes the domain of A_j excluding the overlapping portion with A_i. The scaling factors s_{ji} account for possible overlap of atomic spheres. The integral H_{ji} can be computed exactly because $A_j \backslash A_i$ is a regular domain. Direct evaluation of H_{ji} has $O(N^2)$ complexity. However, linear scaling can be achieved by noting that in the R6 formulation, H_{ji} decays rapidly with increasing r_{ij}. Therefore, a direct cutoff scheme can be used without much loss in accuracy.

We next discuss implementing a tree code algorithm[20] to compute the pairwise interactions in Eq. (5.8). In tree code, the particles are grouped into a hierarchical structure of nested boxes. Starting from the root box, each box is recursively subdivided into eight smaller boxes until the particle number in a child box is less than a specified value N_0. By traversing the hierarchical tree, the total interactions with atom j can be represented by a sum of particle-cluster and particle-particle interactions for far-field and local interactions respectively. A multipole acceptance criterion (MAC) is used to determine whether a particle-cluster interaction is allowed by comparing the ratio of box size h_A to the particle-cluster distance r_{jA} with a predefined error control parameter θ.

Let us consider the interaction between atom j and cluster A, where the reaction field energy is expressed as

$$q_j \Phi_{jA} = q_j \sum_{i \in A} \phi_j(r_i) = \beta q_j \sum_{i \in A} \frac{q_i}{\sqrt{r_{ij}^2 + \alpha_i \alpha_j \exp(-4r_{ij}^2/\alpha_i \alpha_j)}} \tag{5.12}$$

where $\beta = -1/2(1/\varepsilon_i - 1/\varepsilon_0)$. As a first-order approximation $O(h_A)$, the exponential term can be re-expressed in terms of the average over all the particles in the cluster,

$$q_j \Phi_{jA} \approx \beta q_j \sum_{i \in A} \frac{q_i}{\sqrt{r_{ij}^2 + \alpha(r_i)\alpha_j \exp(-4r_{jA}^2/\alpha_A \alpha_j)}} \tag{5.13}$$

where r_{jA} is the distance between $\boldsymbol{r}_j$ and the center of the cluster, $\boldsymbol{r}_A$, α_A is the average Born radius of cluster A, and $\alpha(r)$ is the Born radius function that will be approximated by using $\{\alpha_i,\ i = 1, \ldots, N\}$ and their derivatives at atomic locations. Then the multipole expansion for Φ_{jA} is given by

$$q_j \Phi_{jA} \approx q_j \sum_{\|k\|=0}^{p} \frac{1}{k!} \partial^k \phi_j(r_A) \sum_{i \in A} q_i (r_i - r_A)^k = \beta q_j \sum_{\|k\|=0}^{p} T_k M_A^k \tag{5.14}$$

where $k=(k_1, k_2, k_3)$ is the multi-index notation, T_k and M_A^k represent the kth Taylor expansion coefficient (divided by β) and the moment of the cluster A, respectively. As the moment of each cluster needs to be computed only once, the operations of evaluating the particle-cluster interactions depends only on the number of clusters, thereby leading to an $O(\log N)$ total operations for each atom.

To assess the accuracy of the tGB algorithm, we first computed the electrostatic solvation energies for the entire test set of 25 proteins with a cutoff scheme ($L_c=8$ Å) for the effective Born radii calculation. The solvation energies of the 25 proteins with the P1 (the first order multipole expansion) and P2 (the second order expansion) tree code algorithms are compared to the Poisson (obtained with APBS solver)[2] and direct summation GB results. The statistics for the relative errors are given in Table 5.2. Compared to the direct GB, the average unsigned error for the P1 tree code is 0.2133%, and for P2 is 0.1397%, demonstrating that the tree code algorithm provides a good approximation for the pair wise interactions. It is also shown that the tGB energies approximate the PB energies with an average relative error less than 0.6%.

We also computed the absolute and relative force errors using the P2 expansion with two MAC parameters $\theta=0.4$ and 0.3. The resulting error statistics is summarized in Table 5.3. It is evident that the force calculation is less accurate than the solvation energy calculation, mainly because the force calculation uses one less order of multipole expansions. The average relative errors are about 3.54% when the MAC parameter $\theta=0.4$. The errors are decreased if θ is reduced to 0.3, with the average relative error about 1.15% and the maximum error about 1.73%. The maximum absolute error for $\theta=0.3$ is 0.001254 kcal/mol/Å.

Table 5.2 Statistics for relative errors of electrostatic solvation energies between tGB and direct sum GB $(\Delta GB_{direct}-\Delta tGB)/\Delta GB_{direct}$, and between tGB and PB, $(\Delta PB-\Delta tGB)/\Delta PB$.

	P1, GB	*P2, GB*	*P1, PB*	*P2, PB*
Standard deviation (%)	0.1785	0.0574	0.7854	0.7845
Signed error (%)	−0.1712	−0.1397	−0.1521	−0.1208
Unsigned error (%)	0.2133	0.1397	0.5956	0.5969
Maximum error (%)	0.4586	0.2588	2.1422	2.2112

Table 5.3 Error performance (average and maximum errors) for the force calculation with the P2 tree code with MAC parameters $\theta=0.4$ and $\theta=0.3$.

	$\theta=0.4$	$\theta=0.3$
Average absolute error (kcal/mol/Å)	2.229e-3	6.913e-4
Maximum absolute error (kcal/mol/Å)	4.739e-3	1.254e-3
Average relative error (%)	3.542	1.151
Maximum relative error (%)	5.141	1.734

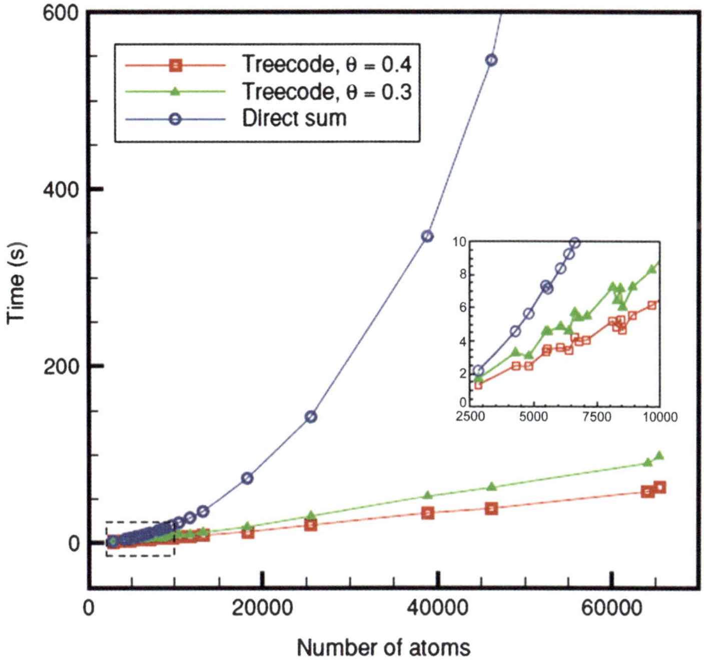

Figure 5.2 CPU times for computing the GB solvation energies of 25 test proteins using the P2 tree code algorithm ($\theta = 0.4$ and 0.3) and direct summation, together with those of the effective Born radii calculations. The inset shows an enlarged view of the dashed portion. Reprinted with permission from Z. Xu, X. Cheng and H. Yang, Treecode-based generalized Born method, *J. Chem. Phys.*, 2011, **134**, 064107.

The CPU times for computing the effective Born radii, the P1 and P2 tGB energies, along with the direct summation GB energies, are shown in Figure 5.2. The CPU times for both the effective Born radii calculation and the tree code algorithms increase almost linearly with the number of particles. The P1 and P2 tree codes outperform the direct sum GB for even the smallest protein (PDB code 1NP4, 2815 atoms) in the test set, suggesting the prefactor for tGB is reasonably small. Including the CPU times for the effective Born radii calculation (with cutoff distance Lc = 8 Å), the total speedups of P1 are 3.10, 5.96 and 10.74, and those of P2 are 2.79, 5.26 and 9.41, respectively for the particle numbers of 10k, 20k and 40k.

5.2.2 Scaling of a Multimillion-atom Lignocellulose Simulation on a Petascale Supercomputer

Atomistic MD simulations of lignocellulosic biomass still face significant challenges. Among these are the characteristic length scales (Å to *μm*) and time scales (*ns* to *μs* and beyond) of events pertinent to the recalcitrance of biomass to hydrolysis into sugars. To access these length and time scales, standard MD protocols that are commonly employed to study systems of ~10–100 Å on time scales of ~100 *ns*, must be modified.

Current supercomputers, such as the Jaguar Cray XT5 at Oak Ridge National Laboratory used for our recent study, are beginning to consist of over 10^5 cores and in this way reach petaflop nominal speeds. However, the challenge for MD and other applications is to achieve efficient scaling up to ~10^4–10^5 cores. Thus, simulations are limited by the parallel efficiency of the MD algorithms, *i.e.* their ability to run in parallel on many thousands of processors.

The most computationally demanding step in MD simulation of biological systems is the treatment of long-range interactions, which in non-polarizable force fields are represented by Coulomb and van der Waals (Lennard-Jones) terms (for a review, see A.D. Mackerell, 2004).[21] Whereas the van der Waals interactions are almost always truncated at a cutoff distance, the electrostatic interactions extend to longer ranges. A common method to treat the electrostatics is to directly calculate the Coulomb interaction for any pair of atoms separated by less than a cutoff distance R_c, and outside this distance to calculate the interactions using the Particle Mesh Ewald (PME) method[12,22] (assuming periodic boundary conditions are applied to the system). By using an Ewald summation to split the Coulomb interactions into a short-range part that converges quickly in real space and a long-range part that converges quickly in reciprocal space, the PME method reduces the computational cost of N particles interacting with each other from O(N) to O($N\log N$).

An alternative method to PME that avoids the electrostatics bottleneck is the reaction field method (RF).[23,24] In RF, it is assumed that any given atom is surrounded by a sphere of radius, R_{rf}, again within which the electrostatic interactions are calculated explicitly. Outside the sphere the system is treated as a dielectric continuum. The occurrence of any net dipole within the sphere induces a polarization in the dielectric continuum, which in turn interacts with the atoms inside the sphere. Due to the approximate treatment of electrostatics outside the cutoff sphere, the RF method is not expected to yield accurate results for systems containing full charges. Nevertheless, the RF model allows the replacement of the infinite Coulomb sum by a finite sum modified by the reaction field, and therefore improves the parallel scaling relative to the PME method.

In a recent publication, we outlined a strategy for fast and accurate all-atom simulation of multi-million atom lignocellulose systems.[25] Two technical problems were addressed. First, we compared the accuracy of MD using PME and RF non-bonded approaches on the benchmark systems, and, second we examined scaling of MD for large systems on a petascale supercomputer.

5.2.3 Comparison of Simulations with Different Electrostatic Methods

First, the accuracy of RF for the biomass test systems was examined. For this analysis, structural and dynamical properties were compared in simulations using different electrostatics methods. The particular choice of properties for

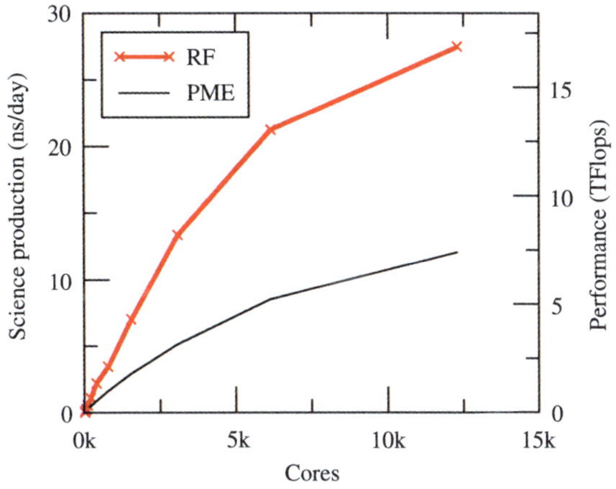

Figure 5.3 Strong scaling (*i.e.* increasing the number of cores with fixed system size) of the 3.3 million-atom lignocellulose model.

comparison was based on their structural importance and their anticipated sensitivity to possible electrostatic artifacts. Six quantities were compared: the total dipole moment of the fiber, the Kirkwood function between dipoles of different chains (providing information on the distance-dependent correlation of molecular dipoles), root mean square fluctuations, and finally three specific dihedral angles of structural importance in cellulose. In all cases, excellent agreement was found between RF and PME.

5.2.4 Scaling

The parallel efficiency of the RF MD simulation was evaluated by considering its strong scaling behavior. To test strong scaling, the system size (number of atoms in the system) is held constant while the number of cores is varied. For the 3.3 million-atom MD simulation of lignocellulose using the RF on the ORNL Jaguar Cray XT5, the strong scaling is shown in Figure 5.3. For this system, RF scales well to 12 288 cores and achieves 27.5 ns/day, running at 16.9 TFlops, a three-fold improvement over the commonly used PME method.

5.3 Applications

The applications section is in two parts: biofuels and bioremediation (subsurface biogeochemistry).

5.3.1 Biofuel Research

Lignocellulose is a complex material found in plant cell walls, composed of cellulose microfibrils embedded in matrix of cross-linked hemicellulose and lignin cell-wall polymers. Although plant biomass has the potential to be a renewable feedstock for biofuel and biomaterial production, its recalcitrance to hydrolysis necessitates expensive pretreatment prior to fermentation.[26] Pretreatment significantly increases the cost of cellulosic biofuel production because of its energy requirements as well as capital and operating costs. Dilute-acid pretreatment that involves soaking biomass in dilute sulfuric acid at $\sim 160\,^{\circ}\mathrm{C}$ depolymerizes and solubilizes hemicellulose, increasing the accessibility of cellulose to cellulolytic enzymes. On the other hand, lignin aggregates and re-precipitates back onto the cellulose after biomass pretreatment, hindering access to cellulose by hydrolytic cellulase enzymes.[27]

Understanding the structural changes that occur at the molecular level during pretreatment provides the fundamental information needed to understand and control this process and improve the efficiency of cellulosic biofuel production. Computer simulation is a powerful tool for complementing experiment in obtaining molecular-level understanding of the structure, dynamics and energetics of lignocellulose. Here, we discuss the technical challenges that must be addressed before performing all-atom MD simulation of lignocellulose. We then discuss results of recent calculations on lignocellulose.

5.3.1.1 A Molecular Mechanics Force Field for Lignin

In recent work, we presented the first essential step toward the accurate computer simulation in which we derived an empirical molecular mechanics (MM) force field for lignin.[28]

5.3.1.2 Parameterization Strategy

Here, we outline the general strategy employed to obtain the lignin force field. The CHARMM potential energy function[29] of a molecule is given by Eq. (5.15):

$$\begin{aligned} E = & \sum_{bonds} K_b(b-b_o)^2 + \sum_{angles} K_\theta(\theta-\theta_o)^2 + \sum_{U-B} K_{ub}(s-s_o)^2 \\ & + \sum_{dihedrals} K_\phi[1+\cos(n\phi-\delta)] + \sum_{impropers} K_\psi(\psi-\psi_o)^2 \\ & + \sum_{non\text{-}bonded} \left\{ \varepsilon_{ij}\left[\left(\frac{R_{ij}}{r_{ij}}\right)^{12} - 2\left(\frac{R_{ij}}{r_{ij}}\right)^{6}\right] + \frac{q_i q_j}{4\varepsilon_0 \pi r_{ij}} \right\} \end{aligned} \tag{5.15}$$

where the contributions to the energy include bonded (bond, angle, Urey-Bradley, dihedral and improper dihedrals) and non-bonded (Lennard-Jones 6-12 potential for the van der Waals interactions and Coulomb electrostatic) terms. The parameters, including the force constants K and partial atomic charges q_i, are molecule-dependent and must be derived prior to performing simulations. The parameterization of the force field for lignin was the main task of L. Petridis and J. C. Smith, 2009.[28]

This parameterization of lignin followed the main procedure of parameterization of proteins. Two model compounds were employed: methoxybenzene and *p*-hydroxyphenyl (PHP), the simplest lignin unit. Parameters were optimized by considering two factors. First, the 'target data' were reproduced as closely as possible, which effectively ensures that the force field describes specific properties of lignin accurately. Second, compatibility with the existing CHARMM force field was ensured by restricting the optimization to include only those parameters that did not already exist in the force field.

The optimization strategy for the new parameters is summarized in Figure 5.4. Equilibrium values for bond lengths, angles and dihedrals were taken from high-level (MP2/6-31G*) QM-optimized geometries and were not further revised. The van der Waals parameters were taken from the existing CHARMM force field.[30] Charges were optimized with respect to the QM interaction energies using a supramolecular approach of methoxybenzene interacting with one water molecule. The partial charges were adjusted to reproduce minimum distances and interaction energies between anisole and water. In order to mimic the effect of electronic polarizability, which is not explicitly taken into account in additive force fields, atomic charges were purposely overestimated. The empirical calculations were found to reproduce well the scaled QM interaction energies, with the error being less than 3%.

After completing the non-bonded interactions, parameters for dihedral rotations were deduced from QM potential energy surfaces. The parameter optimization was based on reproducing quantum chemical adiabatic energy surfaces, in which the selected dihedral is held constant while the remaining

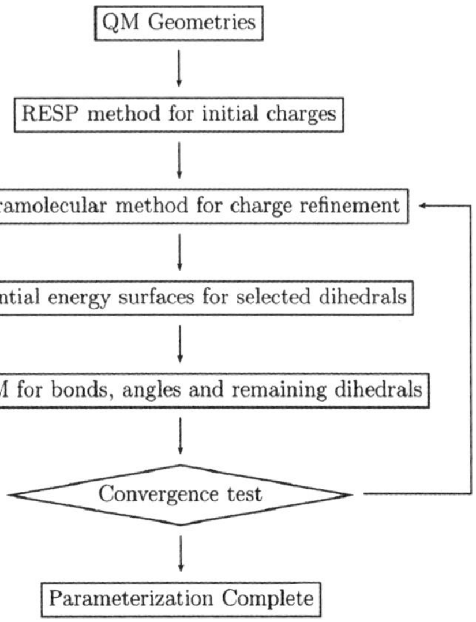

Figure 5.4 Schematic representation of the strategy employed to parameterize the force field for lignin.

degrees of freedom are allowed to relax to a constrained energy minimum. The remaining bonded parameters (bonds and angles) were optimized to reproduce vibrational frequencies and eigenvector projections derived from QM calculations. For this step, the Automated Frequency Matching Method was employed, which automates optimization of the MM parameter set until the best fit with the QM reference set is obtained.[31] Finally, the parameter set was tested without further adjustment against condensed phase experimental properties of lignin that were not used during the parameterization.

5.3.1.3 *Lignin Structure Elucidated by Combination of Simulation and Neutron Scattering*

Lignin is one key source of the recalcitrance of biomass to hydrolysis because it aggregates and re-precipitates back onto the cellulose after biomass pretreatment, hindering hydrolytic cellulases access to cellulose.[32] *In vivo* lignin aggregates have been found to be polydisperse in size, spanning three orders of magnitude, from ~nm to ~μm. However, the detailed structural characteristics of lignin in aqueous solution are largely unknown.[33]

Small-angle neutron scattering (SANS) provides structural information of materials on the ~10–2000 Å length scales. Figure 5.5(a) displays SANS data of loblolly pine lignin.[34] The average radius of gyration R_g, obtained from the low-Q data, is ~1300 Å, where $Q=(4\pi/\lambda)\sin(\theta/2)$, λ is the wavelength of the incident neutrons and θ the scattering angle. A power-law dependence on the scattering vector Q, *i.e.* $I(Q)\propto Q^{-\alpha}$ is found over the Q-range, $0.005\,\text{Å}^{-1}<Q<0.25\,\text{Å}^{-1}$. The power-law exponent, α of 3.38 ± 0.02, lies in the range $3<\alpha<4$, consistent with scattering from surfaces. The surface scan be

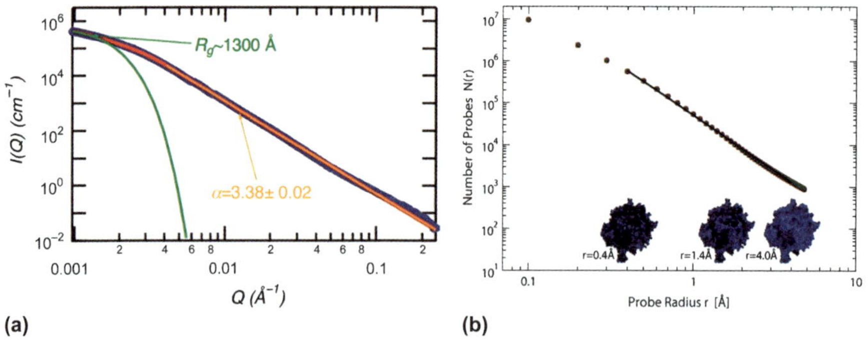

Figure 5.5 (a) SANS profile of lignin in fully deuterated aqueous solution shown in blue. The unified-fit is shown in red, separated into two contributions, from the Guinier region (green) and power-law region (orange). Background contributions, solvent and incoherent (hydrogen) scattering were subtracted. (b) N(r), the number of spherical probes required to cover the surface of the aggregate, as a function of the radius R of the probe. Simulation data of the four models are shown in symbols. The fit of the power-law region of their ensemble average is a solid black line.

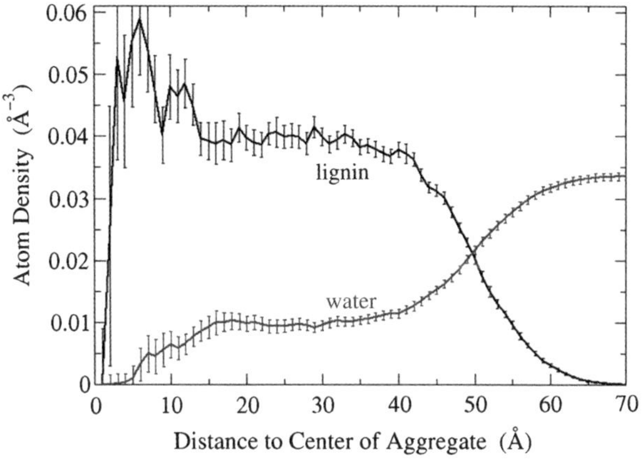

Figure 5.6 Number density of heavy atoms of lignin (black), all water molecules (blue) averaged over the last 50 ns of the 150 ns simulation.

characterized by dsexp, the surface fractal dimension which is related to α by the relation $d_s = 6 - \alpha$. d_s is a measure of the roughness or irregularity of a surface and ranges from a value of between 2 for a smooth surface to a value of 3 for a rough surface. Here, $d_s = 2.62 \pm 0.02$, which corresponds to a highly folded surface.

The surface fractal dimension was obtained from complementary MD simulations using the relation $N(r) \propto r^{-d_s}$, where N(r) is the minimum number of spheres of radius r required to fully cover the surface of the molecule.[34] N(r) plots for the aggregate models Figure 5.5(b) exhibit power-law behavior for probe radii in the range $0.4\,\text{Å} < r < 4.0\,\text{Å}$. The resulting ensemble MD value of $d_s = 2.65 \pm 0.01$, is in excellent agreement with the experiment.

The highly-folded surface and its possible function in water conduction render the hydration structure associated with the lignin aggregates of particular interest. Although hydrophobic, the MD aggregates have considerable water penetration, as shown by the distribution of lignin and water atoms as a function of distance *R* from the center of mass of the MD aggregates (Figure 5.6).[34]

5.3.1.4 Coarse-grained Simulations of Cellulose

Even with the advent of efficient computational power combined using sophisticated routines, atomistic simulations are unable to explore long time and length scales associated with biomass structure and dynamics. The smallest cellulose microfibrils consist of 36 individual chains with a large degree of polymerization.[35] These large system sizes, together with the associated long time dynamics required to examine cell-wall processes, puts simulation studies of cell-wall decomposition beyond the scope of a complete atomistic representation.[36,37] To overcome this difficulty, recent simulation studies have focused on alternative methodologies based on lower resolution, coarse-grained (CG) molecular dynamics (MD).[38,39] CG approaches have been shown to be useful for exploring long time and length scales associated with the dynamics of proteins,[40,41] lipids,[42,43] polymers[44] and other materials.[45] In the following section, a brief overview of CG methodology as applied to cellulose fibrils is described.

In general, a CG approach starts with chemical and structural information from experiments or higher resolution simulations. Based on structure or chemistry, a certain number of atoms is grouped into one or more CG units. For example, each glucose monomer in a cellulose chain can be represented as a single CG unit (Figure 5.7). As a result, a 36-chain fibril contains 2880 CG units instead of 56 000 atoms. Clearly, the CG method involves a reduced number of interacting sites compared to its atomistic counterparts. This reduction, combined with relatively smoother potentials, results in dynamics that are at least two orders faster than atomistic simulations.

The above-mentioned CG model for crystalline cellulose was obtained based on structural information from a 20 ns long atomistic trajectory of a cellulose fibril in water.[25] Similar to atomistic potentials, CG potentials involve bonded and non-bonded contributions,

$$V_{CG}(r) = V_{bond} + V_{angle} + V_{torsion} + V_{nonbond}$$

and all the bonded potentials (distance, angle and torsion) are modeled as harmonic. Because each glucose monomer is modeled as a single CG unit, there is no net charge associated with any CG unit. In determining bonded parameters, distributions between bonded units from atomistic simulations serve as the target observables. The CG bonded parameters are fine-tuned to reproduce the target observables with reasonable accuracy.

The non-bonded interactions are relatively complex and involved. The strategy for developing non-bonded interactions is similar to that of bonded

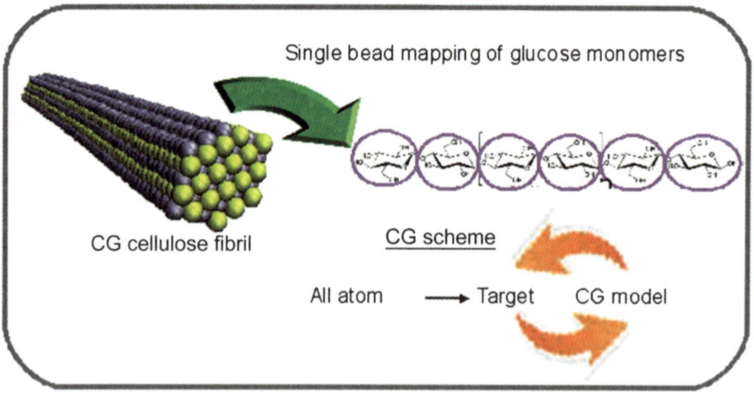

Figure 5.7 Coarse-grained approach for a crystalline cellulose fibril. Each glucose monomer in the cellulose chain is grouped into a single CG unit. Origin (OR-yellow) and center (CE-blue) layers within the fibril are treated distinctly by assigning separate interaction potentials.

interactions, namely obtaining distributions between the non-bonded units of interest from corresponding atomistic simulations and reproducing the distributions using CG simulations by adjusting the underlying potentials. For cellulose fibril, the radial distribution function $g(r)$ is the target observable from the atomistic simulations. Since there are two different CG units, OR and CE, three separate non-bonded potentials must be determined (OR-OR, OR-CE and CE-CE).

$g(r)$ for crystalline cellulose is complex with multiple discrete distinct peaks, and thus contrasts with the simple, less structured functions of typical bulk/liquid systems. Consequently, it is impractical to employ simple Lennard-Jones type potentials in this case, as the discrete nature of $g(r)$ causes potentials to be discontinuous, resulting in force divergence. Using such simplified potentials may also cause instabilities in cellulose crystalline structure, as was observed in previous CG simulation studies of the cellulose fibril.[46]

The non-bonded CG potentials are optimized iteratively using the target observable, $V_{new} = V_{old} + k_B T\ ln(g_{cg(r)}/g_{aa}(r))$, where the updated potential V_{new} between CG units i and j in each step is obtained by modifying the potential from the previous simulation, V_{old}. k_B is the Boltzmann constant, T the absolute temperature and the distributions from the atomistic and CG simulations are $g_{aa}(r)$ and $g_{cg(r)}$, respectively. The iterative procedure begins with a simple Lennard-Jones 6-12 potential as the starting CG potential and is then modified using target and CG distribution functions. At each iteration, the CG simulation is carried out using the updated potentials, and the distribution functions obtained from the updated trajectory are used to update the potentials. Iteration continues until satisfactory agreement with the target observables is obtained.

Comparison (after five or six iterations) of the target and CG observables for the origin chains showed that this method is unable to capture the fine structure of $g_{aa}(r)$ arising from the crystalline nature of cellulose. Such problems are common when dealing with crystalline materials, for example, in the study of crystalline fatty acids.[47] As noted in K. R. Hadley and C. McCabe,[47] a damping factor (δ_{ij}) needs to be incorporated to suppress abrupt changes in the potentials, $V_{new} = V_{old} + \delta_{ij} k_B T\ ln(g_{cg(r)}/g_{aa}(r))$, δ_{ij} is assigned a positive value between 0 and 1. When δ_{ij} is 1, a non-crystalline system is recovered.

The potentials obtained using the iterative procedure did not correspond to any simple analytical form and hence all the non-bonded potentials are tabulated.[48] The CG distribution functions obtained in this fashion reproduce their atomistic counterparts remarkably well. Importantly, the positions of all the peaks corresponding to the crystal structure are reproduced faithfully by the CG simulations. Excellent agreement between the atomistic and CG structures demonstrates the ability of the single-bead model to capture the structural features at the monomer level. Unlike other CG models for cellulose, this model imposes no constraints to retain the crystalline structure. The solvent effects are implicitly incorporated in the model.

The CG force field is used in combination with stochastic dynamics so as to simulate solvent effects implicitly. The dynamics of the individual CG particles are governed by a Langevin equation, $m_i(d\mathbf{v}_i/dt) = F_i(\mathbf{r}_i) - m_i\xi_i\mathbf{v}_i + \boldsymbol{R}_i$. The CG particle mass, velocity and position are represented by m_i, $\mathbf{v}_i$ and $\mathbf{r}_i$ respectively. F_i is the systematic force on particle i calculated using the tabulated potentials and ξ_i is the friction coefficient. The random force $\boldsymbol{R}_i$ is selected from a Gaussian distribution with zero mean. Simulations using the GROMACS package with a 1 fs time step showed that the CG fibril remains stable over a microsecond period. The crystalline fibril yielded average RMSD values (0.11 nm), much smaller than the size of the CG units (0.52 nm), indicating fibril stability.

The cell wall contains both crystalline and amorphous celluloses. To simulate amorphous celluloses, a system of individual cellulose chains dispersed in water was considered as a model for the fully amorphous/noncrystalline form. From this, several non-crystalline forms of cellulose fibrils with different degrees of crystallinity were generated by introducing a scaling factor (λ) between fully crystalline and fully amorphous cellulose systems. The CG potential for any of these structures can be described using a combination of crystalline (V_{cr}) and amorphous (V_{am}) potentials as $V_{cg}(r_{ij}) = \lambda\ V_{cr}(r_{ij}) + (1 - \lambda)\ V_{am}(r_{ij})$. Final snapshots observed in individual simulations with different λ values are shown in Figure 5.8.

The CG model allows the exploration of crystalline and non-crystalline cellulose fibril structures for length and time scales beyond the reach of atomistic simulation studies. In particular, the above example provides a systematic way of generating and representing both crystalline and a range of partially crystalline and amorphous cellulose states. By developing a CG model for lignin, one can efficiently explore lignin-cellulose interactions on reliable time and length scales.

5.3.2 Bioremediation: Mechanisms of Bacterial Mercury Resistance

5.3.2.1 Bacterial Detoxification of Mercury

Mercury is a toxic, highly reactive element with no known biological function. Inorganic Hg(II) and organomercurial species such as methylmercury,

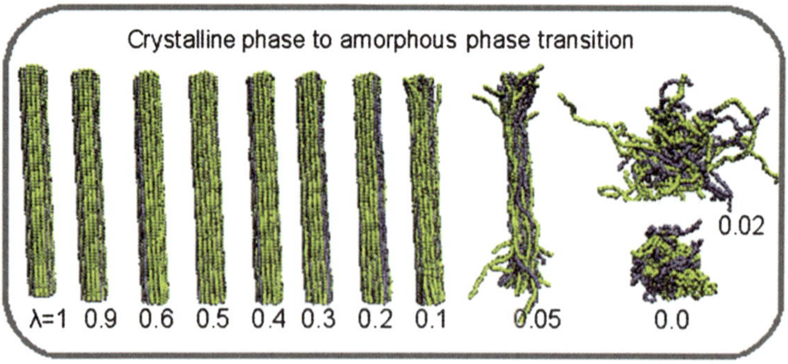

Figure 5.8 Final snapshots of CG cellulose fibrils with different λ values.

$[CH_3Hg(II)]+$, are both toxic due to their extremely high affinity for reduced sulfur ligands, whereas elemental Hg(0) is relatively non-toxic.

Bacteria inhabiting mercury-contaminated environments often possess a set of genes called the *mer* operon. Encoded in the *mer* operon is a set of proteins and enzymes that efficiently convert inorganic and organomercurial species to Hg(0), which diffuses out of the bacterial cell. Specific proteins and enzymes of the *mer* operon vary from species to species, but the typical *mer* operon consists of a periplasmic Hg(II) metallochaperone, MerP, membrane-bound transport proteins (MerC, MerT and others), and the mercuric reductase, MerA. Regulation of the *mer* operon is controlled by MerR. The organomercurial lyase, MerB, which is catalyzes Hg-C bond cleavage, is also present in some, but not all *mer* operons.

In this section, we describe recent work from our laboratory directed toward understanding the molecular mechanisms of bacterial mercury resistance. In particular, we describe progress in understanding the enzymatic degradation of methylmercury catalyzed by the organomercurial lyase, MerB, and transcriptional regulation carried out by the metalloregulator MerR.

5.3.2.2 Hg-C Bond Cleavage Catalyzed by MerB

Here, we provide a brief synopsis of our work on the enzymatic degradation of methylmercury.[49] Demethylation of methylmercury in mercury resistant bacteria is catalyzed by the organomercurial lyase, MerB. The reaction proceeds through protonation of the CH_3- leaving group, with Hg(II) and methane as the products: $[CH_3\text{-}Hg(II)]^+ \rightarrow CH_4 + Hg(II)$.

Lafrance-Vanasse *et al.* solved high-resolution X-ray crystal structures of *apo*- and Hg(II)-product bound MerB.[50] The active site of MerB consists of two cysteines, Cys96 and Cys159, and an aspartate, Asp99, all of which have been found to be essential for catalysis[51] (Figure 5.9).

Several mechanisms have been proposed for the Hg-C bond cleavage catalyzed by MerB.[52] One proposal, which we call Mechanism I, simply involves coordination of methylmercury to one of the two active-site cysteine residues (Cys96 or Cys159), followed by protonation of the CH_3- leaving group by the second cysteine residue. A second mechanistic proposal, which we call Mechanism II, involves methylmercury coordination to *both* of the active-site cysteines, followed by protonation of the CH_3- leaving group by Asp99. To test these two mechanisms, density functional theory (DFT) calculations were performed on a crystallographically derived active-site model of MerB.

The side chains of Cys96, Cys159, Asp99 were included in the model, with alpha carbons held fixed at their crystallographic positions. Hg(II), a crystallographic water molecule, and a single methane molecule, representing the Hg-C cleavage product of methylmercury, were also included. Once the product state was constructed and optimized, other stationary points corresponding to reactants, transition states and intermediates were generated in a similar way. The B3PW91 hybrid exchange-correlation functional[53,54] was used in all calculations. The Stuttgart-Dresden (SDD) relativistic effective core potential and

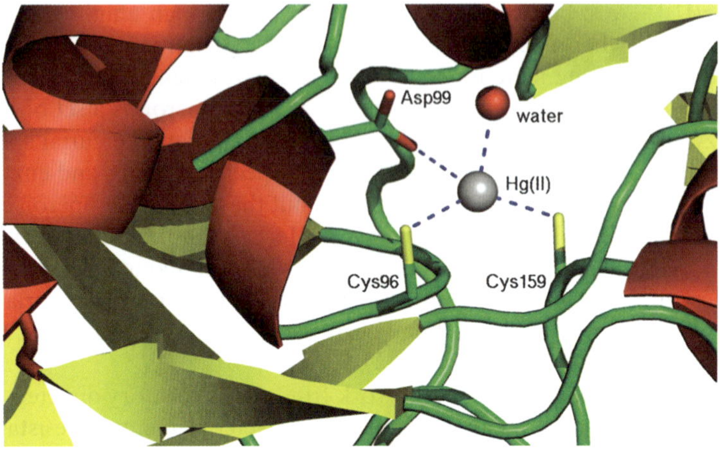

Figure 5.9 Active site of Hg(II)-bound MerB (Lafrance-Vanasse, 2009). Cys96, Cys159, Asp99, Hg(II) and a crystallographic water molecule are labeled. Reprinted with permission from J. M. Parks, H. Guo, C. Momany, L. Liang, S. M. Miller, A. O. Summers, J. C. Smith, *J. Am. Chem. Soc.*, 2009, **131**, 13278–13285. Copyright 2009 American Chemical Society.

basis set[55] was used to describe Hg. Continuum solvation with a dielectric constant of 4.0 was used to represent the enzyme-solvent environment, as is the standard protocol for such calculations.[56–58] Natural Population Analysis[59] was used to compute atomic partial charges of transition-state structures.

Stationary point structures (reactants, transition states, intermediates and products) were generated and optimized for the methylmercury cleavage reaction pathways corresponding to Mechanisms I and II. The experimental and calculated activation barriers are shown in Table 5.4. The calculated barrier for Mechanism I of 33.9 kcal/mol was significantly higher than the experimental value of 20.1 kcal/mol.[60] Mechanism II, in which Cys96 and Cys159 are both coordinated with the substrate and Asp99 protonates the leaving group, yielded a computed activation barrier of 20.4 kcal/mol, which is quite close to the experimental value. Additional tests of Mechanism II were then performed for two other substrates, vinylmercury and *cis*-2-butenyl-2-mercury, and the results were again close to the experimental values (Table 5.4). Together, these data suggested that the active-site model of MerB accurately describes the Hg-C cleavage reaction, and that Mechanism II most likely captures the essence of MerB catalysis.

To understand how MerB lowers the activation barrier for Hg-C cleavage, transition states for the rate-limiting steps in Mechanisms I and II were analyzed (Figure 5.10). Partial atomic charges derived from Natural Population Analyses were computed for each case. In Mechanism I, proton transfer from Cys96 to CH_3- is unfavorable, as evidenced by the significant electron density on the proton. With a charge of +0.20, the catalytic proton is a poor electrophile. On the other hand, in the transition state for Mechanism II,

Table 5.4 Comparison of experimental reaction rates and calculated DFT activation barriers for the Hg-C cleavage of organomercurial substrates by MerB. All energies are in kcal/mol. Adapted with permission from J. M. Parks, H. Guo, C. Momany, L. Liang, S. M. Miller, A. O. Summers, J. C. Smith, *J. Am. Chem. Soc.*, 2009, **131**, 13278–13285. Copyright 2009 American Chemical Society.

Substrate	$\Delta G^{\neq}_{exp}$ [a]	$\Delta E^{\neq}_{Mech\ I}$ [b]	$\Delta E^{\neq}_{Mech\ II}$ [b]
methylmercury	20.1	33.9	20.4
vinylmercury	18.4	—	18.3
cis-2-butenyl-2-mercury	16.6	—	15.6

[a] Kinetics data from T. P. Begley, A. E. Walts, C. T. Walsh, 1986,[60] converted to free energy using Transition State Theory.
[b] Data from J. M. Parks, H. Guo, C. Momany, L. Liang, S. M. Miller, A. O. Summers, J. C. Smith, *J. Am. Chem. Soc.*, 2009, **131**, 13278–13285.[49]

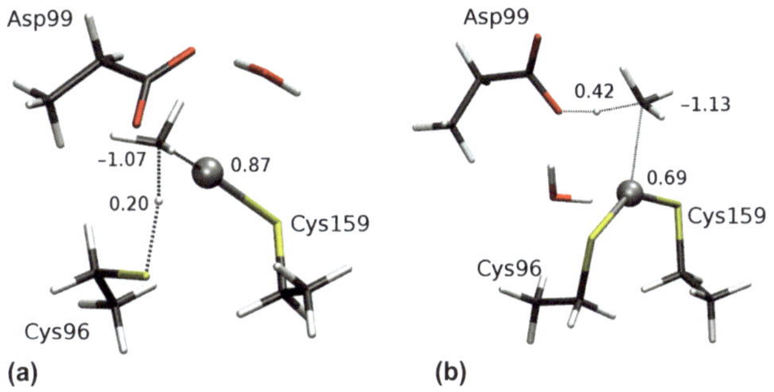

Figure 5.10 NPA charges of selected atoms in the transition state structures for (a) Mechanism I (proton transfer from Cys96 to the CH_3^- leaving group of methylmercury), and (b) Mechanism II (leaving group protonation by Asp99 of the bis-coordinated methylmercury substrate). For Mechanism I, the calculated activation barrier was 33.9 kcal/mol, and the barrier for Mechanism II was 20.4. Adapted with permission from J. M. Parks, H. Guo, C. Momany, L. Liang, S. M. Miller, A. O. Summers, J. C. Smith, *J. Am. Chem. Soc.*, 2009, **131**, 13278–13285. Copyright 2009 American Chemical Society.

bis-coordination of the substrate by Cys96 and Cys159 results in a more electrophilic proton (charge = +0.4), as well as a slightly more nucleophilic CH_3- leaving group (charge = −1.13 versus −1.07 for Mechanism I).

Another important observation is that the S_{Cys96}-Hg(II)-S_{Cys159} moiety in Mechanism II bears a formal neutral charge, making it an excellent leaving group. Quenching of the anionic CH_3- group by a precisely placed catalytic proton from Asp99 affords significant rate enhancement of the enzymatic Hg-C cleavage reaction. The strength of the quantum chemical cluster approach for understanding the mechanism of MerB lies in its simplicity. A system consisting of only 35 atoms was, in this case, sufficient to describe the metal coordination chemistry involved in the enzymatic Hg-C bond cleavage reaction.

5.3.2.3 *Dynamics of Hg(II)-MerR: Combining Low-resolution SAXS Data with MD Simulation*

In this section, we provide a summary of our recent work on the metalloregulator MerR employing small-angle X-ray scattering (SAXS) experiments and MD simulations.[61] MD simulations were performed to determine the conformational dynamics of Hg(II)-bound MerR, and the results were compared to and validated against SAXS experiments.

The metalloregulator MerR is responsible for initiating transcription of *mer* genes by RNA polymerase when cellular levels of Hg(II) reach a certain threshold. In the absence of Hg(II), MerR represses transcription of *mer* genes, but transcription is activated by nanomolar concentrations of Hg(II)[62] Binding of Hg(II) to MerR induces a conformational change that underwinds the DNA and ultimately leads to transcription. *In vivo*, the MerR homodimer is always bound to its operator DNA. As a first step in understanding its regulatory mechanism, we have simulated the protein without its operator DNA.

No NMR or X-ray crystal structures have been determined for MerR. Fortunately, several related proteins from the MerR family of transcriptional regulators have been have been crystallized.[63–70] In addition, MerR has been characterized extensively using a number of genetic, biochemical and biophysical approaches (see H. B. Guo, A. Johs, J. M. Parks, L. Olliff, S. M. Miller, A.O. Summers, L. Liang, J.C. Smith, 2010[61] and refs therein).

MerR functions as a homodimer, and each monomer consists of six α-helices. Based on homology to other MerR family members, helices α1 and α2 form a helix-turn-helix motif that binds DNA. The α2 helices of each monomer insert directly into the major groove of the operator DNA at the 19-bp spacer region.[65,69] Helices α5 and α5′ form a long, antiparallel coiled-coil, which links the two monomers. Helices α3 and α4 are referred to as the *coupling domain*, because they connect the DNA-binding domain to the coiled-coil.

MerR has an extremely high affinity for Hg(II), exhibiting binding at nanomolar concentrations of Hg(II). Three cysteine residues, Cys117 and Cys126 from one monomer and Cys82′ from the other monomer, make up the Hg(II) binding sites of MerR. It is known from experimental studies that only one of the two binding sites in the MerR dimer is occupied by Hg(II) *in vitro*.[71]

To probe the conformational dynamics of MerR in solution, we combined SAXS, which provides low-resolution structural information about the protein in solution, and MD, which yields atomic-detail structural and dynamics. Because there are currently no X-ray crystal structures of MerR, the homologous copper efflux regulator, CueR,[63] was used as a structural template to construct our model.

MD simulations were performed for MerR with and without Hg(II).[61] As MerR is known to bind only one Hg(II) even though it has two metal binding sites, a single Hg(II) was placed at the center-of-mass of the three sulfur atoms of Cys117, Cys126 and Cys82′ (Figure 5.11). CHARMM force field parameters were developed for the trigonally coordinated Hg(II) in MerR (Figure 5.10).[61] MD simulations were performed using NAMD[72] with the CHARMM force

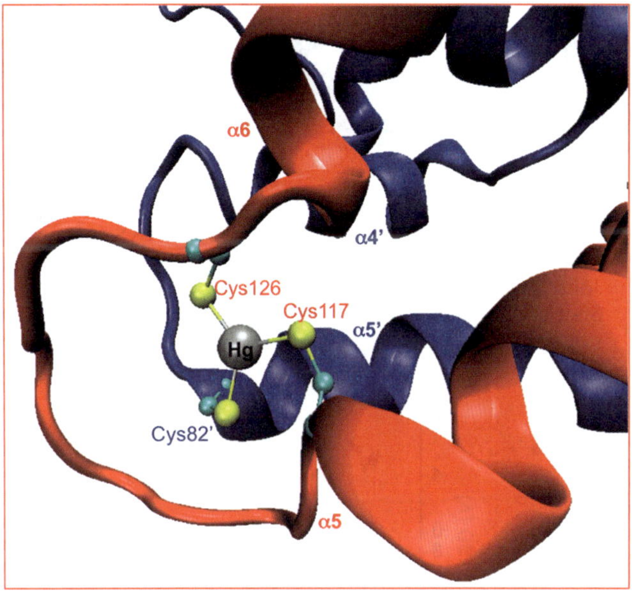

Figure 5.11 Trigonal planar coordination of Hg(II) with Cys117, Cys126 and Cys82′ of the MerR dimer. Colors: Hg, silver; S, yellow; C, green.

field[30] and the TIP3P water model.[73] A series of five 50 ns production runs beginning from different initial coordinates were performed to ensure reproducibility and convergence. Principal component analyses were performed to elucidate large-amplitude conformational changes of Hg(II)-bound MerR.

As a first test of the validity of the MerR model of Hg(II)-MerR, the radius of gyration (R_G) from SAXS experiments was compared with the average R_G from MD simulations. MD gave a systematically smaller value (26.4 ± 0.6 Å) compared to SAXS (28.6 ± 0.5 Å), which is to be expected. The density of the first hydration shell of the protein is greater than that of bulk solvent, yielding larger effective R_G values in SAXS experiments.

The Hg(II)-MerR MD model was then compared with the molecular envelope shape derived from SAXS (Figure 5.12). Encouragingly, the MD model reproduced the overall topology of the protein. The single best-fitting snapshot from the MD trajectory was used to compute a scattering profile (Not shown). Again, the MD structure reproduced the experimental SAXS data quite well. The overall agreement between the MD/homology model of Hg(II)-MerR and SAXS experiments validated the Hg(II)-MerR model for further studies.

The average distance between the α2 and α2′ helices ($R_{\alpha 2\text{-}\alpha 2'}$) in the DNA binding domains of other MerR family proteins,[63–66,68–70] which is similar to the helical repeat distance in canonical B-DNA (34 Å). However, MD simulations revealed large fluctuations in $R_{\alpha 2\text{-}\alpha 2'}$ and R_G. The torsional angle θ between the two DNA binding domains was defined as the angle between the

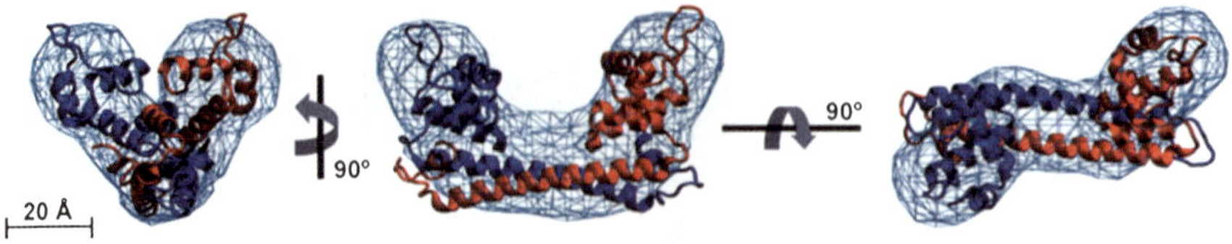

Figure 5.12 Best-fitting conformation of Hg(II)-MerR from MD simulations superimposed on the 3D envelope shape from SAXS (wireframe representation) obtained from SAXS experiments. Reprinted from H. B. Guo, A. Johs, J. M. Parks, L. Olliff, S. M. Miller, A. O. Summers, L. Liang, J. C. Smith, *J. Mol. Biol.*, 2010, **398**, 555–568. Copyright 2010, with permission from Elsevier.

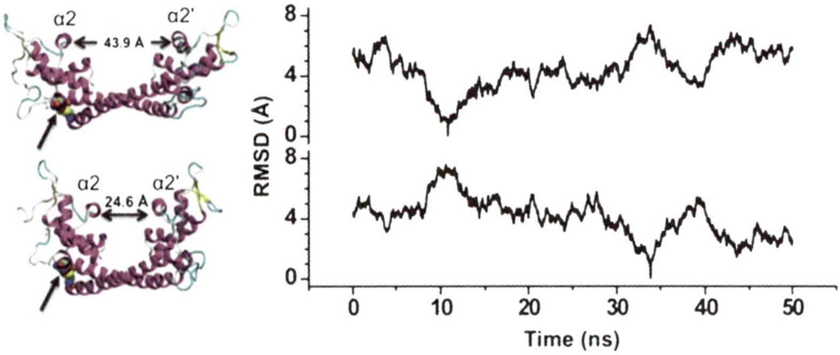

Figure 5.13 Open (top left) and closed (bottom left) states from 50 ns MD simulations of Hg(II)-MerR. The Hg(II) binding site is indicated by black arrow in the lower left, and the distances between the DNA binding domains are labeled. RMSD profiles relative to the open and closed extrema are also shown (right). Reprinted from H. B. Guo, A. Johs, J. M. Parks, L. Olliff, S. M. Miller, A. O. Summers, L. Liang, J. C. Smith, *J. Mol. Biol.*, 2010, **398**, 555–568. Copyright 2010, with permission from Elsevier.

plane of each monomer, and θ was also found to vary significantly during the MD trajectories. Although θ fluctuated between 40 and 90°, its average value of 63°, was in good agreement with the value obtained from the SAXS molecular envelope (65°).

To analyze the dynamics of Hg(II) in greater detail, correlation analyses of the RMSD, $R_{\alpha 2\text{-}\alpha 2'}$, and R_G were performed for Hg(II)-MerR. The correlation coefficient (CC) between R_G and $R_{\alpha 2\text{-}\alpha 2'}$ was 0.96. clearly showing that the opening-and-closing dynamics of the protein are the dominant contributor to changes in R_G. The RMSD of the protein relative to the closed state ($R_{\alpha 2\text{-}\alpha 2'} = 24.6$ Å), open state ($R_{\alpha 2\text{-}\alpha 2'} = 43.9$ Å) yielded correlations of 0.97 and 0.92, respectively. RMSD values for Hg(II)-MerR are shown in Figure 5.13. Principal component analyses also revealed that the largest-amplitude modes of Hg(II)-MerR involve opening and closing of the two DNA-binding domains relative to each other.

An important observation from the MD trajectories was that the opening-and-closing motion in Hg(II)-MerR is not parallel to the coiled coil. As the distance between the two DNA-binding domains increases the tilt angle θ between the planes of the two monomers also increases.

MD simulations of Hg(II)-MerR revealed large-amplitude opening-and-closing motion of the DNA binding domains on an ~10 ns timescale. This motion occurs when the operator DNA is not bound to MerR, but similar motion is expected to occur when DNA is present. The role of this motion is presumed to be in underwinding the DNA strand, which is the first step in transcription by RNA polymerase. Ongoing work in our laboratory is aimed at simulating the dynamics of the Hg(II)-MerR/DNA complex.

5.4 Conclusions

The work in progress involves both methodological development and applications to systems of prime interest in the energy biosciences.

First, we described two scalable implicit solvation models, including FMM accelerated PB solver and tree code based GB method, which not only are capable of extending both the time and length-scales of MD simulations, but also make more efficient use of large-scale supercomputers. By employing the reaction field method on massively-parallel supercomputers simulations of multimillion-atom systems for time scales approaching the microsecond are now within reach. We anticipate that a wealth of structural and dynamical information of biological importance will thus be revealed.

Towards realistic models of lignocellulosic biomass, we developed a CHARMM compatible atomistic force field for lignin molecules and a single bead per monomer level CG models for celluloses with a wide range of crystallinity. The self-similar multiscale structure of lignin revealed by neutron scattering and molecular dynamics simulation provides a basis for understanding the detailed interactions between lignin, cellulose and cellulolytic enzymes, and may therefore aid in rational engineering of plants and/or enzymes aimed at overcoming biomass recalcitrance.

Finally, we described work aimed at understanding the molecular mechanisms of bacterial mercury resistance. We have carried out quantum chemical calculations to illustrate the enzymatic degradation of methylmercury catalyzed by the organomercurial lyase, MerB, and have used extensive MD simulations combined with SAXS experiments to characterize the large-scale conformational dynamics of the metalloregulator MerR.

Acknowledgements

This work was funded by the U.S. Department of Energy, Office of Science, Office of Biological and Environmental Research, by DOE-OBER Scientific Focus Area "Biofuels" grant ERKP704 and by the DOE–OBER Science Focus Area "Biogeochemical and Molecular Mechanisms Controlling Contaminant

Transformation in the Environment", and by DOE's Scientific Discovery through Advanced Computing (SciDAC) program and OBER, under FWP ERKJE84. Computer resources were provided by the National Center for Computational Sciences (NCCS) at Oak Ridge National Laboratory through a DOE INCITE award, by the National Energy Research Scientific Computing Center (NERSC) sponsored by the U.S. Department of Energy, and by the National Science Foundation through TeraGrid resources (Grants TG-MCA08X032 and TG-CHE090035) provided by the National Institute for Computational Science (NICS), and the National Center for Supercomputing Applications (NCSA).

References

1. N. A. Baker, *Methods Enzymol.*, 2004, **383**, 94.
2. N. A. Baker, D. Sept, S. Joseph, M. J. Holst and J. A. McCammon, *Proc. Natl. Acad. Sci. U S A*, 2001, **98**, 10037–10041.
3. A. H. Boschitsch, M. O. Fenley and H. X. Zhou, *J. Phys. Chem. B*, 2002, **106**, 2741–2754.
4. B. Lu, X. Cheng, J. Huang and J. A. McCammon, *Proc. Natl. Acad. Sci. U S A,* 2006, **103**, 19314–9.
5. B. Lu, X. Cheng, J. Huang and J. A. McCammon, *J. Chem. Theory Comput.*, 2009, **5**, 1692–1699.
6. H. Cheng, L. Greengard and V. Rokhlin, *J. Comput. Phys.*, 1999, **155**, 468–498.
7. L. Greengard and J. F. Huang, *J. Comput. Phys.*, 2002, **180**, 642–658.
8. R. Constanciel and R. Contreras, *Theor. Chim. Acta*, 1984, **65**, 1–11.
9. W. C. Still, A. Tempczyk, R. C. Hawley and T. Hendrickson, *J. Am. Chem. Soc.*, 1990, **112**, 6127–6129.
10. D. Bashford and D. A. Case, *Annu. Rev. Phys. Chem.*, 2000, **51**, 129–152.
11. J. Chen, C. L. Brooks and J. Khandogin, *Curr. Opin. Struct. Biol.*, 2008, **18**, 140–148.
12. T. Darden, D. York and L. Pedersen, *J. Chem. Phys.*, 1993, **98**, 10089–10092.
13. W. Cai, Z. Xu and A. Baumketner, *J. Comput. Phys.*, 2008, **227**, 10162–10177.
14. A. Onufriev, D. A. Case and D. Bashford, *J. Comput. Chem.*, 2002, **23**, 1297–1304.
15. T. Grycuk, *J. Chem. Phys.*, 2003, **119**, 4817.
16. H. Tjong and H. X. Zhou, *J. Phys. Chem. B*, 2007, **111**, 3055.
17. G. D. Hawkins, C. J. Cramer and D. G. Truhlar, *Chem. Phys. Lett.*, 1995, **246**, 122–129.
18. G. D. Hawkins, C. J. Cramer and D. G. Truhlar, *J. Phys. Chem.*, 1996, **100**, 19824–19839.
19. E. Gallicchio and R. M. Levy, *J. Comput. Chem.*, 2004, **25**, 479.
20. K. Lindsay and R. Krasny, *J. Comput. Phys.*, 2001, **172**, 879.
21. A. D. Mackerell, *J. Comput. Chem.*, 2004, **25**, 1584.
22. U. Essmann, L. Perera, M. L. Berkowitz, T. Darden, H. Lee and L. G. Pedersen, *J. Chem. Phys.*, 1995, **103**, 8577–8593.

23. W. F. van Gunsteren, H. J. Berendsen and J. A. Rullmann, *Faraday Discuss. Chem. Soc.*, 1978, **66**, 58–70.
24. M. Neumann, *Mol. Phys.*, 1983, **50**, 841–858.
25. R. Schultz, B. Lindner, L. Petridis and J. C. Smith, *J. Chem. Theory. Comput.*, 2009, **5**, 2798–2808.
26. M. E. Himmel, S. Y. Ding, D. K. Johnson, W. S. Adney, M. R. Nimlos, J. W. Brady and T. D. Foust, *Science*, 2007, **315**, 804–807.
27. S. V. Pingali, V. S. Urban, W. T. Heller, J. McGaughey, H. O'Neill, M. Foston, D. A. Myles, A. Ragauskas and B. R. Evans, *Biomacromol.*, 2010, **11**, 2329.
28. L. Petridis and J. C. Smith, *J. Comput. Chem.*, 2009, **30**, 457–467.
29. B. Brooks and M. Karplus, *Proc. Natl. Acad. Sci. U S A.*, 1983, **80**, 6571–6575.
30. A. D. MacKerell, B. Brooks, C. L. Brooks, L. Nilsson, B. Roux, Y. Won, M. Karplus, in *The Encyclopedia of Computational Chemistry*, ed. P.v. R. Schleyer, John Wiley & Sons, 1998, pp. 271–277.
31. A. C. Vaiana, Z. Cournia, I. B. Costescu and J. C. Smith, *Comput. Phys. Commun.*, 2005, **167**, 34.
32. S. V. Pingali, V. S. Urban, W. T. Heller, J. McGaughey, H. O'Neill, M. Foston, D. A. Myles, A. Ragauskas and B. R. Evans, *Biomacromolecules*, 2010, **11**, 2329–2335.
33. B. S. Donohoe, S. R. Decker, M. P. Tucker, M. E. Himmel and T. B. Vinzant, *Biotechnology and Bioengineering*, 2008, **101**, 913–925.
34. L. Petridis, S. V. Pingali, V. S. Urban, W. T. Heller, H. O'Neill, M. Foston, A. Ragauskas and J. C. Smith, *Physical Review E*, 2011, **83**, 061911.
35. J. F. Matthews, C. E. Skopec, P. E. Mason, P. Zuccato, R. W. Torget, J. Sugiyama, M. E. Himmel and J. W. Brady, *Carbohydrate Research*, 2006, **341**, 138–152.
36. L. Zhong, J. F. Matthews, M. F. Crowley, T. Rignall, C. Talon, J. M. Cleary, R. C. Walker, G. Chukkapalli, C. McCabe, M. R. Nimlos, C. L. Brooks, M. E. Himmel and J. W. Brady, *Cellulose*, 2008, **15**, 261–273.
37. L. Zhong, J. F. Matthews, P. I. Hansen, M. F. Crowley, J. M. Cleary, R. C. Walker, M. R. Nimlos, C. L. Brooks, W. S. Adney, M. E. Himmel and J. W. Brady, *Carbohydrate Research*, 2009, **344**, 1984–1992.
38. S. Nielsen, C. Lopez, G. Srinivas and M. L. Klein, *J. Phys. C. (cond mat)*, 2004, **16**, R481.
39. W. G. Noid, J.-W. Chu, G. S. Ayton, V. Krishna, S. Izvekov, G. A. Voth, A. Das and H. C. Andersen, *J. Chem. Phys.*, 2008, **128**, 244114.
40. K. Moritsugu and J. C. Smith, *Biophys. J.*, 2007, **93**, 3460–3469.
41. K. Moritsugu and J. C. Smith, *Biophys. J.*, 2008, **95**, 1639–1648.
42. J. C. Shelley, M. Y. Shelley, R. C. Reeder, S. Bandyopadhyay and M. L. Klein, *J. Phys. Chem. B.*, 2001, **105**, 4464.
43. J. C. Shelley, M. Y. Shelley, R. C. Reeder, S. Bandyopadhyay, P. B. Moore and M. L. Klein, *J. Phys. Chem. B*, 2001, **105**, 9785.
44. G. Srinivas, D. E. Discher and M. L. Klein, *Nature Materials*, 2004, **3**, 638–644.
45. G. Srinivas and J. W. Pitera, *Nanolett*, 2008, **8**, 611–618.

46. L. Bu, G. T. Beckham, C. F. Crowley, C. H. Chang, J. F. Matthews, Y. J. Bomble, W. S. Adney, M. E. Himmel and M. R. Nimlos, *J. Phys. Chem. B*, 2009, **113**, 10994–11002.
47. K. R. Hadley and C. McCabe, *J. Chem. Phys.*, 2010, **132**, 134505.
48. G. Srinivas, X. Cheng and J. C. Smith, *J. Chem. Theory Comput.*, 2011, **7**, 2539–2548.
49. J. M. Parks, H. Guo, C. Momany, L. Liang, S. M. Miller, A. O. Summers and J. C. Smith, *J. Am. Chem. Soc.*, 2009, **131**, 13278–13285.
50. J. Lafrance-Vanasse, M. Lefebvre, P. Di Lello, J. Sygusch and J. G. Omichinski, *J. Biol. Chem.*, 2009, **284**, 938–944.
51. K. E. Pitts and A. O. Summers, *Biochemistry*, 2002, **42**, 10287–10296.
52. S. M. Miller, *Nat. Chem. Biol.*, 2007, **3**, 537–538.
53. A. D. Becke, *J. Chem. Phys.*, 1993, **98**, 5648–5652.
54. J. P. Perdew and Y. Wang, *Phys. Rev. B*, 1992, **45**, 13244–13249.
55. D. Andrae, U. Haeussermann, M. Dolg, H. Stoll and H. Preuss, *Theor. Chem. Acc.*, 1990, **77**, 123–141.
56. F. Himo, *Theor. Chem. Acc.*, 2006, **116**, 232–240.
57. P. E. M. Siegbahn and T. Borowski, *Acc. Chem. Res.*, 2006, **39**, 729–738.
58. P. E. M. Siegbahn and F. Himo, *J. Biol. Inorg. Chem.*, 2009, **14**, 643–651.
59. A. E. Reed, L. A. Curtiss and F. Weinhold, *Chem. Rev.*, 1988, **88**, 899–926.
60. T. P. Begley, A. E. Walts and C. T. Walsh, *Biochemistry*, 1986, **25**, 7192–7200.
61. H. B. Guo, A. Johs, J. M. Parks, L. Olliff, S. M. Miller, A. O. Summers, L. Liang and J. C. Smith, *J. Mol. Biol.*, 2010, **398**, 555–568.
62. T. Barkay, S. M. Miller and A. O. Summers, *FEMS Microbiology Reviews*, 2003, **27**, 355–384.
63. A. Changela, K. Chen, Y. Xue, J. Holschen, C. E. Outten, T. V. O'Halloran and A. Mondragon, *Science*, 2003, **301**, 1383–1387.
64. M. Kumaraswami, K. J. Newberry and R. G. Brennan, *J. Mol. Biol.*, 2010, **398**, 264–275.
65. K. J. Newberry and R. G. Brennan, *J. Biol. Chem.*, 2004, **279**, 20356–20362.
66. K. J. Newberry, J. L. Huffman, M. C. Miller, N. Vazquez-laslop, A. A. Neyfakh and R. G. Brennan, *J. Biol. Chem.*, 2008, **283**, 26795–26804.
67. J. D. Kahmann, H. J. Sass, M. G. Allan, H. Seto, C. J. Thompson and S. Grzesiek, *EMBO J.*, 2003, **22**, 1824–1834.
68. M. H. Godsey, N. N. Baranova, A. A. Neyfakh and R. G. Brennan, *J. Biol. Chem.*, 2001, **276**, 47178–47184.
69. E. E. Z. Heldwein and R. G. Brennan, *Nature*, 2001, **409**, 378–382.
70. S. Watanabe, A. Kita, K. Kobayashi and K. Miki, *Proc. Natl. Acad. Sci. U S A*, 2008, **105**, 4121–4126.
71. L. Y. Song, Q. Teng, R. S. Phillips, J. M. Brewer and A. O. Summers, *J. Mol. Biol.*, 2007, **371**, 79–92.
72. J. C. Phillips, R. Braun, W. Wang, J. Gumbart, E. Tajkhorshid, E. Villa, C. Chipot, R. D. Skeel, L. Kale and K. Schulten, *J. Comput. Chem.*, 2005, **26**, 1781–1802.
73. W. L. Jorgensen, J. Chandrasekhar, J. D. Madura, R. W. Impey and M. L. Klein, *J. Chem. Phys.*, 1983, **79**, 926–935.

Sampling and Rates

CHAPTER 6

Enhancing the Capacity of Molecular Dynamics Simulations with Trajectory Fragments

ALFREDO E. CARDENAS*[a] AND RON ELBER*[a,b]

[a] Institute for Computational Engineering and Sciences; [b] Department of Chemistry and Biochemistry, University of Texas at Austin, 1 University Station C0200, Austin, Texas 78712, USA
*Email: alfredo@ices.utexas.edu; ron@ices.utexas.edu

6.1 Introduction

Molecular Dynamics has become a versatile tool in condensed phase simulations, making it possible to understand molecular mechanisms at the microscopic level and compare simulation results to experiments. Empirical measurements of equilibrium and non-equilibrium observables are defined and computed as ensemble or time averages in statistical mechanics. Therefore, to connect computer simulations to experiments sampling multiple configurations (for equilibrium) and multiple trajectories (for studying kinetics) is necessary. In the present review we focus on kinetics.

Consider a molecular system with two metastable states *A* and *B* (reactants and products). A typical question of kinetics is: What is the probability of the system to reach state *B* at time *t* for the first time after initiation at state *A* at time zero? One trajectory with a single transition event provides zero or one as an answer to this question. An ensemble of trajectories, initiated according to a given distribution at A, provides a statistically converged fraction that better reflects the kinetics of a

RSC Biomolecular Sciences No. 23
Innovations in Biomolecular Modeling and Simulations: Volume 1
Edited by Tamar Schlick

Published by the Royal Society of Chemistry, www.rsc.org

macroscopic number of molecules. In principle, a single long trajectory going forward and backward from A to B, sampling multiple transitional events, can be used for the same purpose, if the transitions are observed in equilibrium. The computational cost of such a long trajectory is at least as high as the computational cost of an ensemble of trajectories initiated at A and propagated until they either hit state B for the first time, or return to A and are terminated.

Calculations of kinetics present a significant computational challenge. Not only do the molecular systems studied by simulations keep growing in size and complexity, but the time scales of kinetic processes of interest are long and necessitate sampling of trajectories with a very large number of integration steps (and of long times). If a clear, identifiable, and dominant free energy barrier is present in the system, then Transition State Theory[1] can be used to effectively investigate the kinetics. If the location of the dominant barrier is not so clear, but is present, we expect transitional trajectories to be short avoiding spending time at high free energy domains. Then Transition Path Sampling and related approaches are appropriate.[2–4] If a highly significant barrier is not present, which is the case in numerous biophysical problems, individual trajectories are not short and pose a major challenge to approaches that compute them explicitly. A discussion on a theoretical approach to bridge the time scale gap between experiments and simulations for a broad range of cases is the focus of the present paper. Before describing the trajectory-fragment approach[2,5,6] to long time processes, we briefly review other approaches to long time dynamics of complex molecular systems.

Consider first a straightforward calculation of a long time trajectory. In the best-case scenario the computational complexity of a trajectory is growing linearly with the system size N. Similarly, the complexity increases linearly with the length of the trajectory L. The number of degrees of freedom of a typical system in a biophysical simulation is $\sim 10^5$. The typical number of steps accessible in current simulations is $\sim 10^8$. A time step of a femtosecond and overall simulation time of tens of nanoseconds are common. The combined complexity of $N \times L$ can be tackled with theory, numerical algorithms, and improved hardware. Some approaches are focused on the N-factor (*e.g.* by parallelization of the forces). However, in the last decade significant advances that were made in theory, enable advances that also reduce the L factor. A combination of approaches that reduce the cost or speed up the calculation per time step and techniques that make it possible to compute kinetic observables with a significantly smaller number of steps are of particular interest.

Improved hardware for simulations is typically based on parallelization. Common approaches reduce the clock time required to compute a single integration step using Initial Value Solvers (IVS). IVS integrate Newton's equations of motion in small time steps. It takes one force evaluation to generate a step of an IVS trajectory. The number of force evaluations is a useful measure of the computational cost, which is roughly proportional to $N \times L$. Parallelization of IVS code can reduce the cost in the best-case scenario to $((N - n)/P + n) \times L$ where P is the number of processes that run in parallel, and n is the code segment that must be run in serial. In practice on commodity computers speedups rarely exceed 100.

An alternative approach to the parallelization of an integration of a step is the consideration of multiple time steps simultaneously, or the parallelization of time. Algorithms to parallelize directly multiple time slices (the factor L) are available, and are based on the optimization of a functional (action). Optimization of an action, a boundary value formulation (BVF) of classical mechanics,[7] and of stochastic processes,[8,9] generates classical trajectories[10–12] or solutions to the Langevin equation.[13] These approaches are radically different from step parallelization and therefore continue to be of considerable interest. However, their cost should be appreciated. The parallel optimization of an action is conducted simultaneously on different time points. With a sufficient number of independent processors, every time slice is optimized on a different processor. Since processes communicate only between neighboring time slices (to estimate velocities and acceleration from coordinates), the communication time is negligible compared to other calculations. The complexity of a single calculation of the action is proportional to $N \times (L/P)$. Since L can be large ($\sim 10^8$ for 1fs time step and ten nanosecond simulation), P (in practice) is only limited by the number of available processors. This is in contrast to IVS in which the number of useful processes is bound by a fraction of the system size N or the size of the non-bonded list. Unfortunately, optimization of actions is more expensive to be conducted than IVS[14,15] (if the same level of accuracy is desired) since all the times of a trajectory are considered simultaneously. It takes many steps to optimize the action and to generate an optimal BVF path. The number of degrees of freedom of an action in BVF formulation is $N \times L$. For a quadratic Hamiltonian, the maximum number of conjugate-gradient steps to find the global minimum of the action is $N^2 \times L^2$. This is an upper bound and heuristic optimizations can be used,[16] however, the calculation remains costly. BVF solutions are effective if the focus is on calculations of approximate trajectories with large time steps (small L)[10,17] or exact calculation of short and rare trajectories[16] that are difficult to sample using IVS. BVF techniques generate stable, approximate solutions that provide qualitative insight into molecular mechanisms. The use of large steps, illustrated in,[14,17–19] cannot be done in typical IVS that lose their stability with step $\sim$5fs or larger. These BVF solutions filter out high frequency motions from the trajectories.[10] The removal of high frequency motions adds to the stability of the calculation but makes the estimates of their statistical weights, approximate and heuristic. While some estimates of weights were promising,[18] it is difficult to prove correctness or to estimate error bars in the general case. Hence, for the calculation of ensembles of trajectories to be used for calculations of rates and thermodynamic properties, other approaches are needed.

Numerical analysis brought to the field multi-time stepping algorithms (such as RESPA)[20] in which slow forces are integrated less frequently. In RESPA a larger time step for integration of slowly varying forces is assigned. For example, while the fast degrees of freedom are integrated with a time step of $\sim$1fs, the larger time step for long range interactions can be $\sim$4fs. The overall computational gain from multi-time stepping algorithms is modest and is typically less than a factor of two.

Expensive special purpose machines for Molecular Dynamics focus only on reducing the factors proportional to N.[21] While it is strikingly successful in producing a few millisecond trajectories, the problem of kinetic at biophysical times (milliseconds) remains prohibitively costly due to the requirement of an ensemble of trajectories. Furthermore, it is desirable to make the calculations less costly and available at a single-researcher laboratory setting.

With the considerable success in speeding up the N-factor calculations, the most significant remaining barrier for routine calculations of kinetic and thermodynamic properties of molecular systems is the factor L – the trajectory length. The computational and numerical approaches discussed so far either successfully attack the N factor and leave the length "as is", or propose an approximate solution to the trajectory, reducing the impact of the length L. The approximations proposed so far are insufficient in the general case for kinetic and thermodynamic calculations at long times since the trajectory weights are uncertain.

All the technical arguments above about efficient calculations of trajectories should not make us forget why these trajectories are computed, how they are used and if there are ways of circumventing the expensive straightforward approaches discussed so far. In the case of thermodynamic calculations configurations are generated by Molecular Dynamics for averaging the values of observables. Averaging using straightforward trajectories is correct for ergodic systems, but correct does not mean efficient. Exact and enhanced sampling techniques are used for a long time in statistical mechanics calculations of thermodynamic variables. For example, the method of umbrella sampling[22] is widely used to probe and estimate probabilities of infrequent events in phase space. Straightforward Molecular Dynamics trajectories should not be used to compute thermodynamic properties that can be estimated much faster with enhanced sampling techniques. Straightforward calculations generate a large number of configurations. Many configurations of little interest are sampled extensively while other structures of significant interest are not. For example, configurations with probability of 10^{-10} will not be sampled effectively in a trajectory of 10^8 steps but can be sampled with high probability with umbrella sampling[22] or "blue-moon" sampling[23] using the same trajectory length.

Similar arguments can be made for kinetics. While straightforward calculations of an ensemble of trajectories from A to B provide the exact answer, it is not the only way of obtaining the exact result. The cost of calculations of kinetics is even higher than simulations of equilibrium and many straightforward calculations face a severe statistics problem. Alternative approaches can provide the desired statistics and overcome the time scale barriers, or a large value of L. It is the reduction of the lengths of the trajectories, breaking them into fragments, running them in parallel, and still computing observables of long time dynamics, which is the focus of the present manuscript.

6.2 Trajectory Fragments: General Overview

One of the theoretical advances of the last ten years that enhances simulation capabilities is the use of trajectory fragments. The fragments are trajectory

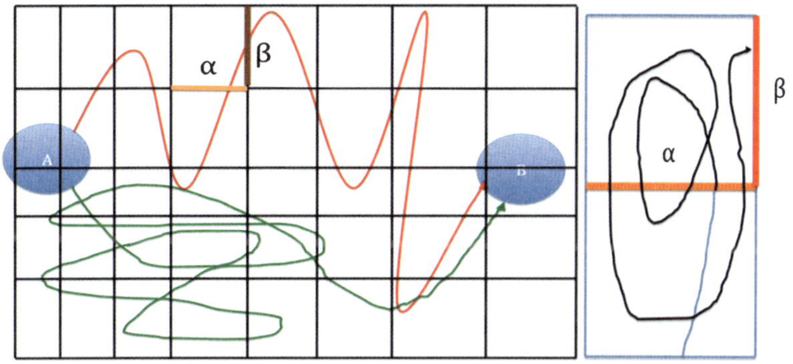

Figure 6.1 An illustration of trajectory fragments in Milestoning and exact trajectories. Left: The edges (interfaces) of the grid (*e.g.* α or β) are the Milestones. Red and green lines represent exact trajectories and the blue circles represent states. Right: An illustration of a partial Milestoning trajectory. The blue trajectory is hitting the interface α for the first time as it arrives from a milestone which is not α. We continue the blue (first hitting) trajectory by a black (Milestoning) trajectory fragment starting from the first hitting point at the interface α until the trajectory hits for the first time a milestone which is not α (in the picture it is β). The black trajectory can re-cross the interface α numerous times. In Milestoning we initiate trajectories from Milestone α by sampling from a first hitting point distribution on α. The distribution is prepared in two steps. First a molecular dynamics trajectory constrained to the interface α samples configurations with Boltzmann weight (the Milestoning calculation we discuss is for systems close to thermal equilibrium). Second, blue trajectories are integrated backward in time following a procedure used in PPTIS (Partial Paths Transition Interface Sampling: a theory similar in its construction of short trajectories to Milestoning).[2] The backward integration checks if the initial condition is a first hitting point. If it is not, this phase point is removed from the set.

pieces defined between portions of phase space. We call these portions cells. The hyper-surfaces that divide cells are called Milestones[6] by the authors and Interfaces by others.[2,3] We will use interchangeably both names. In Figure 6.1 we show a schematic drawing of interfaces in two dimensions. While the interfaces considered in the present manuscript are along a one-dimensional reaction coordinate, generalizations for higher dimensions have been made.[24,25] Transitions between interfaces of cells are marked as passage events and generate trajectory fragments. The lengths of the trajectory fragments are much shorter than the expected length of an exact first passage trajectory (with number of integration steps L). Interestingly, the accumulated length of sets of M fragments representing each of the K individual trajectories is much shorter than KL/M. This is a somewhat unexpected observation, which we discuss below.

Consider first a diffusive process. Diffusive motion is the typical dynamics found in biomolecular motions beyond tens of picoseconds (macromolecular dynamics in the liquid phase). Let the reactant and product be separated by a distance R. The time scale for free diffusion along one dimension is roughly $t \sim R^2$. If we consider M-1 cells between the end interfaces then the time scale for diffusion between a pair of divisions is of order of $(R/M)^2$. In order to complete a trajectory we need to select M pieces of the fragments and hence the time scale using fragments is $t_M \sim M \cdot (R/M)^2 = R^2/M$. This analysis suggests a speed up by a factor of M with respect to a straightforward trajectory. For a moment it may seem that we have gotten something for nothing.

What is the origin of this saving? Diffusive trajectories are going back and forth many times. In contrast, the fragments are computed without explicitly simulating back and forth transitions. In the Milestoning picture we first generate a bank of transitional trajectory fragments, say from cell i to cell j and from cell j to i. We found by experience that adequate sampling of trajectory fragments for estimating transition probabilities can be obtained in some molecular systems[24,26,27] using hundreds or thousands of trajectories. To describe properties of an ensemble of trajectories transitioning from cell i to cell j and back from cell j to i we sample corresponding fragments. The sampling intends to estimate the transition probability between the interfaces and not necessarily to provide a comprehensive picture of the dynamics within the cell. For example, a probability of 10 percent can be estimated quite accurately using 100 events of trajectory fragments of Milestoning. Milestoning is designed to provide uniform sampling of events as the reaction progresses or returns. If the trajectory goes back there is no need to re-compute the trajectory fragments since we re-sample from the prepared pool.

We will obtain similar statistics if we are at a minimum or at the top of the free energy barrier. On the other hand, there is no such guarantee for straightforward Molecular Dynamics and we usually are getting a lot more statistics near the minima, spending unwisely our computational resources. This brings us to another advantage of the trajectory fragments; overcoming a barrier is more efficient compared to a complete solution of IVS. Consider climbing a barrier of height B (Figure 6.2). In the canonical ensemble, the time to reach the top of the barrier is proportional to $\exp(+\beta B)$, where β is the Boltzmann factor. Imagine that the barrier is broken into cells. Each Milestoning transition is required to climb up with an intermediate time of $\exp(+\beta B/M)$. This time is exponentially shorter than $\exp(+\beta B)$. Adding up $\sim M$ Milestones has a small impact on the overall time $\tau_{total} \sim M \exp(+\beta B/M)$ in this case, keeping the rate significantly faster than of a single trajectory. In practice the speedup easily exceeds a factor of millions. For example, in the simulation of the recovery stroke in myosin[28] the actual accumulated length of all the trajectory fragments was of the order of 100 nanoseconds. The predicted mean first passage time of the process (fraction of a millisecond) is within a factor of 10 from experimental numbers[29] and is a million times longer than the simulated time. Hence, the use of trajectory fragments dramatically reduces the collective length of the computed trajectories and increases the computational efficiency.

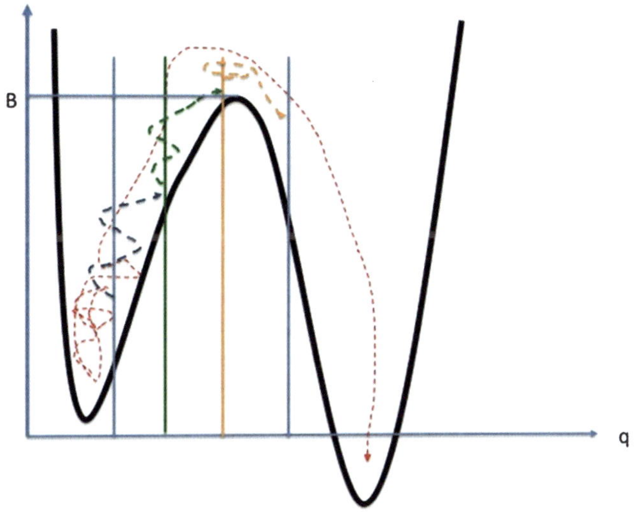

Figure 6.2 A schematic view of overcoming a barrier with trajectory fragments. We consider passage over a barrier of height B along a reaction coordinate q. The black curve represents the double well potential energy. The red dotted line sketches a straightforward trajectory initiated at the left that eventually ends at the right minimum. The "sticks" represent Interfaces or Milestones. Trajectory fragments are in different colors and are climbing up the barrier.

6.3 Challenges in Using Trajectory Fragments

Perhaps the most significant challenge in using trajectory fragments is to combine the pieces of the trajectories using appropriate statistical weights to obtain a comprehensive picture of kinetics (and thermodynamics). How do we start trajectory fragments and how do we terminate them? This is a critical point that requires careful consideration and is at the core of the differences between theories and algorithms that use the trajectory fragment concept. In the present review we consider and compare in detail the methods of Milestoning[6,24,26] and Forward Flux.[30,31] While other important approaches to trajectory fragmentation are available (most notable are the Partial Path Transition Interface Sampling (PPTIS),[2] Markovian Milestoning with Voronoi Tesselation (MMVT),[25] and Transition Interface Sampling (TIS)[3]) for clarity, compactness, and illustration of the main ideas we remain with these two. TIS[32] inspired the Forward Flux approach (FFS) and is similar in a number of ways to the FFS. PPTIS is close to Milestoning along a one-dimensional reaction coordinate. This explains our choice of the two techniques to discuss in detail.

6.3.1 Milestoning

In brief, Milestoning estimates fluxes at interfaces with trajectory fragments. The fluxes are used as input for a non-Markovian theory to extract the kinetics and thermodynamics of the system. The theory assumes that the system is close to equilibrium and loss of correlation between fragments of trajectories. These assumptions help in the derivation of compact and coarse-grained equations for the dynamics, but must be tested carefully as described in reference.[26] If Milestones are placed spatially close to each other (to increase efficiency) the dynamics at the interfaces may be correlated and the accuracy of the results will be poor. A more detailed description of Milestoning follows.

A trajectory fragment of Milestoning starts from a dividing hypersurface and terminates the first time it "touches" another surface (Figure 6.1). A dividing surface is determined with the help of a set of anchors $X_i, i = 1, \ldots, K \ X \in R^{6N}$ and coarse variables $Q_\alpha(X_i)$. An anchor is, in principle, a point in phase space. However, in applications conducted so far it is reduced to a coordinate vector. The set of anchors provides a rough sampling of the most relevant part of conformation space. The anchors are used to assist the selection of trajectory fragments and define interfaces with the coarse variables to be sampled and crossed. The set of anchors can be revisited and expanded as the sampling of trajectories is conducted. Hence, if some of the trajectory fragments are found at phase space domains far from any anchor, anchors could be added to cover the newly found conformations. Of course the choice of the anchor must be made carefully since the anchors need to capture the overall direction of the process. In the past anchors were chosen along a numerically computed reaction coordinates (Figure 6.4);[28,33] a formulation that was only recently extended to higher dimensions in the approaches of MMVT[25] and DiM (Directional Milestoning).[24]

Once a set of anchors is defined, the coarse variables are selected. A coarse variable can be simple, for example, a distance or a torsion of a molecular system. It can be more complex and includes the steepest descent path[28,34] or a minimum free energy coordinate[27] as one of the coarse variables. The requirement from the set of coarse variables is that it is sufficient to make the anchors distinguishable. Defining the distance from a point X_i to another point X_j in the space of the coarse variables as

$$d\left(X_i, X_j\right) = \sqrt{\sum_l \left[Q_l(X_i) - Q_l\left(X_j\right)\right]^2} \tag{6.1}$$

we require for all $i, j \ i \neq j$ that $d\left(X_i, X_j\right) > \varepsilon$, where ε is a minimal separation we set between the anchors. If this criterion is not satisfied it implies that more coarse variables are needed to capture the differences between the anchors, or perhaps that the anchors were placed too densely and some of them can be removed.

With the anchors and the coarse variables at hand we define Milestones, hypersurfaces in phase space that divide it to cells between the reactants and products. A Milestone in DiM has a sense of direction (which explains the name). It is defined as the following set of points

$$M(i \to j) = \left\{ X \middle| d(X, X_i)^2 = d(X, X_j)^2 + \Delta^2 \right\} \text{and } \forall k \ d\left(X, X_k\right) > d\left(X, X_j\right) \tag{6.2}$$

The points at the interface X are closer to the final state j and hence the sense of directionality. The parameter Δ determines the extent of asymmetry between the two end states. The Milestone is closer to the final state. The term Δ is added to minimize the possibility of rapid termination between milestones that crossed each other.

As we further discuss below, the physical assumption of Milestoning theory is of memory loss between hypersurfaces. The coarse variables of the individual trajectories or trajectory fragments, in accord with a statistical mechanics view of dynamics, suffer numerous collision events with other degrees of freedom and their motion is overall diffusive. After a typical time period the coarse variables de-correlate and it is not possible to trace them back to their point of origin or to follow an ensemble (spaghetti) of trajectories to their individual point of origin. A formal statement of this approximation and the profound simplifications it suggests for the calculations of kinetics and thermodynamics are given in the next section. Here we continue to describe the algorithm of fragment generation that uses this assumption.

To define a trajectory fragment we need starting and terminating conditions and obviously an algorithm to propagate the phase space point as a function of time. The choice in Milestoning is arbitrary (Langevin, Newtonian dynamics, *etc.*). However, to reflect true microscopic time scales straightforward mechanics is preferred. A termination point of one fragment (in exact continuation) is the starting point of another fragment. We choose the termination point to be the first hitting point of a trajectory on a Milestone different from the Milestones it started on (Figure 6.1). A starting point (which is not the Milestone of the reactants) must therefore be the same, a first hitting point of a trajectory that reached the present Milestone from yet another Milestone. Since we sample plausible starting points at the interface directly (see below) we need to verify that these points indeed represent a first hitting distribution. We therefore integrate this point back in time and verify that it crosses another Milestone, before re-crossing the Milestone it started from. If it re-crosses the Milestone of initiation, then it is not a first hitting point. This phase space point is removed from the sample set.

In summary, the generation of the trajectory fragments uses the following steps:

(i) Generate a canonical sample of configurations at a Milestone. This is achieved either with constant temperature Molecular Dynamics while restraining the system to the hypersurface,[24] similar to what is done in umbrella sampling,[22] or with constrained dynamics implemented with Lagrange's multipliers.[26] The set of selected configurations is conditioned to start at the interface and is distributed in the interface with weights of $\exp(-\beta U(X))$ where β is the Boltzmann factor and $U(X)$ the potential energy.

(ii) Examine if the phase space points sampled in step (i) are first hitting points. Since our sampling in (i) is of configurations only, sample first atomic velocities from the Maxwell distribution conditioned on the overall velocity directed backward from the hypersurface. Each point is integrated backward in time using Newtonian mechanics (constant energy) until it hits and terminates on a Milestone. The use of the NVE ensemble is important for the calculations of dynamics. Other ensembles provide only phenomenological parameterization of time dependent properties. If the terminating Milestone is different from the interface we started from, accept this initial configuration and velocity as a first hitting point. If not, reject the point.

(iii) Integrate the first hitting points forward in time. The trajectory fragment is terminated when it hits for the first time a Milestone different from the Milestone it was initiated on. Note the important difference between the backward and the forward integrations. During the forward integration we do not terminate trajectories that re-cross the initial Milestone. We continue the forward trajectories until they find a new Milestone to terminate on. All the forward trajectories count, and the removal of some of the sampled phase points at the interface occur only in step (ii).

What do we do with the sampled fragments? The Milestoning theory is built around a kernel or a transition operator, which we denote by $K_{\alpha\beta}(t)$. It is the probability density that a trajectory fragment initiated at interface α will hit interface β at time t. This probability density is normalized: $\sum_{\beta\in\bar{\alpha}}\int_0^\infty K_{\alpha\beta}(t)dt = 1$. The normalization states the obvious that at infinite time the trajectory must terminate on one of the nearby Milestones β. The symbol $\bar{\alpha}$ means Milestones that can be reached from α without crossing other Milestones along the way.

How do we use the trajectory fragments to estimate the value of the kernel (or time moments of it)? We compute the kernel (or moments of it) by binning. For example, let the number of first hitting point trajectories initiated at hypersurface α be n_α. Let the number of trajectories that hit the new milestone β between time t and time $t+\Delta t$ be $n_{\alpha\beta}(t)$. The kernel element $K_{\alpha\beta}(t)$ is therefore estimated as $n_{\alpha\beta}(t)/n_\alpha\Delta t$. We will be mostly interested in the moments of the kernel. For example, the probability that a trajectory fragment will make it from α to β (at any time) is the zero moment (in time) of $K_{\alpha\beta}(t)$, $p_{\alpha\beta} = \int_0^\infty K_{\alpha\beta}(t)dt \approx \sum_i n_{\alpha\beta}(t_i)/n_\alpha$. Computing the moments is more stable statistically since less sampling is required to compute moments compared to accurate estimates of many bins of K. The use of K in rate calculation is described in 6.4.

6.3.2 Forward Flux Sampling (FFS)

The Forward Flux Sampling method is an algorithm that was first developed to simulate kinetics of rare events in nonequilibrium systems and then was extended to study equilibrium properties as well.[5,35,36] This methodology allows the computation of rate constants and the sampling of complete and stochastic transitional trajectories. Similar to Milestoning, FFS divides the

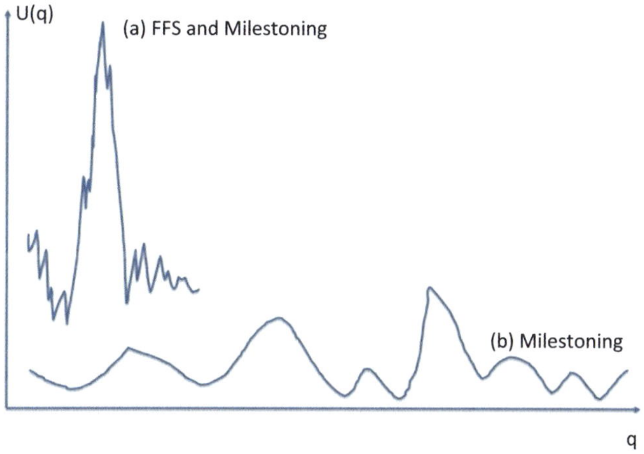

Figure 6.3 A schematic drawing of one dimensional energy landscape which is activated and can be studied with FFS (exactly) and Milestoning (with memory loss assumption) (a), and a diffusive energy landscape that requires individually long trajectories that can be studied with Milestoning (b).

phase space between the initial and final state into a series of milestones (called interfaces in the FFS literature). Then it uses stochastic dynamics to compute trajectories crossing these interfaces in the forward direction (from reactant to product). Trajectories starting from a given interface are free to visit previous interfaces several times. They are only stopped when the forward interface is hit. In this method, the assumption of memory loss during interface transitions is not required. The calculation is conducted serially. The prime difference between FFS and Milestoning is that in FFS complete trajectories are eventually computed, while in Milestoning they are not. The good news of complete trajectories is that the calculation of individual trajectories is exact. The speedup in FFS is obtained by selection of rare trajectories, not by computations of trajectory fragments. The bad news is the inherent assumption of FFS that long time dynamics is observed because the reactive trajectories are rare (but short in time) and not because individual trajectories are long (Figure 6.3). FFS is also restricted to calculation of rate constants while Milestoning can address non-exponential kinetics. Finally, FFS is limited to one-dimensional order parameter while DiM is not.

In FFS $n+1$ non-intersecting interfaces separating the initial state A and the final state B can be labeled by an order parameter λ with $\lambda < \lambda_0 \equiv \lambda_A$ describing the initial state and $\lambda > \lambda_n \equiv \lambda_B$ the final state. As we will see in the next section, the two key quantities that need to be evaluated to compute the rate constant for the forward transition from A to B are the flux $\langle \Phi_{A,0} \rangle$ of trajectories leaving A and reaching λ_0 and the conditional probabilities $P(\lambda_{i+1}|\lambda_i)$ that a trajectory that started at A and crosses λ_i for the first time will reach λ_{i+1} before going back to A.

In the original version of FFS (called direct-FFS) the determination of the initial flux $\langle \Phi_{A,0} \rangle$ and conditional probabilities $P(\lambda_{i+1}|\lambda_i)$ for interface transitions follows four steps:

(i) Starting with a configuration from the initial state A, use a free simulation of total time t to generate a collection of N_0 configurations representing crossings of interface λ_0. This set of configurations should only count crossings if the system arrives to the interface from A. The initial flux is the ratio of the number of λ_0 crossing to the total time of the run (N_0/t). The set of N_0 configurations is stored for the next step. For the present estimate to be effective, it is important that the initial state A will be meta-stable. As we shall see below the first interface is better placed somewhat far from the state boundaries.
(ii) Choose one of the N_0 configurations at random and use it as a starting point of a trajectory that is continued until either it reaches λ_1 (a success attempt) or returns to A (a failure). If λ_1 is reached the configuration at the crossing is stored. This procedure is repeated M_0 times, each time starting with a randomly chosen configuration from the N_0 set. In this way, if N_1 is the number of successful crossings of λ_1, $P(\lambda_{i+1}|\lambda_i)$ can be computed as the ratio N_1/M_0.
(iii) Repeat step (ii) for interface j (with $1 \leq j \leq n-1$) using the N_j stored configurations from the previous step. Specifically, M_j trial runs are attempted, a trajectory is successful and configurations are stored if is reached, or a failure if λ_0 is reached first. If N_{j+1} is the number of successful attempts, the conditional probability is $P(\lambda_{j+1}|\lambda_j) = N_{j+1}/M_j$.
(iv) Repeat step (iii) until λ_n is reached.

In this version of the algorithm, many transition paths are generated simultaneously.[5,35] In a variation called the branched growth method[35] branched paths from A to B are generated one by one. In a second variation, called the Rosenbluth method,[35] unbranched transitions are generated one by one. (For a more detailed description of these variations see the original papers introducing them or two recent reviews.)[37,38]

An advantage of the FFS methodology is that ensembles of complete transitional trajectories with correct weights can be extracted in cases with a single reactive channel. From the direct FFS procedure we can trace back the paths by starting with the collection of successful attempts that cross the last interface λ_n from the configurations in λ_{n-1}; then from these configurations in λ_{n-1} obtain the partial paths coming from the configurations in λ_{n-2} and continue these steps until the initial state A is reached. Complete transitions are generated from the connection of these partial paths. The disadvantage of FFS is that the complete trajectories must be relatively short compared to the overall kinetic times. If long individual trajectories are making significant contributions to the rate the FFS results are unlikely to find them. This is because order parameters are usually chosen in preference for short reactive trajectories.

6.4 Applications of Trajectory Fragments: Equilibrium and Kinetics

Milestoning: Assuming that we have at hand the ensemble of trajectory fragments, and as a prime result an estimate of the kernel $K_{\alpha\beta}(t)$ at hand, how do we proceed to obtain kinetics and thermodynamics? At the core of the Milestoning theory one finds an equation for the flux through Milestones. A flux is defined as the number of trajectories fragments that pass through a Milestone at time t. We write a general and exact equation for the flux (irrespective of the dynamics used to generate the trajectory fragments):

$$q_\alpha(t, X_\alpha) = p_\alpha(0, X_\alpha)\delta(t^+) + \sum_{\beta\epsilon\bar{\alpha}} \int_0^t q_\beta(t', X_\beta) K_{\beta\alpha}(t - t', X_\beta, X_\alpha) dt' dX_\beta \quad (6.3)$$

where the indices α, β are used to denote Milestones, and $p_\alpha(t, X_\alpha)$ is the probability that the last Milestone that was crossed at time t is α. The coordinate vectors, X_α and X_β, are at the interfaces, and $q_\alpha(t, X_\alpha)$is the flux at time t through the Milestone point X_α. Eq. (6.3) is difficult to solve directly. The flux is a function of the position in the hypersurface, which means a function of N-k dimensions of all degrees of freedom (where N is the number of degrees of freedom, and k the number of coarse variables). The kernel itself depends on position vectors in two Milestones. This exact equation is therefore not useful for simulation of large molecular systems with a number of coarse variables that easily exceeds hundreds. What comes to our rescue is the memory loss assumption mentioned in the previous section. In the kernel language it means that a trajectory fragment depends only on the label of the Milestone it started from, but is independent of the exact location within the Milestone. Hence

$$K_{\alpha\beta}(t, X_a, X_\beta) \cong K_{\alpha\beta}(t, X_\beta) \quad (6.4)$$

The approximation in Eq. (6.4) is a critical difference between FFS and Milestoning. FFS continues a prior trajectory, and therefore produces an exact path. Milestoning is using fragments to estimate the kernel. With the approximation of Eq. (6.4) at hand we define

$$\begin{aligned} K_{\alpha\beta}(t) &= \int K_{\alpha\beta}(t, X_\beta) dX_\beta \\ q_\alpha(t) &= \int q_\alpha(t, X_\alpha) dX_\alpha \\ p_\alpha(t) &= \int p_\alpha(t, X_\alpha) dX_\alpha \end{aligned} \quad (6.5)$$

Integrating Eq. (6.3) with respect to X_α (and also integrating over X_β on the right-hand-side equation) we obtain the basic formula of the Milestoning theory[6]

$$q_\alpha(t) = p_\alpha(0)\delta(t^+) + \sum_{\beta\epsilon\bar{\alpha}} \int_0^t q_\beta(t')K_{\beta\alpha}(t-t')dt' \tag{6.6}$$

Eq. (6.6) can be solved analytically with the help of Laplace transform tricks to provide the stationary distribution, $p_\alpha(t \to \infty)$ and the mean first passage time $\langle\tau\rangle$ (and higher moments of it) as was shown in a number of publications.[26,27,39] In the absence of external forces and (or) fluxes in and out the system, $p_\alpha(t \to \infty)$ is the equilibrium distribution. The overall mean first passage time, $\langle\tau\rangle$, is computed for a system with an absorbing boundary at the product state. Every trajectory which makes it to the product state is terminated. The final expressions for the stationary flux and distribution are

$$\begin{aligned} \mathbf{q}_{stat}(\mathbf{I}-\mathbf{K}) &= 0 \\ p_{\alpha,stat} &= q_{\alpha,stat} \cdot \langle t_\alpha \rangle \end{aligned} \tag{6.7}$$

The vector $\mathbf{q}$ is of length L – the number of Milestones. We have $(\mathbf{q})_{\alpha,stat} = q_\alpha(t \to \infty)$. Similarly $\mathbf{K}$ is a matrix such that $(\mathbf{K})_{\alpha\beta} = \int_0^\infty K_{\alpha\beta}(t)dt$ and $\mathbf{I}$ is the identity matrix. The average $\langle t_\alpha \rangle$ is the life-time of Milestone α, *i.e.*, the average time that it takes a trajectory fragment initiated at Milestone α to terminate on any other Milestone. It is given by

$$\langle t_\alpha \rangle = \sum_{\beta\epsilon\bar{\alpha}} \int_0^\infty t \cdot K_{\alpha\beta}(t)dt \tag{6.8}$$

From the first line of Eq. (6.7) we realize that $\mathbf{q}$ is an eigenvector of the matrix $(\mathbf{I}-\mathbf{K})$ with an eigenvalue of zero – a straightforward problem in linear algebra.

The calculation of the Mean First Passage Time (MFPT) follows another analytical expression

$$\langle\tau\rangle = \mathbf{p} \cdot (\mathbf{I}-\mathbf{K})^{-1}\langle t \rangle \tag{6.9}$$

where $\mathbf{p}$ is the vector of initial conditions $\left((\mathbf{p})_\alpha = p_\alpha(0)\right)$, and $\langle t \rangle$ is a vector with components $\langle t \rangle_\alpha \equiv \langle t_\alpha \rangle$. Higher moments of the first passage time can be computed as well using moments of the kernel.[26,39]

Forward Flux Sampling: In Forward Flux Sampling the rate constant is computed using the effective positive flux formalism for transition between an initial state A and final state B as described in the context of transition interface sampling (TIS) theory.[3,32,35] TIS defines a set of $n+1$ non-intersecting interfaces $\{i = 0, 1, \ldots, n\}$. Those interfaces can be chosen such as $\lambda_{i-1} < \lambda_i$ where λ (as introduced in 6.3.2) is a reaction coordinate for the transition and the boundaries of the stable states A and B are λ_0 and λ_n, respectively. If x_0 is a phase space point in the system, the backward and forward trajectories (backward and forward following the direction of the order parameter) starting from that point are fully determined (assuming deterministic dynamics). Then, we can introduce $t_i^b(x_0)$ and $t_i^f(x_0)$ to measure the time it takes the backward or forward trajectories to hit interface λ_i for the first time from x_0. These time variables are used to define two-interface theta functions indicating if the backward or forward trajectories starting from x reach interface i before j or *vice versa*:

$$h_{i,j}^b(x) = \theta\left(t_j^b(x) - t_i^b(x)\right)$$
$$h_{i,j}^f(x) = \theta\left(t_j^f(x) - t_i^f(x)\right)$$

If $i<j$, both functions are unity for $\lambda(x)<\lambda_i$ and zero for $\lambda(x)>\lambda_j$. The functions need to be evaluated for values $\lambda_i<\lambda(x)<\lambda_j$. Those theta functions are used to define overall regions $\mathcal{A}$ and $\mathcal{B}$ which satisfy the following two characteristic functions

$$\begin{aligned} h_{\mathcal{A}}(x) &\equiv h_{0,n}^b(x) \\ h_{\mathcal{B}}(x) &\equiv h_{n,0}^b(x) \end{aligned} \tag{6.10}$$

The phase space point x belongs to $\mathcal{A}$ ($h_{\mathcal{A}}(x) = 1$) if it came from A without visiting B, and belongs to $\mathcal{B}$ ($h_{\mathcal{B}}(x) = 1$) if it came from B without visiting A. These overall states encompass the entire phase space of the system. Based on the characteristic functions $h_{\mathcal{A}}$ and $h_{\mathcal{B}}$ we can write the correlation function

$$C(t) = \frac{\langle h_{\mathcal{A}}(x_0) h_{\mathcal{B}}(x_t)\rangle}{\langle h_{\mathcal{A}}(x_0)\rangle} \tag{6.11}$$

The equation is similar to the correlation function used in transition path sampling $C(t) = \langle h_A(x_0)h_B(x_t)\rangle / \langle h_A(x_0)\rangle$ but the former is linear ($\sim k_{AB}t$) from the start $t=0$ while the latter is linear only after a typical molecular transition time. The correlation function remains linear until the reaction time $\tau_{rxn} = (k_{AB} + k_{BA})^{-1}$ has been reached. Then at this linear regime the time derivative of the correlation function (Eq. (6.11)) gives the rate:

$$\begin{aligned} \dot{C}(t) = k_{AB} &= \frac{\langle h_{\mathcal{A}}(x_0)\dot{h}_{\mathcal{B}}(x_t)\rangle}{\langle h_{\mathcal{A}}(x_0)\rangle} = \frac{\langle h_{\mathcal{A}}(x_0)\dot{h}_{\mathcal{B}}(x_0)\rangle}{\langle h_{\mathcal{A}}(x_0)\rangle} \\ &= \lim_{\Delta t \to 0} \frac{1}{\Delta t} \frac{\left\langle h_{0,n}^b(x_0)\theta(\lambda_{\mathcal{B}} - \lambda(x_0))\theta(\lambda(x_{\Delta t}) - \lambda_B)\right\rangle}{\langle h_{\mathcal{A}}(x_0)\rangle} \\ &= \frac{\langle h_{0,n}^b(x_0)\dot{\lambda}(x_0)\delta(\lambda(x_0) - \lambda_B)\rangle}{\langle h_{\mathcal{A}}(x_0)\rangle} \end{aligned} \tag{6.12}$$

where only positive terms contribute to the rate. This equation is difficult to compute as it is because the number of phase points x_0 very close to interface λ_B

that belong to $\mathcal{A}$ can be very small. That justifies the introduction of a set of interfaces between A and B. Defining $\Omega_{\lambda_i} \equiv \{x|\lambda(x) > \lambda_i\}$ (for example, Ω_{λ_A} represents the phase space region outside state A), and introducing the flux

$$\Phi_{A,\lambda_i}(x_0) \equiv \lim_{\Delta t \to 0} \frac{1}{\Delta t} h^b_{A,\Omega_{\lambda_i}}(x_0)\theta(\lambda_i - \lambda(x_0))\theta(\lambda(x_{\Delta t}) - \lambda_i)$$

The equation for the rate constant (Eq. 6.12) is reduced to

$$k_{AB} = \frac{\langle \Phi_{A,\lambda_n} \rangle}{\langle h_{\mathcal{A}} \rangle} = \frac{\langle \Phi_{A,\lambda_0} \rangle}{\langle h_{\mathcal{A}} \rangle} P(\lambda_n|\lambda_0) \tag{6.13}$$

$\langle \Phi_{A,\lambda_n} \rangle$ is the positive flux of trajectories leaving state A and reaching B (interface λ_n). The right-hand side of the equation states that the flux of trajectories leaving state A and reaching B is equal to the flux of trajectories leaving A and crossing λ_0 times the probability that a trajectory that leaves A by crossing λ_0 will reach B (λ_n) before returning to A. The flux $\langle \Phi_{A,\lambda_0} \rangle$ is easy to compute if the interface λ_0 is close to the basin A. The probability $P(\lambda_n|\lambda_0)$ can be computed with the interfacial conditional probabilities:

$$P(\lambda_n|\lambda_0) = \prod_{i=0}^{n-1} P(\lambda_{i+1}|\lambda_i) \tag{6.14}$$

These conditional probabilities are easier to compute with trajectory fragmenting than the original probability $P(\lambda_n|\lambda_0)$, and with Eqs (6.13) and (6.14) we compute the forward rate.

FFS is especially suited to determine kinetic data for the transition rather than the stationary probability distribution of the system. With additional work it is possible to compute the stationary distribution along the order parameter λ, for example, by estimating the backward rate besides the forward rate or with a combination with umbrella sampling.[40]

6.5 Examples

As a pedagogical illustration of the speedup gains of these two methods we considered transition rates on the Mueller potential (Figure 6.4). We computed the dynamics on this 2D model system with the Langevin equation:

$$\dot{x}(t) = v(t),$$
$$m\dot{v}(t) = -\nabla V(x(t)) - \gamma v(t) + \sqrt{2\beta^{-1}\gamma}\eta(t)$$

where $x(t)$ and $v(t)$ are the two-dimensional position and velocity, respectively, $V(x(t))$ is the potential energy, β is the inverse temperature, γ is the friction coefficient and $\eta(t)$ is a white noise with mean zero and $\langle \eta_i(t)\eta_j(t') \rangle = \delta_{ij}\delta(t-t')$. For simplicity, the mass m is assumed to be one.

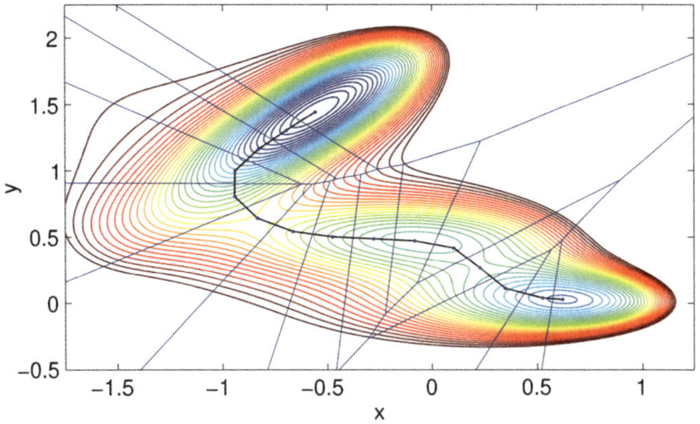

Figure 6.4 Contour plot of the Mueller potential, $U(x,y) = \sum_{i=1}^{4} A_i \exp[a_i(x-x_i)^2 + b_i(x-x_i)(y-y_i) + c_i(y-y_i)^2]$, with $A = (-200, -100, -170, 15)$, $a = (-1, -1, -6.5, 0.7)$, $b = (0, 0, 11, 0.6)$, $c = (-10, -10, -6.5, 0.7)$, $x = (1, 0, -0.5, -1)$ and, $y = (0, 0.5, 1.5, 1)$. It also shows a minimum energy path connecting the two lowest minima energy basins and Voronoi cells generated from the 15 conformations describing the curve. The 14 edges dividing the curve were used as milestones in the Milestoning calculation. For the FFS method only 11 edges were used in the calculation starting from the horizontal edge at $y = 0.9$.

We used the algorithm of reference[41] to integrate the equations of motion with a time step $\Delta t = 10^{-4}$, $\beta^{-1} = 10$ and $\gamma = 100$. With these values of temperature and friction the transition between the two lowest minima basins is a rare event and the mean first passage time MFPT computed by brute force of 2000 trajectories with the Langevin equation is $(2.07 \pm 0.07) \times 10^4$. Hence it is an example of an activated system in which reactive trajectories are short but infrequent and therefore accessible to both Milestoning and FFS.

To estimate the MFPT with FFS and Milestoning, we need to define interfaces. Therefore, we first computed a discrete approximation of the steepest descent path connecting the two lowest minima using the LUP algorithm.[42] Fifteen conformations from this minimum energy path were used as centers of Voronoi cells.[25] The edges of the Voronoi cells are interfaces/milestones for the calculation (see Figure 6.4).

The simplest version of FFS, the direct scheme algorithm as described in Section 6.3.2, was used to obtain the forward flux and MFPT estimates by FFS. The choice of the first interface separating the initial state from the rest was found to be critical in FFS calculation. A bad choice of the initial interface leads to errors that propagate to the rest of the interfaces in the calculation. We did not find a simple general procedure for accurate selection of this interface. By trial and error we found that if the first interface in FFS is the fourth milestone in Figure 6.4, the results converge to a reasonable value (we did not find this problem with later interfaces). Therefore, eleven interfaces were used in FFS separating the two lowest minima energy basins. To estimate the initial flux from A to the first interface a free Langevin simulation was initiated at the lowest basin minima. Every time the system leaves the basin and crosses interface 4 the coordinates were stored and the trajectory was continued. The number of times the interface was hit (from trajectories coming from A) and the time of the simulation were used to compute the initial flux $\langle \Phi_{A,\lambda_0} \rangle$. Then beginning with the configurations phase points at the interface, a number of trial trajectories were attempted to estimate the forward probability of moving to interface 5 before hitting the initial basin. If the trajectory was successful reaching the following interface, the trajectory was stopped and the final configuration was saved. If the trajectory reached the initial basin, it was stopped and a new trial was attempted by choosing randomly a configuration in interface 4. A similar procedure was followed to obtain the forward probability from interface 5 to 6, starting with the configurations saved in the previous step, and the process was continued until the final basin state was reached. These data are used in the calculation of the conditional interfacial probabilities $P(\lambda_{i+1}|\lambda_i)$.

For the Milestoning calculation we also used its simplest version[26] where milestones are described as hyperplanes (in the present case they are lines) perpendicular to the reaction coordinate. To avoid plane crossing (which is possible when the reaction coordinate is curved and the fluctuations in the normal direction to the path are large) we adjust our description to that of Voronoi cells.[25] A center of a Voronoi cell is an image along the steepest descent path. First, we sampled conformations constrained to each of the milestones and then launched backward and forward unconstrained trajectories from these milestone conformations. The backward trajectories are run first to determine if the conformation (including the initial velocities) is a first hitting point. If the backward trajectory was successful then a forward trajectory was launched. These trajectories were stopped when they hit a neighboring interface and the time of the trajectory was saved. These data are all that are needed to compute the transition kernel and the MFPT according to Eq. 6.9.

Figure 6.5 shows the estimates for the MFPT using FFS and Milestoning methods compared to the results coming from a straightforward determination of the transition times using the Langevin equation. In the abscissa, we plot the number of calls to the force subroutine in each case. In FFS, the variation in the number of calls comes from changes in the length of the trajectory to compute the initial flux and the number of attempted transitions between interfaces. In Milestoning, the variations come from changes in the number of initial configurations in each milestone and the number of trajectories launched from each milestone. Clearly the introduction of interfaces/milestones makes the evaluation of the rate about two orders of magnitude more efficient than the brute force estimate. In Milestoning the calculation can be made even more efficient by using trivial parallelism (the interfaces are independent). The major source of errors in the MFPT estimates when the number of force calls is low, is the limited statistics to describe the interface transition at the barrier. But for this simple case, a sample of 3–5 transitions between the milestones at the barrier gives good estimates for the rate. With the largest number of sampling points considered in both methods, the MFPT obtained with Milestoning was

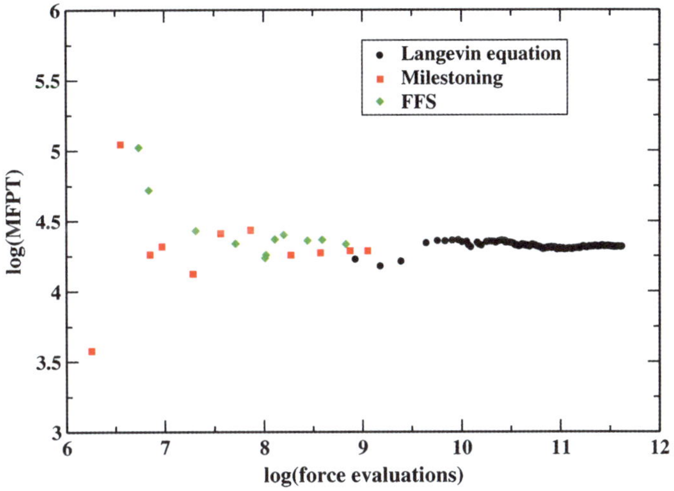

Figure 6.5 A log-log plot of the mean first passage time versus the number of force evaluations used to estimate it. In each of the methods the algorithm of reference[41] was used to compute the dynamics with a time step $\Delta t = 10^{-4}$, $\beta^{-1} = 10$, and $\gamma = 100$. To complete a trajectory with an average transition time $10^{8.3}$ force evaluations are required.

$(1.93 \pm 0.06) \times 10^4$ and with FFS was $(2.15 \pm 0.28) \times 10^4$. (A reminder: the rate computed with straightforward trajectories was $2.07 \pm 0.07 \times 10^4$).

At higher temperatures, the transition region will broaden and multiple additional channels can be open. In that case, different interfaces can intersect and cross and both algorithms used in the example will fail. Versions of these trajectory fragment methods, appropriate for higher dimensions, can be used, like Markovian Milestoning with Voronoi Tessellation,[25] or Directional Milestoning.[24]

6.6 Conclusions

We presented pedagogical descriptions and examples for two leading methods that exploit trajectory fragments to compute rate in complex systems: Forward Flux and Milestoning.

Forward Flux Sampling (FFS) has the advantage that it is in principle exact for activated processes, since complete trajectories are calculated, while Milestoning assumes loss of correlation between interfaces. A disadvantage of FFS is a limitation to activated processes. Diffusive trajectories that are long individually cannot be produced exactly in reasonable computer time. Milestoning is able to describe the kinetics in these systems since a kernel is estimated using trajectory fragments and complete trajectories are not integrated.

Milestoning has the advantage that it is trivially parallelized. Each interface is sampled and evaluated independently. It is similar in this regard to the

PPTIS method.[2] Since the number of interfaces for complex processes (*e.g.* the conformational transition of myosin)[28] is in the hundreds, this trivial parallelization can reduce computational clock time by orders of magnitude compared to FFS that is run serially. The disadvantages of Milestoning are the assumption of correlation loss and calculations that are restricted at present to near equilibrium for non-collective variables.

For the two-dimensional system investigated the efficiency and the accuracy of both techniques is comparable, while Milestoning seems somewhat more robust. Both are significantly more efficient than a single long time trajectory, as was measured by the number of required force evaluation to reach an average value. Both techniques can be improved further as was discussed in the text (by optimizing the initial interfaces of FFS and extending Milestoning to more than one dimension) and we anticipate that the efficiency and accuracy of these methodologies will increase as they are used further in the field.

References

1. D. G. Truhlar, B. C. Garrett and S. J. Klippenstein, *J. Phys. Chem.*, 1996, **100**, 12771–12800.
2. D. Moroni, P. G. Bolhuis and T. S. van Erp, *J. Chem. Phys.*, 2004, **120**, 4055–4065.
3. T. S. van Erp, D. Moroni and P. G. Bolhuis, *J. Chem. Phys.*, 2003, **118**, 7762–7774.
4. P. G. Bolhuis, D. Chandler, C. Dellago and P. L. Geissler, *Annu. Rev. Phys. Chem.*, 2002, **53**, 291–318.
5. R. J. Allen, P. B. Warren and P. R. ten Wolde, *Phys. Rev. Lett.*, 2005, **94**, 018104.
6. A. K. Faradjian and R. Elber, *J. Chem. Phys.*, 2004, **120**, 10880–10889.
7. L. D. Landau and E. M. Lifshitz, *Mechanics*, Pergamon Oxford, 1976.
8. S. Machlup and L. Onsager, *Phys. Rev.*, 1953, **91**, 1512–1515.
9. L. Onsager and S. Machlup, *Phys. Rev.*, 1953, **91**, 1505–1512.
10. R. Olender and R. Elber, *J. Chem. Phys.*, 1996, **105**, 9299–9315.
11. R. Elber, A. Ghosh and A. Cardenas, *Acc. Chem. Res.*, 2002, **35**, 396–403.
12. R. Elber, A. Cardenas, A. Ghosh and H. A. Stern, in *Adv. Chem. Phys.*, eds. I. Prigogine and S. A. Rice, John Wiley & Sons, Inc., NJ, 2003, Vol. 126, pp. 93–129.
13. P. Faccioli, M. Sega, F. Pederiva and H. Orland, *Phys. Rev. Lett.*, 2006, **97**, 108101.
14. A. E. Cardenas and R. Elber, *Proteins: Struct., Funct., Bioinf.*, 2003, **51**, 245–257.
15. A. E. Cardenas and R. Elber, *Biophys. J.*, 2003, **85**, 2919–2939.
16. D. Bai and R. Elber, *J. Chem. Theory Comput.*, 2006, **2**, 484–494.
17. R. Elber, J. Meller and R. Olender, *J. Phys. Chem. B*, 1999, **103**, 899–911.
18. K. Siva and R. Elber, *Proteins: Struct., Funct., Bioinf.*, 2003, **50**, 63–80.

19. A. Ghosh, R. Elber and H. A. Scheraga, *Proc. Natl. Acad. Sci. U. S. A.*, 2002, **99**, 10394–10398.
20. M. Tuckerman, B. J. Berne and G. J. Martyna, *J. Chem. Phys.*, 1992, **97**, 1990–2001.
21. D. E. Shaw, P. Maragakis, K. Lindorff-Larsen, S. Piana, R. O. Dror, M. P. Eastwood, J. A. Bank, J. M. Jumper, J. K. Salmon, Y. B. Shan and W. Wriggers, *Science*, 2010, **330**, 341–346.
22. J. Valleau, in *Classical and Quantum Dynamics in Condensed Phase Simulations*, ed. G. C. Bruce J. Berne and David F. Coker, World Scientific, Singapore, 1998.
23. E. A. Carter, G. Ciccotti, J. T. Hynes and R. Kapral, *Chem. Phys. Lett.*, 1989, **156**, 472–477.
24. P. Majek and R. Elber, *J. Chem. Theory Comput.*, 2010, **6**, 1805–1817.
25. E. Vanden-Eijnden and M. Venturoli, *J. Chem. Phys.*, 2009, **130**, 13.
26. A. M. A. West, R. Elber and D. Shalloway, *J. Chem. Phys.*, 2007, **126**, 145104.
27. S. Kirmizialtin and R. Elber, *J. Phys. Chem. A*, 2011, **115**, 6137–6148.
28. R. Elber and A. West, *Proc. Natl. Acad. Sci. U. S. A.*, 2010, **107**, 5001–5005.
29. A. Malnasi-Csizmadia, J. Toth, D. S. Pearson, C. Hetenyi, L. Nyitray, M. A. Geeves, C. R. Bagshaw and M. Kovacs, *J. Biol. Chem.*, 2007, **282**, 17658–17664.
30. A. Dickson, A. Warmflash and A. R. Dinner, *J. Chem. Phys.*, 2009, **131**, 154104.
31. J. Juraszek and P. G. Bolhuis, *Biophys. J.*, 2008, **95**, 4246–4257.
32. T. S. van Erp and P. G. Bolhuis, *J. Comput. Phys.*, 2005, **205**, 157–181.
33. R. Elber, *Biophys. J.*, 2007, **92**, L85–L87.
34. K. Kuczera, G. S. Jas and R. Elber, *J. Phys. Chem. A*, 2009, **113**, 7461–7473.
35. R. J. Allen, D. Frenkel and P. R. ten Wolde, *J. Chem. Phys.*, 2006, **124**, 024102.
36. R. J. Allen, D. Frenkel and P. R. ten Wolde, *J. Chem. Phys.*, 2006, **124**, 018104.
37. R. J. Allen, C. Valeriani and P. R. ten Wolde, *J. Phys.: Condens. Matter*, 2009, **21**, 463102.
38. F. A. Escobedo, E. E. Borrero and J. C. Araque, *J. Phys.: Condens. Matter*, 2009, **21**, 333101.
39. D. Shalloway and A. K. Faradjian, *J. Chem. Phys.*, 2006, **124**, 054112.
40. C. Valeriani, R. J. Allen, M. J. Morelli, D. Frenkel and P. R. ten Wolde, *J. Chem. Phys.*, 2007, **127**, 114109.
41. A. Brunger, C. L. Brooks and M. Karplus, *Chem. Phys. Lett.*, 1984, **105**, 495–500.
42. A. Ulitsky and R. Elber, *J. Chem. Phys.*, 1990, **92**, 1510–1511.

CHAPTER 7

Computing Reaction Rates in Bio-molecular Systems Using Discrete Macro-states

ERIC DARVE*[a,b] AND ERNEST RYU[a]

[a] Institute for Computational and Mathematical Engineering; [b] Mechanical Engineering Department, Stanford University, Lomita Mall, Stanford CA 94305, USA
*Email: darve@stanford.edu

7.1 Introduction

Computing reaction rates in biomolecular systems is a common goal of molecular dynamics simulations. The reactions considered often involve conformational changes in the molecule, either changes in the structure of a protein or the relative position of two molecules, for example when modeling the binding of a protein and ligand. Here we will consider the general problem of computing the rate of transfer from a subset A of the conformational space Ω to a subset $B \subset \Omega$. It is assumed that A and B are associated with minimum energy basins and are long-lived states. See Table 7.1 for the notations used in this chapter.

Rates can be obtained using many different methods. In this chapter, we will review some of the most popular approaches. We organize the different approaches roughly in chronological order and under four main categories: reactive flux, transition path sampling, and conformation dynamics. The fourth class of methods, to which we do not give any specific name, in some sense attempts to combine

RSC Biomolecular Sciences No. 23
Innovations in Biomolecular Modeling and Simulations: Volume 1
Edited by Tamar Schlick

Published by the Royal Society of Chemistry, www.rsc.org

Table 7.1 Notations used in this chapter. Not all the notations used in this chapter can be found here. Notations that are local to a page or paragraph have been omitted.

TIS	transition interface sampling
FFS	forward flux sampling
MSM	Markov state models
WEB	weighted ensemble Brownian dynamics
RTS	reactive trajectory sampling
MFEP	minimum free energy pathway
Ω	conformational space
x_k	coordinates of atom k
m_k	mass of atom k
$U(x)$	potential energy
t	time
T	temperature
β	$\beta=(kT)^{-1}$
$\langle\ \rangle$	statistical average
$\rho(x)$	$\rho(x)=e^{-\beta U(x)}/Z$
Z	partition function, $Z=\int e^{-\beta U(x)}\,dx$
A	subset of Ω; reactant states
B	subset of Ω; product states
k_{AB}	rate from A to B
τ_{AB}	mean passage time from A to B
k_{TST}	rate as predicted from transition state theory
k_{Kramers}	rate as predicted from Kramers' theory
$\dot{f}(t)$	time derivative of $f(t)$
τ	Lag-time in Markov state models
$P_{ij}(\tau)$	probability to be in state j when starting from i after time τ
ξ	reaction coordinate or order parameter that monotonically increase from A to B
$(\xi_1, \ldots, \xi_p)$	set of generalized coordinates
$A(\xi)$	free energy
D	diffusion tensor
χ_A	characteristic function of some set, A
$C(t)$	conditional probability to find the system in B at t provided it was in A at time 0
$x(\mathscr{T})$	a discrete trajectory in Ω
$\mathscr{P}(x(\mathscr{T}))$	probability density function of trajectories
$\mathscr{P}_{AB}[x(\mathscr{T})]$	probability density function for the transition path ensemble
S_i	hypersurface in Ω; used in transition interface sampling, forward flux sampling, *etc.*
V_i	cell or macro-state in Markov state models and related methods In many cases S_i is the set of points such that $\xi(x)=\xi_i$
$\Phi_{A,1}$	flux from region A to S_1
$P_A(\xi_{i+1}\|\xi_i)$	for trajectories coming from A, probability to reach S_{i+1} starting from S_i
$\pi(x)$	committor function
μ_i	eigenvalue of the transition matrix
λ_i	$\lambda_i=-(\ln\mu_i)/\tau$; they are often an approximation of the eigenvalues of the Fokker-Planck equation
$\mathbf{P}(\ )$	used to denote the probability of some event happening
X_n, X_t	homogeneous Markov process indexed by n or t
N_{ij}	number of observed crossings from macro-state i to j
T_i	length of simulation in macro-state i
W_i	statistical weight of macro-state i

Table 7.1 (*Continued*)

w_i	statistical weight of walkers in weighted ensemble Brownian dynamics or reactive trajectory sampling
$\rho(x,t\|x_0, 0)$	probability to be at x at time t if the system was at x_0 at time 0
ρ_k	eigenfunction of the forward Fokker-Planck equation
ψ_k	eigenfunction of the backward Fokker-Planck equation
λ_k	corresponding eigenvalues

features from transition path sampling and conformation dynamics. They include non-equilibrium umbrella sampling,[1,2] and weighted ensemble dynamics.[3]

Reactive flux. We start with methods that were derived around 1930s,[4–8] were revisited later by, for example D. Chandler, 1978,[9] and are based on the concept of reactive flux. In these methods, the rate is derived from the free energy, and it is assumed that it is controlled by the flux at a saddle point at the top of the energy barrier separating A and B. The advantages of this approach is that it involves quantities which are relatively easy to calculate. However, it makes relatively strong assumptions about the system, and in practice assumes that a lot of information is already available regarding the transition mechanism and important pathways between A and B. We relate some of the results to Kramers' method, which applies to systems modeled using Langevin dynamics and overdamped dynamics.[10,11]

Transition path sampling. Many of the ideas developed in that context were used to develop another class of methods based on sampling transition pathways between A and B.[12–14] From the ensemble of pathways, rates and other properties can be obtained. The advantage of some of these approaches is that they do not require determining the saddle point separating A and B, and they apply to more general situations, for example, when multiple pathways contribute to the rate. See Vanden-Eijnden[15] for a discussion of transition-path theory, which proposes a mathematical framework to study transition pathways and the associated probability current. To address the shortcomings of some of these approaches, other methods were pursued along similar lines, including transition interface sampling,[16] and forward flux sampling.[17–19] We will present in the same category the milestoning technique, Faradjian and Elber,[20] which, although different in spirit, shares some similarities with transition interface sampling. This approach requires that the system "loses" memory when moving from a milestone to the next, for example by ensuring that the milestones are sufficiently separated from one another. Recent advances will be presented based on the work of Vanden-Eijnden *et al.*,[21] who introduced the concept of optimal milestones using the committor function. Using these milestones, the rate can be obtained exactly (in the absence of statistical errors), even when the milestones are close to one another. Májek and Elber[22] attempted to define the milestones in a way that would be computationally more general and advantageous compared to the original formulation.

Conformation dynamics. This is a large class of methods that can be traced back to Deuflhard and Schütte,[23,24] and are based on the concept of metastable states and transfer operator (or transition matrix). Broadly speaking, Ω is decomposed into metastable sets, which are sets that are long-lived and in which the system gets trapped. Then a transition matrix $P_{ij}(\tau)$ is defined using the probability to reach a metastable set j if one starts a trajectory of length τ (the lag-time) in set i. The analysis of the eigenvalues leads to the concept of Perron cluster. From the eigenvectors and eigenvalues, one can derive the rate and other kinetic information.

Although derived apparently independently and at a later date, some groups started exploring how one could model molecular systems using Markov state models, a well-known theory but which has been only (relatively) recently applied to modeling bio-molecular systems.[25–31] See related work by Shalloway (1996). Many of the theory and results for Markov state models can be found in the literature on conformation dynamics. In some sense, Markov state models can be viewed as a practical implementation of conformation dynamics, that attacks the high-dimensionality of Ω by subdividing the space into a "small" number of cells, also called macro-states.

Specifically, the transition matrix $P_{ij}(\tau)$ giving the probability of reaching macro-state j when starting from state i after some lag time τ is used to compute the rate. Its first eigenvalue is equal to 1 and corresponds to the equilibrium distribution. The second eigenvalue is very close to 1 and can be used to estimate the relaxation rate, as well as the forward (reactant to product) and backward (product to reactant) rates. An important issue is the effect of the lag time τ. At short times, non-Markovian effects, or memory, are present, that is the Markov state model is not accurate and the estimated rate suffers from a systematic bias.

Reactive trajectory sampling. The last class of methods groups two separate approaches that in some sense combine ideas from transition path sampling and a subdivision of space similar to Markov state models. One such method, called weighted ensemble Brownian dynamics, originates in Huber and Kim.[3] Although this paper is similar in spirit to transition interface sampling or milestoning, it can be easily extended to a general partitioning of space, using for example Voronoi cells. This is an important extension since, as a result, the method remains efficient in cases where multiple pathways contribute to the rate or when the most important pathway is not known. This approach leads to a sampling of transition pathways between A and B and therefore does not rely on the Markovian assumption made in Markov state models. However, like Markov state models, the efficiency of the sampling is improved by partitioning space into macro-states. A large number of walkers (simulations) are run in each macro-state. In order to maintain the population of walkers in each macro-state, a procedure was created to kill walkers in macro-states that are too crowded, and to split walkers when the number of walkers becomes too low. This method was recently revisited by Zhang *et al.* and Bhatt *et al.*,[32,33] who showed how the original approach could be extended.

The technique of non-equilibrium umbrella sampling of Warmflash *et al.*,[1,2] is similar in spirit. It applies, like weighted ensemble Brownian dynamics, to non-equilibrium systems and systems with memory (the Markovian approximation is not required) and uses a partitioning of Ω into macro-states. In each macro-state, a large number of walkers are simulated. Each time a walker attempts to leave a macro-state, its position (and velocity if needed) is recorded. Then in order to restart (continue) the simulation, a random position is chosen from the set of walkers who attempted to enter this macro-state from other macro-states.

In the second half of the paper, we will discuss in more detail weighted ensemble Brownian dynamics, renamed Reactive Trajectory Sampling (RTS) to reflect its broader application. The original method, Huber and Kim,[3] involves a procedure to split and kill walkers. We will revisit this method and propose an optimal procedure which leads to walkers with identical weights in each macro-state, a strategy which minimizes the statistical errors. We will discuss how the choice of macro-states affects the statistical errors and what the optimal choice is. We will present a new "multi-colored" algorithm that allows computing multiple rates (eigenvalues of the transfer operator from conformation dynamics) and accelerate convergence. This approach shares some similarities with the technique of core sets, Schütte *et al.*[34]

We will propose a novel error analysis of Markov state models, by considering the sensitivity of the eigenvalues to perturbations in the transition matrix. This will lead to estimates of the systematic errors (non-Markovity) and statistical errors, and their dependence on the lag-time τ (length of trajectories used to calculate the transition matrix). The choice of macro-states influences the decay of the non-Markovian effects; we will discuss what the optimal choice is. This optimal choice is, as can be expected, difficult to realize in practice, but this provides nonetheless a guideline to improve and refine an initial guess. We will make an argument showing that in the general case, statistical errors increase with τ, showing that an optimal tradeoff must be found between memory effects (small τ) and statistical errors (large τ). Several papers discuss the numerical errors in Markov state models.[35–39]

Some numerical results on simple benchmark problems in 1D and 2D are given at the end to illustrate the numerical properties of Markov state models and RTS.

Since there is a large number of methods to choose from, with different strengths and weaknesses, we attempt to summarize their main features for the purpose of comparing these methods together. We considered three axes in our evaluation: generality, computational cost, and parallel scalability. Each characteristic is ranked low, medium, or high. Generality relates to the number of approximations or assumptions that are required by the method to be accurate. For example, the method of reactive flux focusses its analysis on the transition region, typically a saddle point. When the assumptions are satisfied, the calculation may be very accurate. However, in some instances the prediction may not be satisfactory and need to be improved for example using a method based on transition path sampling. The computational cost should also be taken as a general guideline since it will vary tremendously depending on the system. However, some overall conclusions can be made regarding computational cost. Typically, the situation is that generality is traded for computational cost. Finally, since all large scale calculations require a parallel computer (multicore processors, graphics processing units, and parallel clusters), we also rank methods according to the amount of parallelism they offer, although in all cases, the amount of parallelism is very large and scalability is typically not an issue. Codes that run on slow networks (grid computing, cloud computing) will be more sensitive to these issues. Finally, the information in the list below should be taken merely as a guideline since all conclusions are in general strongly system and implementation dependent.

Reactive Flux

Generality: low. These approaches are typically the ones that require the greatest amount of knowledge about the system and relatively strict assumptions, in particular regarding the energy landscape near the transition region.

Computational cost: low. A free energy calculation is required. Although this can be difficult in some instances, this type of calculation is typically easier than with the other methods.

Parallel scalability: medium. Many methods are available to calculate free energy and are quite scalable. In most cases, it may become difficult to increase the number of processors compared to the other techniques described here.

Transition Path Sampling

Generality: high. These approaches are among the most "direct" and require little or no assumption.

Computational cost: high. As a result of being very general, they lead to extensive sampling and typically long simulation times.

Parallel scalability: high. In most cases, one can sample paths independently.

Transition Interface Sampling, Forward Flux Sampling

Generality: high. These approaches also require little or not assumption. Forward flux sampling in addition only requires the ability to run forward simulations making them applicable in situations where transition interface sampling fails.

Computational cost: high. A large number of pathways must be sampled. Forward flux sampling may in some cases converge a little slower, in particular when the sampling of the initial interfaces is poor or turns out to be insufficient for later interfaces. These methods typically perform better than transition path sampling.

Parallel scalability: medium. Processing the interfaces is sequential, making the method less parallel.

Milestoning

Generality: medium. Strong assumptions are required regarding loss of memory between milestones. As discussed in the main text, the original method has been extended and made more general.

Computational cost: medium. Only short pathways between adjacent milestones are required making this approach less expensive than the two previous methods.

Parallel scalability: high. The milestones can be processed independently.

Markov State Models

Generality: medium. The Markovian assumption must apply, which depends on the choice of Markov (or macro) states and the lag time between observations.
Computational cost: medium. The sampling is mostly local within each macro state, leading to efficient sampling.
Parallel scalability: high. Macro states are processed independently and require running a large number of short trajectories (in some implementations).

Weighted Ensemble Brownian Dynamics and Non-equilibrium Umbrella Sampling

Generality: high. The Markovian assumption is not required. The level of generality is similar to transition path sampling and related methods. The convergence of weighted ensemble Brownian dynamics is relatively easy to monitor.
Computational cost: high. The fact that the Markovian assumption does not apply typically leads to higher computational cost compared to Markov state models.
Parallel scalability: high. It is similar to Markov state models, although some communication is required to update the weight of walkers. This involves a small amount of communication, but one that occurs at regular intervals.

Weighted ensemble Brownian dynamics is in many respects similar to transition path sampling and differs "mostly" in the technique used to generate paths joining A and B and enhance the sampling. In this paper, the method of weighted ensemble Brownian dynamics will be referred to as reactive trajectory sampling to indicate that more general formulations have been created since the original paper of Huber and Kim (1996).[3]

Many theoretical results for these methods have been proved in the context of Langevin dynamics or Brownian dynamics. Results are often derived in the context of Brownian dynamics (over-damped dynamics). Extensions to Langevin are in most cases possible, although the proofs become more technical. In some cases, a method may depend only on some stochastic process with minimal assumptions. The theoretical derivations at the end of this paper are done in the context of Brownian dynamics but results can be extended to Langevin dynamics. Extensions to Newtonian dynamics (deterministic) are much more difficult and in most cases these extensions do not exist yet.

7.2 Transition Path Sampling

7.2.1 Reactive Flux and Transition State Theory

We start the discussion with the method of reactive flux which is a long standing approach to computing reaction rates. The idea goes back to Marcelin (1915), Eyring and Polanyi (1931), Eyring (1935), Horiuti (1938), Wigner (1938), who developed the initial theory of chemical reaction kinetics.[4–8] We outline the main ingredients in this type of approach. We assume that region

A is a subset of the conformational space of the molecular system and that it represents in the system in its reactant state. Similarly, B denotes the region defining the product states. Analytical approximation for the rate can be obtained if one assumes that a coordinate ξ can be defined which describes the reaction, a reaction coordinate. It is assumed that when $\xi = 0$ the system is in A and when $\xi = 1$ the system is in B. The value $\xi = \xi^*$ corresponds to the transition region or barrier between A and B.

We define the characteristic function χ_A (resp. B) which is 1 in the set A and 0 outside. Then using these functions, we can express the conditional probability to find the system in state B at time t provided it was in A at time 0:

$$C(t) = \frac{\langle \chi_A[\xi(0)]\chi_B(\xi(t)) \rangle}{\langle \chi_A \rangle} \tag{7.1}$$

Brackets $\langle\ \rangle$ are used to denote a statistical average. Regions A and B are separated by a transition region and the rate is determined by the rate at which this transition or barrier is crossed. At the molecular scale, there is some correlation time τ_{mol} associated with this crossing. That is for times larger than τ_{mol}, the system has forgotten how it went from A to B. Then for times t between τ_{mol} and the reaction time τ_{rxn}, $\tau_{\mathrm{mol}} < t \ll \tau_{\mathrm{rxn}}$, the time derivative of $C(t)$, called the reactive flux, reaches a plateau,[9] and

$$\dot{C}(t) \approx k_{AB} \tag{7.2}$$

The symbol $\dot{}$ denotes a time derivative.

Using transition state theory (TST), under the assumption that the recrossing of the barrier between A and B can be neglected, one can derive an expression for k_{AB} using Eq. (7.1) (Chandler [1978, 1987]):[9,133]

$$k_{\mathrm{TST}} = \frac{1}{2}\langle|\dot{\xi}|\rangle_{\xi=\xi^*} \frac{e^{-\beta A(\xi^*)}}{\int_{-\infty}^{\xi^*} e^{-\beta A(\xi)} d\xi} \tag{7.3}$$

where $A(\xi)$ is the free energy, and $\langle\rangle_{\xi=\xi^*}$ denotes an ensemble average with ξ constrained at ξ^*. This approach has some drawbacks. It always overestimates the rate. It requires a good reaction coordinate and a precise determination of the free energy maximum to locate the barrier. Nevertheless the method is computationally efficient and involves only quantities that can be computed with relatively low computational cost. Among the many methods to calculate the free energy (in this context the potential of mean force).[40–46]

Related approaches include Kramers' rate theory,[10,11] which was developed in the context of Langevin equations and overdamped dynamics. There are many connections between transition state theory and Kramers' theory, Hänggi and Borkovec (1990).[11] In particular Kramers' rate can be related to the "simple" TST rate through:

$$k_{\mathrm{Kramers}} = \frac{\lambda_+}{\omega_{\mathrm{bar}}} k_{\mathrm{TST}} \tag{7.4}$$

In this expression the potential at the transition point is assumed to be locally quadratic with stiffness $\omega_{\text{bar}}^2 = -m^{-1}U''(x_{\text{bar}})$ (m is the mass of the particle in a 1D model), and λ_+ is a function of the friction in the Langevin model and ω_{bar}. It can be shown that k_{Kramers} is equal to the multidimensional TST rate for a heat bath describing strict Ohmic friction, Hänggi and Borkovec (1990), pp. 268 & 272.[11] As the friction in the Langevin model goes to zero $\lambda_+ \to \omega_{\text{bar}}$ and $k_{\text{Kramers}} \to k_{\text{TST}}$. Moreover we always have $k_{\text{Kramers}} < k_{\text{TST}}$. The rate k_{Kramers} is itself an upper bound on the true rate given by

$$k(t) = \frac{\langle \dot{\xi}(0)\theta(\xi(t) - \xi^*)\rangle_{\xi(0)=\xi^*}}{\langle \theta(\xi^* - \xi(0))\rangle} \tag{7.5}$$

where θ is the Heaviside function.

This basic approach using TST has been improved in many ways including the use of harmonic approximations to model the minimum energy basins and transition region.[10,11,47] In variational TST, one attempts to improve the predicted rate by finding a dividing surface between A and B that minimizes the rate.[48,49] The overestimation of the rate by TST is a result of neglecting the re-crossing of the dividing surface. Several authors have proposed corrections to the basic TST approach to account for these effects.[9,50]

7.2.2 Transition Path Sampling

The method of reactive flux is attractive as its computational cost is often tractable. The primary calculation is obtaining the free energy profile along the reaction coordinate ξ. The accuracy is very dependent on the choice of coordinate. Specifically, the transmission coefficient, which measures the amount of re-crossing taking place, is dependent on the definition of ξ and a low transmission coefficient leads to inaccuracies or inefficiencies. Transition path sampling were in part proposed to alleviate the need to define this coordinate as they rely primarily on sampling trajectories going from A to B with no knowledge of ξ required. However we will see that later versions, again, are dependent on some knowledge of ξ.

Transition path sampling[12–14] is a Monte-Carlo method that allows sampling the ensemble of trajectories. For each discrete trajectory

$$x(\mathscr{T}) = \{x_0, x_{\Delta t}, \ldots, x_{\mathscr{T}}\} \tag{7.6}$$

one can define a probability $\mathscr{P}(x(\mathscr{T}))$ to observe such a trajectory. Then the transition path ensemble defines a probability density in the space of trajectories that is non-zero only for trajectories that connect A and B, and therefore its probability density function is defined as:

$$\mathscr{P}_{AB}[x(\mathscr{T})] = \chi_A(x_0)\chi_B(x_{\mathscr{T}})\mathscr{P}(x(\mathscr{T}))/Z_{AB}(\mathscr{T}) \tag{7.7}$$

where Z_{AB} is the appropriate normalization factor.

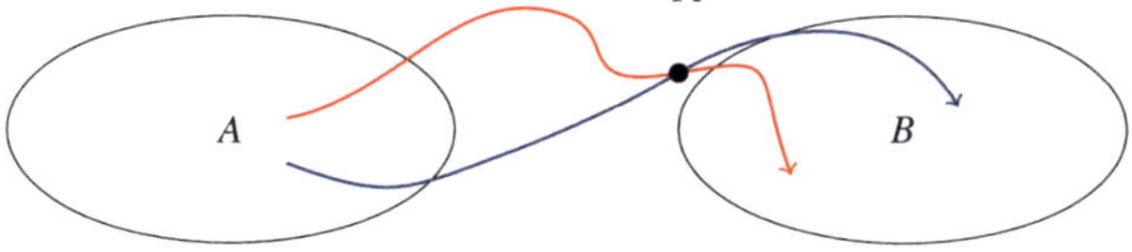

Figure 7.1 Shooting method of transition path sampling. The red curve is the old trajectory. The blue curve is obtained by perturbing the momentum at some point and regenerating a trajectory by using a forward and backward time integrator.

Trajectories in this ensemble can be generated using a Monte-Carlo procedure. A popular method is the so-called shooting method, in which a point is selected along the trajectory, the momentum is modified and a new trajectory is generated using a forward and backward time integration. If a stochastic dynamics is used then no perturbation is necessary since the random number generator will lead to a different trajectory. Then the appropriate acceptance probability is used (Metropolis-Hastings algorithm) to accept or reject this new trajectory. See Figure 7.1. Trajectories that do not start in A and end up in B are always rejected. This procedure can be improved by adding extra moves like shifting moves or path reversal moves.[14]

Many thermodynamic and kinetic properties can be determined from transition path sampling. As before the rate is related to $C(t)$ through $k_{AB}(t) = \dot{C}(t)$ and[14,51]

$$C(t) = \frac{\int \mathscr{D}x(t)\mathscr{P}(x(t))\chi_A(x_0)\chi_B(x_t)}{\int \mathscr{D}x(t)\mathscr{P}(x(t))\chi_A(x_0)} \tag{7.8}$$

where $\int \mathscr{D}x(t)$ is an integration over all possible paths (precisely, all possible points along the approximation at discrete time steps of the continuous trajectory).

There is a numerical difficulty in computing

$$\int \mathscr{D}x(t)\mathscr{P}(x(t))\chi_A(x_0)\chi_B(x_t) \tag{7.9}$$

since for most trajectories of length t, $\chi_A(x_0) = 0$ or $\chi_B(x_t) = 0$. A common procedure to make this tractable is to use umbrella sampling and an order parameter ξ which approximates the reaction coordinate. Denote $P_A(\xi', t)$ the probability that a trajectory started in A is such that $\xi = \xi'$ at time t, then:[51]

$$C(t) = \int_{\xi_{\min}^B}^{\xi_{\max}^B} P_A(\xi, t)\, d\xi \tag{7.10}$$

where $\xi_{\min}^B \leq \xi \leq \xi_{\max}^B$ defines region B. The quantity $P_A(\xi,t)$ will be difficult to obtain for values of ξ that are close to 1/2 (barrier) or beyond. The method of umbrella sampling[52] can then be conveniently used. For a window W_i in the interval [0, 1] define:

$$P_A^{W_i}(\xi,t) \stackrel{\text{def}}{=} \frac{\int \mathscr{D}x(t)\mathscr{P}(x(t))\chi_A(x_0)\chi_{W_i}(x_t)\delta[\xi-\xi(x_t)]}{\int \mathscr{D}x(t)\mathscr{P}(x(t))\chi_A(x_0)\chi_{W_i}(x_t)} \tag{7.11}$$

Then each $P_A^{W_i}(\xi,t)$ can be efficiently computed for small enough windows, while $P_A(\xi,t)$ over the entire [0, 1] interval is obtained by patching together the different $P_A^{W_i}(\xi,t)$ and matching the curves to produce a single profile.

Transition path theory. Transition path theory (TPT) is a theoretical framework to study transition path ensembles. It considers so-called reactive trajectories, which are the trajectories sampled by transition-path sampling. The theory is derived in the context of Langevin and over-damped dynamics. TPT provides definitions for the probability density of reactive trajectories. Most results involve the forward q_+ or backward q_- committor functions. The forward committor function for example is defined as the probability that, starting from some point outside of $A \cup B$, the system first reaches B before A.[15] As an example, the probability density of reactive trajectories is defined as:

$$\rho_{\text{reactive}}(x) = \rho(x)q_+(x)q_-(x)$$

where $\rho(x)$ is the equilibrium distribution for the process. More importantly expressions are provided for the probability current of reactive trajectories. This allows identifying important transition pathways and transition tubes. The example of the maze is helpful. The path in the maze with the largest reactive flux corresponds to the shortest path from A to B. Actual transition pathways take many detours and visit many dead ends before back-tracking and ending up in B. Analyzing the reactive flux allows bypassing these detours and take a straight route to B. In practice this allows identifying important reaction mechanisms joining A and B. Expressions are also given for the reaction rate, basically by integrating the probability flux over a subdividing surface.

Metzner *et al.*[53] presents an application to discrete Markov processes. In this case, the discrete probability current for reactive trajectories is given by:

$$f_{ij}^{A\to B} = \begin{cases} \rho_i\, q_j^-\, l_{ij}\, q_j^+, & \text{if } i \neq j, \\ 0, & \text{otherwise.} \end{cases}$$

where l_{ij} is the infinitesimal generator (rate matrix) of a continuous-time Markov chain.

See E and Vanden-Eijnden[54] for an earlier work. A series of illustrative examples are proposed in Metzner *et al.*[55]

7.2.3 Transition Interface Sampling

One issue in the previous approach is the fact that the trajectories need to have a fixed length t. See Dellago *et al.*[14] (Section IV D "A Convenient

Factorization") for an algorithm that relaxes this requirement. This length must be chosen carefully. A short time will lead to inaccuracies while a long time leads to a larger computational cost to update the trajectories. The method of transition interface sampling (TIS) introduces a novel concept, somewhat related to the previous algorithm with umbrella sampling, in which $n+1$ multi-dimensional surface in the space outside of A and B are defined. These surface are such that they can be used roughly to measure the progress of the reaction. For example, for some order parameter ξ and choosing a value ξ_i, we can define a surface S_i by the equation $\xi(x) = \xi_i$. We assume that $\xi_0 < \xi_1 < \ldots < \xi_n$; S_0 is the boundary of A and S_n is the boundary of B. Several methods (transition interface sampling and forward flux sampling) start from a reinterpretation of the rate as:

$$k_{AB} = \Phi_{A,1}\, P_A(\xi_n|\xi_1) \tag{7.12}$$

The first term, $\Phi_{A,1}$, is the flux of trajectories going out of A and crossing the first surface S_1. This quantity can be obtained by running a long trajectory (ignoring parts of the trajectory that last visited B) and counting the number of times the surface S_1 is crossed, per unit time. Only positive crossing are counted (that is moving away from A towards B) and re-crossings are ignored until A is entered again (that is crossings for which the previous crossing was S_1, not A, are ignored). The second term, $P_A(\xi_n|\xi_1)$ is the probability to reach S_n, assuming that the trajectory has crossed S_1 and that it does not cross A before S_n. See van Erp *et al.*[16]

The probability $P_A(\xi_n|\xi_1)$ is difficult to calculate since it is very small. This can be remedied using the intermediate surfaces S_i, $2 \leq i \leq n-1$. By construction, the function ξ being continuous and since $\xi_0 < \xi_1 < \ldots < \xi_n$, it is not possible to last come from A and cross ξ_i without having crossed first ξ_{i-1}. Using this result, it is possible to show that:[16]

$$P_A(\xi_n|\xi_1) = \prod_{i=1}^{n-1} P_A(\xi_{i+1}|\xi_i) \tag{7.13}$$

where $P_A(\xi_{i+1}|\xi_i)$ is the probability to cross S_{i+1} before A assuming the trajectory had previously crossed A then S_i. Although $P_A(\xi_n|\xi_1)$ can be very small, the quantities $P_A(\xi_{i+1}|\xi_i)$ are much larger and can be reliably obtained through direct sampling. The procedure to calculate $P_A(\xi_{i+1}|\xi_i)$ is similar to the procedure for TPS. It is illustrated in Figure 7.2(a).

7.2.4 Forward Flux Sampling

The forward flux sampling method (FFS) was conceived by Allen *et al.*[17–19] An earlier paper by Harvey and Gabb[57] bears some conceptual similarities. In TIS, one generates new trajectories by perturbing a point and then integrating forward and backward. In the forward flux sampling method, only forward integration is used. This can be essential in cases where backward integration is not possible. For example, in non-equilibrium systems, the lack of detailed

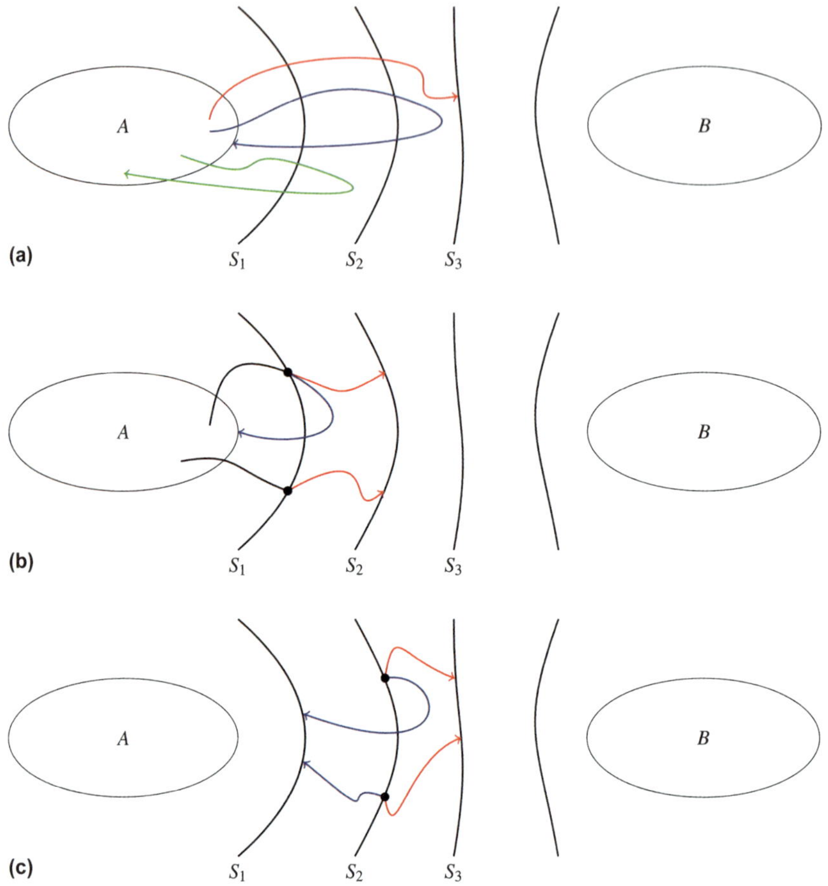

Figure 7.2 These schematic figures illustrate three different schemes: transition interface sampling, forward flux sampling, and milestoning. (a) Transition interface sampling algorithm. Calculation of $P_A(\xi_3|\xi_2)$. The sampling is similar to TPS. The main difference is that the only paths that are considered are those that last come from A, cross S_2 and then cross A or S_3. A new path is generated by perturbing a given path and then integrating forward or backward. The integration can be stopped as soon as A or S_3 are reached. The new path is then accepted only if there is a segment crossing S_2 that last come from A. Green trajectory: it has not reached the surface S_2 and therefore is discarded. Red trajectory: it has reached S_3 before A and therefore counts as 1. Blue trajectory: it has reached A before S_3 and therefore counts as 0. (b) Forward Flux Sampling. The black trajectories are used to generate starting points on S_1. The red trajectories are trajectories started from S_1 who made it to S_2, while the blue trajectories reached A first. The red trajectories are used to generate starting points on S_2. (c) Milestoning algorithm. Trajectories are initiated from S_2 using the equilibrium Boltzmann distributions. The exit times are recorded when the system reaches S_3 (red trajectories, function $K_2^+(t)$) or S_1 (blue trajectories, function $K_2^-(t)$).

balance and absence of time-reversal symmetry means that TIS (or milestoning which will be described in the next section) is not applicable. FFS is one of the few methods applicable to such systems.

FFS uses the same basic framework as TIS expressing the rate as

$$k_{AB} = \Phi_{A,1} \prod_{i=1}^{n-1} P_A(\xi_{i+1}|\xi_i) \tag{7.14}$$

FFS start by calculating $P_A(\xi_2|\xi_1)$ using trajectories initiated from S_1. Some of these trajectories may fail to reach S_2 (*i.e.* reach A before S_2) while others will reach S_2 successfully (before reaching A). For those that reach S_2, the first hitting point (first point where the trajectory crosses S_2) is saved. Those points are then used to calculate the next conditional probability, $P_A(\xi_3|\xi_2)$. FFS therefore only requires forward integration of trajectories. The starting points on S_i are in some sense fixed and are produced solely as a result of the sampling during the calculation of $P_A(\xi_i|\xi_{i-1})$. Typically, the computation for $P_A(\xi_i|\xi_{i-1})$ is continued until satisfactory accuracy is achieved and enough points have been generated on S_i. We note that the forward flux approach not only yields the rate constant, but also the complete transition pathways, which can be reconstructed by "gluing" the successful trajectory segments together. The method is depicted in Figure 7.2(b).

A limitation of this method is that the accuracy at later interfaces depends on the sampling at earlier interfaces. For example, if the first interface is relatively poorly sampled the error will propagate throughout the next interfaces. In addition, it is very possible that initial trajectories with low probabilities, in the end, make large contribution to the flux. Such a situation would lead to a large standard deviation and statistical errors.

Example applications are discussed for example in Valeriani *et al.* (2007), Borrero and Escobedo (2007), Allen *et al.* (2005).[17,19,58,59] Allen *et al.* (2006b)[19] describe FFS along with two other methods, the branched growth method and the Rosenbluth method. Allen *et al.* (2006a)[18] proposes an analysis of the efficiency of these methods.

7.2.5 Milestoning

Similar to the previous method, the milestoning technique of Faradjian and Elber (2004),[20] West *et al.* (2007)[134] is based on a set of separating hyper-surfaces that are used to measure the progress of the system from A to B. The advantage of this method is that trajectories need only to be run from one interface S_i (or milestone) to the next S_{i+1} or previous interface S_{i-1}. In contrast, the previous methods required running trajectories until they reach the next milestone or set A. Milestoning is inherently more parallel since each interface S_i can be processed in parallel with the others, whereas the other methods require a sequential processing of the interfaces. A drawback is that the interfaces need to be sufficiently separated so that we can assume that the system loses memory in the time it takes to reach the next interface.

The milestoning method proceeds by initiating trajectories at S_i, using the equilibrium distribution. Then it records the time it takes to reach S_{i-1} or S_{i+1}. This leads to two time distribution functions $K_i^+(t)$ (to reach S_{i+1}) and $K_i^-(t)$ (to reach S_{i-1}). We define $K_i(t) = K_i^+(t) + K_i^-(t)$. See Figure 7.2(c). Then we can calculate two functions: $P_s(t)$, which is the probability of being at milestone s (that is the last interface that was crossed was S_s), and $Q_s(t)$, the probability to transition to milestone s at time t (cross S_s at t). Then these two functions satisfy the following set of integral equations:

$$P_s(t) = \int_0^t \left[1 - \int_0^{t-t'} K_s(\tau)d\tau \right] Q_s(t')dt' \tag{7.15}$$

$$Q_s(t) = \eta_s \delta(t - 0^+) + \int_0^t \left[K_{s+1}^-(t - t')Q_{s+1}(t') + K_{s-1}^+(t - t')Q_{s-1}(t') \right] dt' \tag{7.16}$$

In these equations η_s is the initial milestone probability distribution. The first equation is saying that in order to be at s the system must first transition to s $[Q_s(t')]$ and then stay at s until time $t > t'$. The second equation says that to reach s one must first get to $s-1$ and then transition from $s-1$ to s (and similarly with $s+1$). The first term $\eta_s\ \delta(t-0^+)$ accounts for the initial distribution at $t=0$ (δ is the Dirac delta function). From these equations, the free energy and reaction rate can be obtained.

This approach is accurate once we assume that the system loses memory between milestones. With this, it becomes justified to independently generate initial conditions on each milestone following the Boltzmann distribution. These assumptions allow running independent calculations at each milestone and make it possible to run trajectories that stop as soon as the previous or next milestone is reached.

See a discussion of this approach in Elber (2005)[61] as well as an application example to an allosteric transition with *deoxy Scapharca* hemoglobin in Elber (2007).[62]

7.2.6 Milestoning Using Optimal Milestones

The method of milestoning was recently revisited, see *e.g.*, Vanden-Eijnden *et al.* (2008).[21] In this paper, it is shown that the assumption that the milestones need to be sufficiently far apart is not required provided that the interfaces S_i are iso-surfaces of the committor function. The committor function, denoted $\pi(x)$, is the probability to reach B before A starting from x. In Vanden-Eijnden and Venturoli (2009a),[63] the milestoning method is extended to the case of a tessellation of the conformational space using Voronoi cells V_i. See Figure 7.3. See a more complete discussion about Voronoi cells in Section 7.3.3. In a more recent paper, Májek and Elber (2010)[22] modify the original milestoning method using an approach that does not require a reaction coordinate (in that sense similar to Vanden-Eijnden and Venturoli [2009a]).[63] However, instead of considering the committor function to guarantee the accuracy of the method, this approach focusses in guaranteeing a minimum separation between the milestones. This leads to greater accuracy since the assumption that memory is lost between crossing milestones is automatically satisfied.

We now review some of the mathematical underpinnings discussed in Vanden-Eijnden *et al.* (2008). In particular, we recall the main arguments to establish that the use of the committor function $\pi(x)$ to define the milestones leads to an exact rate prediction with milestoning, irrespective of whether the milestones are close or well separated. For this reason these milestones are called optimal milestones.

In Vanden-Eijnden *et al.* (2008), one is concerned with computing the mean passage time τ_{AB} from A to B.

Consider, for example, the case of over-damped dynamics and the associated Fokker-Planck equation. The eigen values of the Fokker-Planck equations are denoted λ_i. We assume the λ_i are ranked by magnitude so that $\lambda_1 = 0 < \lambda_2 < \lambda_3 < \ldots$. If there is a single eigenvalue λ_2 close to 0 ($\lambda_2 \ll \lambda_3$), then the mean passage time is related to λ_2 through $\tau_{AB} = \lambda_2^{-1}\rho(B)$ (where $\rho(B) = \int_B \rho(x)dx$).

Milestones, denoted by S_i, are defined as hypersurfaces in Ω. The mean time to go from A to B can be obtained from the mean times to go from a milestone to another milestone. This has important consequences in terms of which assumptions need to be made to get an exact rate (in the absence of statistical errors).

The key property is the following one. Let us assume we consider all the trajectories that go from milestone S_i to S_j ($j = i + 1$ or $j = i - 1$). Specifically we initialize trajectories on milestone S_i with density:

$$\frac{|\nabla\pi(x)|e^{-\beta U(x)}}{\int_{S_i}|\nabla\pi(x)|e^{-\beta U(x)}d\sigma_i(x)} \tag{7.17}$$

where $d\sigma_i(x)$ is the surface element on milestone S_i. Then we can follow all the trajectories and record where they hit milestone S_j (we discard trajectories that hit other milestones first). The density of points on S_j is called the density of first hitting points.

In Vanden-Eijnden *et al.* (2008),[21] it was shown that, if the milestones $\{S_k\}$ are iso-surfaces of π (optimal milestones), then the density on S_j is given by:

$$\frac{|\nabla\pi(x)|e^{-\beta U(x)}}{\int_{S_j}|\nabla\pi(x)|e^{-\beta U(x)}d\sigma_j(x)} \tag{7.18}$$

In particular, this density is independent of the fact that the trajectories were started from S_i. This can be proved using the definition of the committor function and the forward Chapman-Kolmogorov equation.

The implication is that if we want to calculate the mean time to go from i to j we do not need to consider where the system is coming from. It is sufficient to initiate trajectories on S_i with the density given above and calculate the average time required to reach S_j (assuming this is the next milestone crossed).

We now discuss the calculation of the mean passage times and show that the property above is essential to derive expressions for the exact rate. Let us consider an absorbing boundary condition on the milestone, denoted S_n, that surrounds region B (cemetery milestone Vanden-Eijnden *et al.* [2008]).[21] We assume that trajectories are initialized on S_i with a probability density given by Eq. (7.17). We define T_i the mean length of a trajectory that start from S_i and goes to S_n (mean exit time). We denote: p_{ij} the probability that S_j is crossed after S_i, and τ_i the mean time to hit any other milestone S_j, $j \neq i$. Using the result above regarding the first hitting point density, we have:

$$T_i = \tau_i + \sum_{j \neq i,\, j \neq n} p_{ij} T_j \tag{7.19}$$

This is true because the mean escape time for some S_j, $j \neq i$, is the same irrespective of whether the system comes from S_i or not. This is a weaker property than saying that the times to go from S_i to S_j, t_{ij}, and S_j to S_k, t_{jk}, are independent. One can construct examples where t_{ij} and t_{jk} are strongly correlated, while the density of first hitting points at a milestone is independent of the last milestone that was crossed.

This equation can be written in matrix form as:

$$(I - p^n)T^n = \tau^n \tag{7.20}$$

where $p^n_{ii} = 0$ and p^n does not have the row or column corresponding to S_n.

This equation requires computing p_{ij} and τ_i for all milestones. This can be further simplified by observing that p_{ij}/τ_i has a simple interpretation. A formal proof can be given but here we simply outline the main points. We consider a very long trajectory. We will not discuss this further but limits must be taken as the trajectory length goes to infinity. Then:

$$p_{ij} = \frac{\text{Number of times the system crossed } S_j \text{ after } S_i}{\text{Number of times the system crossed any milestone after last crossing } S_i} \tag{7.21}$$

$$\tau_i = \frac{\text{Total length of time during which the last milestone crossed was } S_i}{\text{Number of times the system crossed any milestone after last crossing } S_i} \tag{7.22}$$

So that:

$$\frac{p_{ij}}{\tau_i} = \frac{\text{Number of times the system crossed } S_j \text{ after } S_i}{\text{Total length of time during which the last milestone crossed was } S_i} \tag{7.23}$$

Assume that we now use a discrete integrator to integrate the dynamics (which may be Langevin or overdamped dynamics), with time step Δt. We define:

$$P_{ij} = \mathbf{P}\big(\text{system crosses } S_j \text{ during the next step assuming it last crossed } S_i\big) \tag{7.24}$$

$$P_{ii} = \mathbf{P}(\text{system crosses no milestones during the next step assuming it last crossed } S_i) \tag{7.25}$$

Then from Eq. (7.23):

$$P_{ij} = \Delta t \frac{p_{ij}}{\tau_i} \tag{7.26}$$

if Δt is small compared to the time required to go from a milestone to the next. This shows that:†

$$(P - I)\big(\Delta t^{-1} T\big) = -1 \tag{7.27}$$

The quantity $\Delta t^{-1} T_i$ is the mean number of steps required to go from S_i to S_n.

The advantage of this equation is that it requires computing P_{ij} only, which can be relatively easily computed. Assume that we have defined a partition of the conformational space Ω into cells V_i such that the milestones S_i form the boundary of these cells. See Figure 7.3. Then the approach requires simply running independent simulations in all the cells V_i. For this, we need to use boundary conditions such that the system remains in the cell it started in, during the simulation. Let us assume that the trajectory hits a cell boundary with velocity $\dot{x}$. Then we know from the equilibrium probability density that there is another trajectory in the past or future, with a one-to-one mapping, which re-enters the cell through the same point. Its velocity can be chosen equal to $-\dot{x}$ (using the fact that the equilibrium probability density is even with respect to the momenta), or obtained using a hard wall reflection (now using the fact that the reflection conserves the equilibrium probability density) with

$$\dot{x}_{\text{re-entering particle}} = \dot{x} - 2(\dot{x} \cdot n)n \tag{7.28}$$

and n is the normal to the interface.

With this approach, one can generate a large number of samples in each cell, from which we can estimate P_{ij}:

$$P_{ij} = \frac{\rho(V_a) N_{ij}/n_a}{\rho(V_a) N_i^a/n_a + \rho(V_b) N_i^b/n_b} \tag{7.29}$$

where N_{ij} is the number of times the system was found to cross S_j after S_i, a is the cell bordered by S_i and S_j, while b is the cell on the other side of S_i, N_i^a (resp.

† The equation $\text{diag}(\tau)^{-1}(I - p) = \Delta t^{-1}(I - P)$ is also true along the diagonal since $p_{ii} = 0$ and $\Delta t/\tau_i = 1 - P_{ii}$.

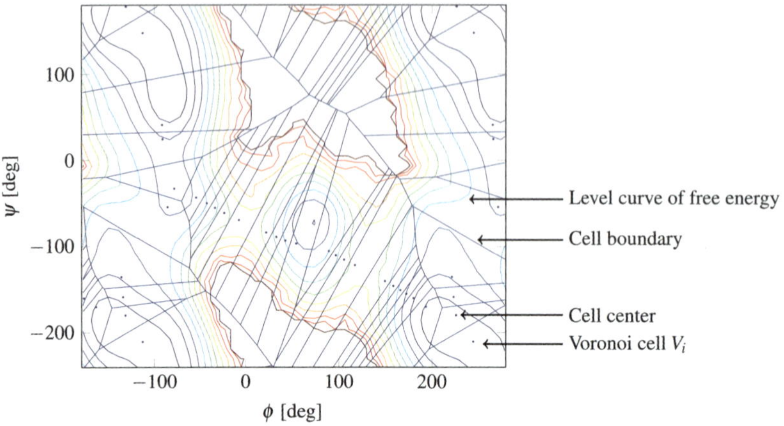

Figure 7.3 Example of Voronoi cells V_i for alanine dipeptide. The angles ϕ, ψ are torsional angles along the backbone of alanine dipeptide. The cell centers are shown using small blue dots. The blue lines denote cell boundaries (milestones S_i). In this case, cell boundaries are by construction equidistant to two centers. The colored curves are level curves of the free energy. Low energy regions are dark blue while high-energy regions are dark red.

N_i^b) is the number of steps for which the last milestone crossed was S_i in cell V_a (resp. V_b), and n_a, n_b are the number of steps computed in each cell. This is basically a direct calculation of Eq. (7.23). Note that by construction, the density of first hitting points on the milestone is the exact one so that, up to statistical errors, Eq. (7.29) is exact.

The advantage of this approach is its efficiency and the fact that the accuracy is more or less independent of the energy barrier between A and B. It does not have any systematic error, unlike the previous approaches. The main drawback is the requirement that the optimal milestones are iso-surfaces of the committor function, which again is difficult to realize in practice.

This requirement was relaxed in Májek and Elber (2010).[22]

7.3 Conformation Dynamics and Markov State Models

7.3.1 Conformation Dynamics

All the methods discussed above attempt in a sense to do a direct calculation of the rate, either by computing a reactive flux at the transition barrier or by sampling reactive trajectories going from A to B. We now discuss another class of methods that also attempt to calculate the free energy and the rate but, indirectly, by calculating the rate of transition between metastable basins. If one derives a statistical model of the system in terms of hops or transition between states, then an eigenvalue analysis can be used to calculate reaction rates, metastable states, and extract many relevant kinetic and thermodynamic information.

The idea goes back to Deuflhard and Schütte, who realized that computing time averages of physical observables or relaxation times of conformational changes (using molecular dynamics, for example) was largely determined by the existence and properties of invariant sets, called metastable sets. These are by definition sets (subsets of Ω, the conformational space of the molecular system) such that the system stays trapped in these sets for extended periods of times and with very rare transitions between sets. This has led to the conformation dynamics approach which aims at identifying these sets, and computing the transition rates between these sets. The first paper goes back to Deuflhard *et al.* (1996),[23] although the term "essential dynamics" can be found in Amadei and Linssen (1993);[64] Grubmüller and Tavan (1994)[65] had introduced the concept of conformational substates in 1994. Deuflhard (2003),[66] Schütte and Huisinga (2003)[24] provided some surveys on this topic.

Central to this model is the concept of transfer operator and the study of its eigenvectors and eigenvalues. Relevant reaction rates (as well as mean passage times, mean exit times,...) can then be extracted from these eigenvalues. We provide a brief account of the key mathematical objects. The definitions can be made in a relatively general context.[24] We consider a stochastic transition kernel $p(t, x, A)$ such that:

$$p(t, x, A) = \mathbf{P}[X_{t+s} \subset A | X_s = x] \tag{7.30}$$

where the family $\{X_t\}$ is a homogeneous Markov process indexed by a time variable t; $p(t, x, A)$ is therefore the probability that a Markov process started at x is in A after a time span t. This allows defining the Perron-Frobenius operator P_t (propagator or forward transfer operator). See Schütte and Huisinga (2003),[24] Schütte *et al.* (2001)[67] for a definition that does not assume that the transition kernel $p(t, x, y)$ is continuous with respect to a probability measure $\mu(dx)$. Then:

$$P_t u(y) = \frac{\int p(t, x, y) u(x) \rho(x) dx}{\rho(y)} \tag{7.31}$$

The density ρ is assumed to be invariant. In a similar fashion, ensemble transition probabilities can be defined as:

$$p(s, C, D) = \frac{1}{\rho(C)} \int_C p(s, x, D) \rho(x) dx \tag{7.32}$$

7.3.2 Perron Cluster Cluster Analysis

The metastable sets alluded to earlier are defined as sets that are almost invariant under the Markov process. With our definition, the set C is almost invariant if:

$$p(s, C, C) \approx 1 \tag{7.33}$$

These almost invariant sets can also be identified by considering the eigenvalues and eigenvectors of P_t. For example, the density ρ, being invariant, satisfies $P_t\chi_\Omega = \chi_\Omega$, where Ω is the space of all conformations of the molecule and χ_Ω is the characteristic function of Ω (in that case simply the function 1). The eigenvalue 1 therefore corresponds to the equilibrium distribution ρ.

Eigenvalues close to one form the so-called Perron cluster. The largest of these eigenvalues (not equal to 1) can then be associated with the slowest rate in the system. See the thesis of Huisinga (2001)[68] for an extensive discussion of transfer operators and metastability. Huisinga and Schmidt (2006)[69] provides some upper and lower bounds on these eigenvalues. There is a significant body of literature on the identification of these sets, in particular using an approach called Perron cluster cluster analysis (PCCA). The premise is that if one has uncoupled Markov chains (that is P_t is a block diagonal matrix assuming discrete states) then, assuming k separate aggregates or clusters, we will have k eigenvectors with eigenvalue 1. Each eigenvector is constant over each cluster and changes sign (positive, negative or 0) between clusters.[70] In PCCA, the sign has been used to identify these clusters or aggregates.

In a real application however, we are not dealing with uncoupled Markov chains but rather weakly coupled chains, resulting in a single eigenvector with eigenvalue 1 and a cluster of eigenvalues near 1, Schütte and Huisinga (2000b).[71] The identification of the clusters based on the sign structure[70] is then more difficult as the sign change is more progressive with a smooth change of value across the transition region, and the determination of the sign is more difficult as the eigenvectors may assume very small values whose sign $(+1, -1, 0)$ is then difficult to determine.

One remedy to this is to recognize, as will be further discussed later on, that the sign of an eigenvector entry is not the right quantity to look at. For example, the second eigenvector ρ_2 with eigenvalue $1 - \varepsilon$ is typically nearly constant with value say $\rho_2^A > 0$ in one cluster and is constant with value $\rho_2^B > 0$ in another cluster. The transition region is not associated with the point where ρ_2 is zero (or changes sign) but rather where ρ_2 assumes the value $1/2\left(\rho_2^A + \rho_2^B\right)$ (midpoint between the two plateaus). This midpoint value can be used in a robust fashion to determine the cluster boundaries.

Another approach pioneered by Deuflhard[72,73] called robust Perron cluster cluster analysis (PCCA+) proposes a fuzzy decomposition where instead of a strict partitioning into clusters one calculates of partition of unity $\sum_i \bar{\chi}(l) = 1$ (at a discrete state indexed by l) where each function $\bar{\chi}_i$ is called an almost characteristic function which smoothly transitions from 1 to 0 outside out a cluster. In effect, Weber and Rungsarityotin (2004)[73] assign a grade of membership between 0 and 1 to each state (in a discrete setting). Therefore, each state l may correspond to different clusters with a different grade of membership, defined by $\bar{\chi}_i(l)$. This approach was used to study a relatively long polyalanine (Ala_8 and Ala_{12}) in Noé *et al.* (2007).[30] The backbone torsion rotamer pattern was used to define the microstates.

The thesis of Weber (2007)[60] reviews these methods and discusses a meshless approach in which the membership functions are used to construct a Galerkin approximation of P_t. The fact that the basis functions are non-orthogonal (in contrast with partition functions, for example) leads to a non-diagonal mass matrix in the Galerkin formulation.

We mention an alternative method based on a singular value decomposition, see Fritzsche *et al.* (2007).[74] The main drawbacks of the approaches mentioned above (PCCA and variants) are the difficulty of identifying the Perron cluster if the transition matrix of the Markov chain has no significant spectral gaps; in addition, the calculation of the eigenvectors may be badly conditioned if the Perron cluster contains many eigenvalues very close to 1. The SVD approach attempts to mitigates some of these issues.

Kube *et al.*[75,76] have used this decomposition into metastable sets with PCCA+ to construct a coarse grained matrix that approximates the exact fine grained propagator. Limitations of the resulting propagator are discussed.

An application of this approach to systems described by a Langevin equation is given in Schütte and Huisinga (2000a).[77] An issue in this approach is the fact that the conformational space needs to be discretized appropriately in order to form a discretized approximation of P_t (called transition matrix in Cordes *et al.* [2002]).[78] This is difficult in practice since molecular systems live in high dimensional space. Cordes *et al.* (2002)[78] propose to focus on dihedral angles to reduce the dimension and further improve their approach by considering a recursive decomposition in which space is first decomposed using the dihedral angle with the longest auto-correlation (which is shown to correlate with metastability). This leads to a first partitioning of space. Then, these metastable subspaces are further decomposed by applying the same strategy recursively (see Cordes *et al.* [2002]).[78] PCCA is finally applied to the resulting coarse decomposition. Galliat *et al.* [2000][79] proposes an approach based on self-organized neural networks, also to attack this problem of dimensionality. Schultheis *et al.* (2005),[80] Kloppenburg and Tavan (1997)[81] approach the problem of dimensionality through the use of density-oriented discretizations that represent the probability density using a mixture of normal distributions.

The concept of metastability also allows viewing the Markov chains as a collection of rapidly mixing chains within certain parts of the state space, that are weakly coupled together. This gives rise to the concept of uncoupling wherein uncoupled Markov chains, which resemble the original chain, are formulated for each metastable states. Then these k chains can be coupled again by a $k \times k$ coupling matrix. The resulting system contains all the important information from the original chain. This is discussed in Meerbach *et al.* (2005),[82] where in addition upper bounds are provided for the 2nd eigenvalue of the uncoupled Markov chains to establish that they are indeed rapidly mixing.

In a related approach to building coarse models without resorting to the transition matrix or transfer operator, Horenko *et al.* (2006)[83] attempts a direct construction of a coarse grained model using a hidden Markov model with output given by stochastic differential equations (SDE). The model is then formulated as the combination of a set of SDEs (Langevin equations) and a rate matrix which determines how one jumps from an SDE to the next.[83,84]

The idea of hidden Markov model is reused in Fischer *et al.* (2007)[85] but this time the output of a hidden Markov state is a probability density function in the observed variables (in this case torsion angles along the protein backbone). Von Mises output distributions are used since the observed variables are angles.

Although many clustering methods have been proposed, we mention the work of Coifman *et al.* (2008)[86] who proposed to build a coarse grained representation based on the eigenvectors of the diffusion map. The algorithm is based on the definition of a weighted graph on the simulated points and the subsequent computation of the first few eigenvalues and eigenvectors of a random walk on this graph. Connections are made to the backward Fokker-Planck operator.

A related approach was developed by Shalloway and his group.[31,87–91] Church *et al.* (1999) review these methods. In Orešič and Shalloway (1994), Gaussian packets are used to characterize metastable basins and equations are provided to calculate and evolve packets. Packets are Gaussian functions in Ω. These packets can be numerically obtained by computing the average value of x locally in some metastable basin, and the variance of x:

$$K_i^{-1} = 2\beta\langle(x - x_i^0)(x - x_i^0)^T\rangle_i \tag{7.34}$$

where x_i^0 is such that $\langle K_i\,(x - x_i^0)\rangle_i = 0$. See Orešič and Shalloway (1994)[87] for the notation $\langle\,\rangle_i$ (this is a suitable local averaging in basin i). Then the Gaussian packet has the form:

$$p_i(x) = \exp\left(-\beta\left[V_i + (x - x_i^0)^T K_i(x - x_i^0)\right]\right) \tag{7.35}$$

In Shalloway (1996), the eigenfunction expansion of the Smoluchowski equation is used to optimize the macrostate expansion (using Gaussian packets) by satisfying a minimum uncertainty condition. Ulitsky and Shalloway (1998)[89] builds macrostates based on a variational principle, from which the transition region (separating the macrostates) can be identified and analyzed. Connections with the variational transition state theory.[92]

7.3.3 Markov State Models

Another broad class of methods are Markov state models (MSM). Markov chains have a long history, however their application to bio-molecular modeling and protein modeling is relatively recent.[25–27,93] Two landmark papers are Chodera *et al.* (2007) and Chodera *et al.* (2006). See Pande *et al.* (2010) for a review and discussion of this model. This approach is closely related to conformation dynamics and many theoretical results from conformation dynamics directly apply to MSM. To build a discrete model of the continuous underlying conformational space Ω, the method starts by partitioning Ω into cells. The advantage is that this decomposition is often easier to obtain in practice than defining a reaction coordinate. In this respect, this approach has some of the advantages of transition path sampling, which also does not require a precise knowledge of the reaction coordinate. This is a significant departure from TIS, FFS, and milestoning who rely primarily on a single order parameter that measures the progress of the reaction from A to B in an essentially sequential way (progress along a single dimensional variable or order parameter). MSM remains easy to construct even in the presence of multiple pathways with no obvious reaction coordinate.

These macro-states can be constructed in different ways, for example from pathway sampling information[25] or by partitioning the Ramachandran map.[28] Schütte *et al.* (1999)[56] uses a decomposition based on torsion angles. In this paper, a method is also proposed to define generalized angle coordinates, in an attempt to reduce the number of coordinates in the problem. It is based on ideas from Amadei and Linssen (1993),[64] and statistical analysis of circular data.[95,96] Chodera *et al.* (2007)[29] proposes an automatic procedure to create macro-states using an iterative procedure and the k-medoids algorithm (a partitioning algorithm similar to the k-means algorithm). Microstates are iteratively lumped into macrostates using the k-medoids algorithm, and split again into microstates to iteratively refine the definition of the macrostates.

Voronoi tesselation has also been proposed since it allows a simple construction of macro-states based only on the definition of the cell centers.[63] See Figure 7.3. Given a set of points $\{x_i\}$, a Voronoi cell V_i is defined as:

$$V_i = \{x|\ |x - x_i| \leq |x - x_j|, \quad j \neq i\} \tag{7.36}$$

Such cells are convenient since there is a straightforward equation to determine in which cell a given point x is. The centers x_i can be obtained in different ways including simulations at high temperature, using nudge elastic bands or the string method, techniques to sample transition pathways, or computing minimum (free) energy pathways. We will show later on that placing centers along minimum energy pathways is often a good choice.[97]

We note that in Schultheis *et al.* (2005)[80] this issue is altogether avoided by constructing a density-oriented discretization (an attempt to circumvent the curse of dimensionality) based on approximating the probability density in conformational space using a mixture of univariate normal distributions.

Once the macro-states have been defined, one calculates the transition matrix:

$$P_{ij}(\tau) = \mathbf{P}(\text{particle in cell } i \text{ at time 0 is in cell } j \text{ at } \tau) \tag{7.37}$$

where τ is the so-called lag-time. The eigenvalues and eigenvectors of this matrix allow computing the different rates in the system, along with mean passage times.[26,28]

Park and Pande (2006)[98] use the concept of Shannon entropy to measure and identify non-Markovity. Shannon entropy measures the amount of uncertainty associated with a random variable. Non-markovity can be measured by evaluating the change in uncertainty (Shannon entropy) for a Markov variable X_n if one prescribes the previous state X_{n-1} vs. the last two states X_{n-1} and X_{n-2}.

Markov chains are such that the uncertainty is the same in both situations. From the definition of the Shannon entropy.

If the previous state is known:

$$H(X_n|X_{n-1}) = -\sum_{x,y} \mathbf{P}(X_n = x, X_{n-1} = y) \ln \mathbf{P}(X_n = x|X_{n-1} = y) \quad (7.38)$$

If the last two states are known:

$$H(X_n|X_{n-1}, X_{n-2}) = \quad (7.39)$$

$$-\sum_{x,y,z} \mathbf{P}(X_n = x, X_{n-1} = y, X_{n-2} = z) \ln \mathbf{P}(X_n = x|X_{n-1} = y, X_{n-2} = z) \quad (7.40)$$

Measure of non-Markovity:

$$R = \frac{H(X_n|X_{n-1}) - H(X_n|X_{n-1}, X_{n-2})}{H(X_n|X_{n-1})} \quad (7.41)$$

From this definition, $R = 0$ for Markov chain[‡] and $R = 1$ if the knowledge of X_{n-2} and X_{n-1} completely determines X_n. This procedure can also be used to refine the definition of macro-states. Another approach to identify and correct non-Markovity is proposed in Nerukh *et al.* (2010)[99] using the concepts of "computational mechanics" (a term coined by Crutchfield [1994])[100] and ε-machines.[101,102]

Huang *et al.* (2009)[103] discusses the use of generalized ensemble algorithms, *e.g.* the replica exchange method, parallel tempering or simulated tempering[104–107] and how they can be combined with Markov State models to provide an efficient conformational sampling algorithm. See Bowman (2009)[108] for a discussion of similar ideas and how they have been implemented in the piece of software MSMBUILDER.

Chiang *et al.* (2010)[109] proposes to construct Markov models with hidden states as a way to construct more accurate models with fewer states (hidden Markov models). In such models, the hidden Markov states do not represent individual protein conformations but rather overlapping probabilistic distributions over the conformational space.

Applications of this approach are discussed in several papers including:

- a polyphenylacetylene (pPA) 12-mer in explicit solvent for four common organic and aqueous solvents (acetonitrile, chloroform, methanol, and water): Elmer *et al.* (2005a,b).[110,111]
- lipid membrane fusion: Kasson *et al.* (2006).[112]
- the villin headpiece: Jayachandran *et al.* (2006). Bowman *et al.* (2009) discusses the application of the software MSMBUILDER to the villin headpiece.[113,114]
- polyalanines: Noé *et al.* (2007).[30] See also Noé and Fischer (2008)[115] with a review and discussion of MSM.

‡ $H(X_n|X_{n-1}, X_{n-2}) \leq H(X_n|X_{n-1})$ so that $R \geq 0$. Note that $R = 0$ does not imply that the system is Markovian since it is possible that $R = 0$ and $H(X_n|X_{n-1}, X_{n-2}, X_{n-3}) < H(X_n|X_{n-1})$.

- PinWW domain: Noé *et al.* (2009),[116] Morcos *et al.* (2010).[117] In addition, in this paper, it is shown how folding pathways can be reconstructed from the MSM using transition-path theory.[53,54]
- the millisecond folder NTL9 (1–39): Voelz *et al.* (2010).[118]

7.4 Non-equilibrium Umbrella Sampling and Reactive Trajectory Sampling

7.4.1 Non-equilibrium Umbrella Sampling

In recent years, some approaches have in some sense tried to combine ideas from the previous sections, for example by calculating reactive trajectories or transition pathways from A to B (thereby being exact, contrary to Markov state models whose accuracy depends on the Markov assumption), while using a decomposition of the conformational space Ω into cells. These methods therefore combine the advantages of transition path sampling with the sampling efficiency of MSM, which requires only "local" sampling.

The first approach we will discuss is the one proposed by Dinner and co-workers.[1,2] The method originates from the method of umbrella sampling, Torrie and Valleau (1977),[52] in the sense that it tries to enhance sampling in poorly sampled region. The latter was broadly speaking adapted to allow modeling non-equilibrium systems. In this case, when computing a reaction rate from A to B, particles are removed from the system each time they reach B and are reinjected in A, thereby creating a steady-state but out of equilibrium situation.

This technique constructs two staggered lattices (using boxes[1] or following minimum energy pathways).[2] Simulations are run inside each macro-state (a box in Warmflash *et al.* [2007]).[1] Then one records when the system attempts to leave a box. At that point, two strategies are applied.

First, we keep track of the number of particles going from box i to j. Each time a particle attempts to go from i to j, the weight of box i, W_i, and j, W_j are adjusted according to:

$$-\Delta W_i = \Delta W_j = sW_i \frac{T^*}{T_i} \tag{7.42}$$

where T_i is the time elapsed in region i (to account for situations in which longer simulations are run in some boxes), T^* is some arbitrary time scale to make the equation dimensionally correct, and s is a small parameter used to adjust the rate at which the weights W_is vary. With this equation, the weight of each box converges to its correct steady-state value.

Second, one needs to determine which point should be used to reinsert a walker that left box i back into box i. For this walkers that leave another box j and attempt to enter i are saved and, among those, one is picked according to the probability rule

$$p_a = \frac{\sum_b N_{ba} W_{j(b)}/T_{j(b)}}{Z_i} \tag{7.43}$$

$$Z_i = \sum_{ab} N_{ba} W_{j(b)} / T_{j(b)}$$

where p_a is the probability of choosing state a in box i, N_{ba} is the number of crossings from state b in box $j(b) \neq i$ to state a.

In Warmflash *et al.* (2007),[1] page 154112–4, end of section G, it is argued that two lattices are needed. Simulations are run in both but walkers that re-enter box i in lattice 1 are chosen with Eq. (7.43) using data from lattice 2. Otherwise it is claimed that the method is unstable and convergence may not be achieved. The argument put forward is as follows:

> Suppose, for example, that the weight of a box (B) fluctuates upward. By Eqs (3) and (4) (Eq. [7.43] in this manuscript), walkers in neighboring boxes will then be reset to boundary states accessible from B more often. However, if transitions from those states to ones in B are allowed, with some probability, the reset walkers will immediately attempt to enter B and increase its weight further according to Eq. (5) (Eq. [42] in this manuscript). This positive feedback loop causes the single-lattice scheme to be unstable in simulations to obtain the steady-state probability distribution as a function of multiple variables. The use of two lattices enables boundary states on one lattice to be chosen using the fluxes from the other lattice, which breaks the feedback loop and enables convergence.

This argument is not so clear unfortunately. As a particle leaves i to enter j, the weight W_i is reduced by $-sW_iT^*/T_i$. If the particle reenters i, the weight is increased again by sW_jT^*/T_j. In general if W_i has an upward fluctuation, the net result is a reduction in W_i. In some cases, walkers do not re-enter i and move to some other box, further reducing W_i. As a result, in the scenario mentioned above of a temporary fluctuation upward of the weight, on average, Eq. (7.42) will slowly reduce the weight of box i and return it to its correct steady-state value.

Following Warmflash *et al.* (2007) and Dickson *et al.* (2009b),[1,2] Vanden-Eijnden and Venturoli (2009b)[119] developed a similar approach but that uses a single partition of space, based on Voronoi cells. The approach is similar with the following differences:

- Weights are adjusted based on fluxes between cells. These fluxes, N_{ij}/T_i (where N_{ij} is the number of crossings from i to j), are used to solve a linear system that provides an approximation to the steady-state weights W_i:

$$\sum_{j,i\neq j} W_i \frac{N_{ij}}{T_i} = \sum_{j,i\neq j} W_j \frac{N_{ji}}{T_j} \tag{7.44}$$

  Another "global" scheme to adjust the weights (by contrast with the local scheme [7.42]) is given in Dickson *et al.* (2009a).[120]

- To pick a re-entry point, one of the *boundaries* α of cell i is randomly picked using a probability law obtained from the cell flux and steady-state probabilities:

$$p_{\text{boundary } \alpha \text{ of cell } i} = \frac{W_j N_{ji}/T_j}{\sum_{k,k \neq i} W_k N_{ki}/T_k} \tag{7.45}$$

  where boundary α is the boundary between cell i and j. Although the implementation is different, this is similar to Warmflash *et al.* (2007).[1]

Based on this, it appears that using two lattices is not necessary and that the scheme correctly works with a single lattice. Also the idea of lattice is no longer discussed in a more recent publication.[121] See also Dickson *et al.* (2011)[122] for an application of this method to unfolding and refolding of RNA. In this paper as well, a single lattice is used. Dickson and Dinner (2010)[121] present some theoretical results regarding non-equilibrium umbrella sampling (an analysis of the convergence of the weights using the local scheme), a comparison with and discussion of forward flux sampling, and recent applications of these methods.

7.4.2 Reactive Trajectory Sampling

The second method, which is related in some fashion to the previous class of techniques, can be attributed to Huber and Kim (1996).[3] In this reference, the method is developed assuming that an approximate reaction coordinate has been chosen. However, it is not difficult to extend this approach to a general decomposition of the conformational space Ω in a manner similar to, for example, Vanden-Eijnden and Venturoli (2009a)[63] with Voronoi cells. This method will be discussed in more details below. It consists in running a large number of simulations (or "walkers") in parallel in such a way that a given number of walkers are maintained in each cell or macro-state. Macro-states that are near an energy barrier will tend to be depleted and therefore a strategy is applied to duplicate walkers in this macro-state, in a statistically correct way. This is done by assigning statistical weights to each walker. For example a walker with weight w can be split into two walkers, starting from the same location in Ω, with weights $w/2$. Conversely, macro-states that are at low energy will tend to become overcrowded and walkers are then removed. If for example we have two walkers with weights w_1 and w_2, we randomly select one with probabilities $(w_1/(w_1+w_2), w_2/(w_1+w_2))$ and assign to it the weight w_1+w_2. This approach ensures an efficient sampling of phase space.

In order to calculate a reaction rate, the macro-state corresponding to region B is transformed into a cemetery state, that is any walker that enters this macro-state is removed from the simulation and re-inserted in region A. In this fashion, although the simulation is effectively out of equilibrium, the population of walkers is kept constant. This method allows computing all the relevant quantities of interest, such as reaction rates, free energy, metastable states, *etc.* We note that contrary to Markov state models, this approach does not suffer from non-Markovity errors and that in the limit of infinite sampling it provides an exact answer.

In Zhang *et al.* (2007),[123] this technique was applied to explore the transition paths ensemble in a united-residue model of calmodulin.[33,124] In Zhang *et al.* (2010),[32] it is shown that the method initially developed in Huber and Kim (1996)[3] is really applicable to a much wider class of problems and proposes some generalizations of this procedure.

We mention that a similar technique has been applied to simulated annealing to find minima of rough (or even fractal) functions (see Huber and McCammon [1997]).[125]

Detailed discussion of reactive trajectory sampling. We now discuss in more details the method of Huber and Kim (1996),[3] Zhang *et al.* (2007, 2010)[32,123] which we rename reactive trajectory sampling method (RTS), in the broader context of macro-state models (*e.g.* Voronoi cell partitioning). In this approach, systematic errors arising from non-Markovian effects are avoided by directly calculating reactive trajectories from A to B and obtaining the probability flux entering B (or A for the backward rate), Metzner *et al.* (2006).[55] When the energy barrier is high, this can be very inefficient since very few trajectories (if any) will make it to B when started from A. However, a simple trick allows improving the efficiency of the calculation to the extent that the decay of the statistical errors becomes essentially independent of the energy barrier height.

As before we split the space of possible configurations into cells. Then a large number of random "walkers" are initialized and advanced forward in time. The basic idea is to use a strategy whereby, in cells that get overcrowded (too many walkers), we merge walkers, thereby reducing their numbers, while in cells that are depleted (near transition regions), we split walkers to increase their number. The end goal is to maintain a given target number of walkers in each cell. With such an approach we are able to observe a constant stream of walkers going from A to B (and vice versa) irrespective of the height of the energy barrier. We now explain the details of the method.

Assume we have n_w walkers whose position gets updated at each time step. It is possible to resample from these walkers without introducing any bias in the calculation using the following procedure. Each walker, whose position is denoted x_i, is assigned a probabilistic weight w_i, for example initially equal to $1/n_w$. A walker can be split into p walkers with weight w_i/p. After the split, each walker can be advanced independently. Averages can then be computed using:

$$\langle f \rangle = \lim_{n_w \to \infty} \frac{1}{\sum_j W_j} \sum_i w_i f(x_i) \tag{7.46}$$

This equation is always true, irrespective of how many times the splitting procedure is applied, or how many steps are performed, as long as the initial position of the walkers is drawn from the equilibrium distribution. This is proved from the fact that the equilibrium distribution is by definition invariant under the dynamics under consideration for $x(t)$.

The reverse operation is possible. Assume we have a group of walkers with weights $w_1, \ldots, w_{n_w}$. Suppose we randomly pick a walker with probabilities $w_1/\Sigma_k w_k, \ldots, w_{n_w}/\Sigma_k w_k$, and assign to it a weight of $\Sigma_i w_i$. Since the average weight of walker j is $(\Sigma_i w_i)(w_j/\Sigma_k w_k)$, Eq. (7.46) remains true. This procedure can be used to reduce the number of walkers in a cell.

The algorithm below, called `resample`, explains the procedure for resampling walkers. Before the procedure `resample` is called we have several walkers in each cell with varying weights. The procedure `resample` loops over the cells and select walkers in a way such that the new set of walkers all have the same weight, equal to the average weight of walkers in the cell. This is an important aspect of the method as assigning a constant weight can be proved to be optimal in terms of minimizing the variance and therefore statistical errors. This approach is different from Huber and Kim (1996),[3] which leads to walkers with varying weights and results in a somewhat larger variance. The proof is given below.

```
# This algorithm, called resample, is written in the programming language
# python. It was written by Eric Darve and Jesus A. Izaguirre (University
# of Notre-Dame).
# Input: list of walkers (list0) and list of walker weights (weights).
# Input: target number of walkers (ntargetwalkers).
# Output: list of walkers (list1) and their weights (newweights).
# weights[x] must be the weight of the walker with ID x.


from numpy import floor, argsort, random, sum
list1 = []                # new list of walkers
newweights = []           # weights of new walkers
nwalkerlist1 = 0          # number of walkers in list 1

wi = # Initialize the list of weights for walkers in the current cell.
ind = argsort(-wi)
# Sort the walkers in descending order based on their weights.
list0 = list(list0[ind])

W = sum(wi)
tw = W / ntargetwalkers
# ntargetwalkers is the target number of walkers in cell

x = list0.pop( ) # We assume that there is at least one walker in the cell

while True : # while loop exits using a break.
    Wx = weights[x]
    if (Wx >= tw or len (list0) == 0):
        r = max (1, int(floor(Wx / tw)))
        # max is required because of round-off errors
        r = min (r, ntargetwalkers-nwalkerlist1)
```

```
31          # required because of round-off errors
            nwalkerlist1 += r          # update the number of walkers in list1
33          for item in repeat (x, r): #insert r copies of walkers in list1
              list1.append(item)
35            newweights.append(tw)
            if nwalkerlist1<ntargetwalkers and Wx - r*tw>0.0:
37            list0.append (x)
              weights [x] = Wx - r*tw
39          if len(list0)>0:
              x = list0.pop( )
41          else:
              break
43      else:
            y = list0.pop( )
45          Wy = weights[y]
            Wxy = Wx + Wy
47          p = random.random( )   # randomly select a walker
            if p < Wy / Wxy:
49            x = y
            weights [x] = Wxy
```

We note that this algorithm terminates since when the last walker is removed from list0 we exactly have W = tw (target weight) so that the while loop does terminate. The maximum number of iterations in the while loop is bounded by n_w + tn: the number of walkers in the cell before the procedure starts + the target number of walkers. (Lines 26 through 40 are executed at most tn times, while the lines 42 to 48 are executed at most n_w times.) In addition it is apparent from line 33 that the walkers have the same weight at the end. The sorting of the weights on line 13 helps reduce data correlation. Indeed when a walker is split, samples become correlated for some time. The initial sorting makes sure that only walkers whose weight is greater than tw are split. As soon as we have processed all the walkers with weight greater than tw, r stays equal to 1. The reinsertion on line 35 is required to ensure a constant total weight W. The weight that is used, W–r*tw, ensures that the resampling is correct and that the total weight in each cell is unchanged by `resample`.

Proof of optimality. To simplify the discussion we assume that we have n_w particles with weights w_i such that $\Sigma_i w_i = 1$. We evolve the system in time such that the walker locations $\{x_i\}$ become uncorrelated. As a consequence, the weights w_i are independent from the positions $\{x_i\}_{i=1}^{n_w}$. We also assume that the walkers do not have any particular order so that the statistics of x_i and w_i are the same as x_j and w_j, $j \neq i$. The estimator of a particular quantity f is

$$\hat{f} = \sum_{i=1}^{n_w} w_i f(x_i)$$

Then we have

$$\langle\hat{f}\rangle = \sum_{i=1}^{n_w} \langle w_i\rangle\langle f(x_i)\rangle = \left\langle \sum_{i=1}^{n_w} w_i \right\rangle \langle f\rangle = \langle f\rangle \tag{7.47}$$

which tells us that $\hat{f}$ is indeed an unbiased estimator of f. The statistical error can be estimated from the variance of $\hat{f}$:

$$\langle(\hat{f}-\mu)^2\rangle = \left\langle \left(\sum_{i=1}^{n_w} w_i(f(x_i)-\mu) \right)^2 \right\rangle \tag{7.48}$$

$$= \sum_{i=1}^{n_w}\sum_{j=1}^{n_w} \langle w_i w_j (f(x_i)-\mu)(f(x_j)-\mu)\rangle \tag{7.49}$$

$$= \sum_{i=1}^{n_w} \langle w_i^2\rangle\langle(f(x_i)-\mu)^2\rangle \tag{7.50}$$

$$= n_w\langle w^2\rangle\langle(f-\mu)^2\rangle \tag{7.51}$$

Therefore

$$\left\langle(\hat{f}-\mu)^2\right\rangle = \left\langle(f-\mu)^2\right\rangle\left(\frac{1}{n_w} + n_w \mathrm{Var}(w)\right) \tag{7.52}$$

since $\langle w\rangle^2 = 1/n_w^2$. In our algorithm the weights are kept constant (if tn in Algorithm `resample` is constant) so that $\mathrm{Var}(w) = 0$. The statistical error is therefore minimized. In Huber and Kim (1996), the weights are not constant resulting in a larger statistical error.

□

An important difference with the Markov model based on cells (coarse states) is that MSM must be run by construction with a known lag time τ. Post-processing is then required to determine whether τ is large enough. After this, the simulation may need to be run again with a larger τ if it is found that memory effects are important. On the contrary, the convergence of RTS is easier to monitor. One simply needs to record the particles that reach B and calculate the average flux. If the error is found to be too large, one can simply continue the simulation to accumulate more statistics, without losing the data already gathered.

In Huber and Kim (1996),[3] walkers that enter B are re-inserted in A thereby ensuring a steady-state system. It is possible to use a slightly different procedure where colors are given to walkers such that when a walker last entered A its color is blue, while it is red if it last entered B. In effect the color changes from blue to red the first time the walker enters B (similarly with A). The population for both colors is kept constant in each macro-state according to the algorithm `resample`. This algorithm allows computing both the forward and backward rates, the free energy, and the equilibrium distribution of particles (by considering all particles, of any color).

7.4.3 Optimal Cells

Even though the method was shown to always converge to the correct answer, the rate of convergence, which depends on the rate at which particles transition from *A* to *B*, depends on the choice of cells. We provide some guidelines to help make a good choice of cells. We note that, as explained previously, RTS is always unbiased and is exact in the absence of statistical errors. This is in contrast with milestoning which is exact only when the milestones are isosurfaces of the committor function. In that case the milestones are called optimal since they minimize systematic errors. Here, RTS is unbiased. The optimal milestones in this context are the ones that minimize the statistical error. Even though the terminology is the same, the meaning is therefore quite different.

RTS ultimately amounts to sampling reactive trajectories from *A* to *B*. Reactive trajectories are defined as trajectories that leave *A* and reach *B* without reentering *A* at any point. These trajectories cluster around the minimum (free) energy paths, Maragliano *et al.* (2006).[126] These paths correspond to reactive trajectories going from *A* to *B* with maximum likelihood, that is the probability density associated with this trajectory is maximum.

To discuss properties of minimum free energy pathways (MFEP), it is convenient to use generalized coordinates $(\xi_1, \ldots, \xi_p)$ to describe the system and use the free energy $A(\xi_1, \ldots, \xi_p)$. This is practically a more useful description and it removes degeneracies such as translation and rotation invariance. For example, these generalized coordinates can be chosen as a set of internal coordinates describing the shape or structure of a molecule. If some information is available about the system we can reduce the number of such variables to focus on the variables of interest for the reaction at hand.

The generalized coordinates can have any units, for example Å or deg. This indicates that some kind of non-dimensionalization procedure is required to work with ξ. This non-dimensionalization can be derived in different ways. We shortly describe how this can be done.

If one assumes for example a Brownian model for these variables:

$$\frac{d\xi}{dt} = \nabla D - \beta D \nabla A(\xi) + R\eta(t) \tag{7.53}$$

where R is such that $RR^T = 2D$, $\beta = (kT)^{-1}$, and $\eta(t)$ is a random term with a normal distribution and variance 1, and A is the free energy. The tensor D is the diffusion tensor. Under simplifying assumptions this tensor can be approximated by:

$$D \approx \beta^{-1} \tau_D M_\xi^{-1} \tag{7.54}$$

$$[M_\xi^{-1}]_{ij} = \sum_k \frac{1}{m_k} \frac{\partial \xi_i}{\partial x_k} \frac{\partial \xi_j}{\partial x_k} \tag{7.55}$$

where m_k is the mass of atom k; τ_D is a time scale associated with the rate of decay of the auto-correlation function for $d\xi/dt$. The tensor M_ξ is non constant. However, we will assume that it can be approximated by its statistical average and that its fluctuations can be neglected.

From Eq. (7.53), we can conclude that along the MFEP we must have that the tangent $d\xi/ds$ (where s is some parameterization such as the arc length) is parallel to D∇A or equivalently:

$$\frac{d\xi}{ds} \propto M_\xi^{-1} \nabla A \tag{7.56}$$

This result is somewhat counter-intuitive since we would expect $d\xi/ds \propto \nabla A$ but is a result of the metric associated with ξ and defined by M_ξ. This suggests normalizing ξ using:

$$\tilde{\xi} = M_\xi^{1/2} \xi \tag{7.57}$$

With the variables $\tilde{\xi}$ we have the expected relation, along the MFEP:

$$\frac{d\tilde{\xi}}{ds} \propto \tilde{\nabla} A \tag{7.58}$$

where $\tilde{\nabla}$ involves derivatives with respect to $\tilde{\xi}$.

Similarly when defining Voronoi cells, the correct distance to use should be consistent with the rate of diffusion and therefore the following distance must be used:

$$\|\Delta\xi\|_{M_\xi} = \left(\Delta\xi^T M_\xi \Delta\xi\right)^{1/2} = \|\Delta\tilde{\xi}\|_2 \tag{7.59}$$

We will now assume that we are using $\tilde{\xi}$ instead of ξ, but we will keep the notation ξ for simplicity. Returning to the issue of reactive trajectories and optimal cells, we note that in each cell, walkers tend to accumulate in low energy regions. Consequently, we can expect the method to be efficient (practically the statistical errors are small) whenever the regions around the MFEPs are well sampled, that is the low energy region in a cell should overlap as much as possible with the MFEP.

Consider a cell and assume that A is not singular in that cell. This implies that A is minimum on one of the boundaries, which we denote S. Let us assume that the MFEP crosses S at ξ^0. In order to minimize the statistical errors, we impose the condition that ξ^0 is the point with the highest probability density in the cell. In that case, the boundary must be orthogonal to ∇A at that point, that is the tangent to the MFEP should be orthogonal to the cell boundary. As a note, we point out that along the MFEP, the gradient of the committor function is also

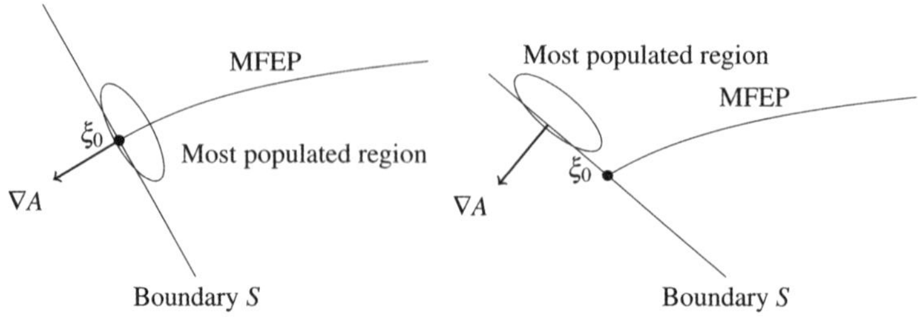

Figure 7.4 Schematic representation of a cell crossed by an MFEP. On the left panel, the boundary of the cell is normal to the MFEP, ensuring a good sampling along the MFEP. This is illustrated by the most populated region containing ξ_0. On the right panel, the orientation of the boundary was changed so that the most populated region is now shifted and no longer overlaps with the MFEP. In that case statistical errors are expected to be somewhat larger. In the extreme case where the boundary is parallel with ∇A the simulation converges very slowly.

parallel to ∇A, which implies that locally the boundary S is an iso-surface of the committor function. See Figure 7.4.

As pointed out previously the method always converges to the correct answer. However, the statistical errors are expected to increase as the cells degrade, that is the cell boundaries are no longer orthogonal to ∇A. Then the walkers in a cell start accumulating in regions that are far away from the MFEP. As a result the probability of seeing a walker reach B becomes correspondingly smaller, leading to larger statistical errors. Instead of observing a steady (with small fluctuations) stream of particles with small weights reaching B, we see a more sporadic stream of particles with larger weights. See Figure 7.4.

7.4.4 Metastability, Convergence, and the Multi-colored Algorithm

In RTS, walkers initially start from region A and are moved out of the simulation when they reach region B. If nothing is done, the total weight slowly diminishes. When the reaction rate is slow, the simulation remains accurate. However, when multiple rates are comparable, this may lead to biased results. One option to prevent the total weight from diminishing is to reinsert a walker in basin A after it enters basin B. Another possible strategy, as discussed earlier, with similar efficiency but somewhat easier to implement, is to use walkers with two different colors, for example blue and red. By definition, red (resp. blue) walkers are those that have last visited A (resp. B). Whenever a blue walker enters B, its color changes to blue, and vice versa. The resampling algorithm is applied to each color separately so that each bin contains the same amount of walkers of each color. This approach allows computing the forward and backward rates, and the free energy.

The efficiency of RTS degrades when there are other slow reactions rates (values of λ_k that are small) in the system, that is other metastable regions in $\bar{A}$

or $\bar{B}$.[30] In that case, the convergence of the algorithm is limited by the rate at which walkers equilibrate in region $\bar{A}$ and $\bar{B}$, which is given by λ_3^{-1}. This leads to long correlation times for the measured fluxes, and therefore overall slow decay of the statistical errors.

This can be addressed by assigning appropriate weights for the walkers at $t=0$. One possibility is to start running RTS for a while and calculate fluxes between cells, given by the weights of walkers moving from cell i to j after a time step divided by the total weight of walkers in cell i. From the flux matrix, the steady-state weights of cells can be computed. These weights can then be used to adjust the weights of walkers in each cell. In principle, this needs to be iterated until convergence. However, only one or two iterations are typically needed. The remaining deviations from steady-state can be later on reduced by the production RTS run. The goal of this procedure is simply to improve the initial weights in each cell in order to bypass the initial slow convergence.

The key property of this procedure is that the accuracy with which fluxes can be computed is independent of the presence of metastability and depends primarily on the sampling inside each cell. Overall convergence is a global property, which is why a few iterations may be required, but since this is followed by the production RTS run, this part of the calculation only requires low accuracy and few iterations. This approach is similar to a method described in Bhatt *et al.* (2010),[127] called "Enhanced weighted ensemble attainment of steady state."

Another approach can be applied to address this shortcoming. It consists in using more than 2 colors. To explain this in a simple fashion we return to the milestoning framework of Section 7.2.6. RTS can be reinterpreted in terms of the milestoning framework. The difficulty in the optimal milestoning approach described in Section 7.2.6 is that the cells must be such that their boundaries are iso-surfaces of the committor function. Even though this is in general difficult to realize, there is a case for which the problem is simplified. Consider the following three cells: cell V_0 enclosing A, cell V_1 enclosing B. Let us assume that we have another metastable basin C. The same reasoning can be extended to an arbitrary number of basins. We define V_2 as a cell enclosing C and V_3 as the complement: $V_3 = \left(\bigcup_{i=0}^{2} V_i\right)^c$. We define S_0 as the interface between V_0 and V_3, and similarly for S_1, and S_2. In the multi-coloring algorithm, each cell is assigned a color, say 0 is blue, 1 is red, and 2 is green. Each time a blue particle enters cell V_2, its color changes to green, and similarly for the other colors. The total weight of all walkers is therefore constant. We use the resampling algorithm to maintain a constant number of walkers in each cell, separately for each color. We then measure the mean flux of particles turning blue to green, *etc.*

Eq. (7.27) still holds for this approach. The fluxes F_{ij}, from cell i to j, are simply related to the probability matrix P through: $F=\Delta t^{-1}P$. Eq. (29) is not needed for this approach. We can directly obtain P from the flux values F, computed using RTS.

In general the milestone S_2 is not an iso-surface of the committor. However, following the proof that optimal milestoning gives exact rates, we will have proved that the rate with multi-coloring is exact if we show that: p_{ij}, the probability to cross j after i, and τ_i, the mean time before crossing another milestone, are independent of the previous milestone k that was crossed (see page 14 for the proof in the optimal milestoning case). This property is in fact true for the multi-coloring approach because of our choice for S_2 and the fact that the equilibration time for V_2 is very small. The key assumption is that V_2 must be associated with a metastable state (minimum energy basin) so that the relaxation time in V_2 is small compared to the mean escape time.

This shows that the mean passage time predicted using Eq. (7.27) in the multi-coloring framework is for all practical purposes exact, with no systematic bias. This approach allows considering the case of multiple slow rates, with no significant degradation in efficiency. The computational cost merely grows with the number of colors, but is independent of the degree of metastability (the values of the first λ_k, $k = 1, 2,$ *etc.*).

This approach has some conceptual similarities with the technique of core sets of Schütte *et al.* (2011).[34] Their analysis of accuracy (section E, pp. 204105–7) carries over to RTS with multiple colors. In addition, if ones applies the Galerkin discretization approach from Schütte *et al.* (2011)[34] to RTS with $m + 1$ colors, one can calculate the slowest m rates in the system (or phenomenological rates, see Schütte *et al.* [2011][34] pp. 204105–4, section F) which may be of interest for certain applications.

As a final note, we point out that the method is embarrassingly parallel and can be easily implemented on a parallel machine. This is important as this allows making only few changes to a serial (sequential) molecular dynamics code to make it run efficiently on a parallel cluster, with RTS, without having to parallelize the core of the code.

7.5 Analysis of Statistical Errors in Markov State Models

We now discuss some mathematical results for Markov state models. In particular, we will analyze the sensitivity of the eigenvalues to perturbations in the transition matrix. This analysis will lead to estimates for the statistical errors in the method. This will also lead to an analysis of the systematic errors, due to the finite lag time τ (the length of the short trajectories used to build the Markov state model).

Several papers have discussed error analysis in the context of conformation dynamics.[35–39,128] Metzner *et al.* (2009a) takes a different approach to error analysis by considering a method that generates random transition matrices and as a result can estimate errors in various quantities computed from the transition matrix. In Metzner *et al.* (2009a),[39] it is argued that this approach is more accurate since it does not rely on Taylor expansions to approximate the impact of small variations of the numerical transition matrix from the exact matrix. In that sense this is a more direct estimate of the statistical errors and how they impact various quantities of interest (stationary distribution, eigenvalues, committor function, *etc.*). Darve *et al.* (2009)[47] discusses the effect of memory in building coarse grained models in the context of the Mori-Zwanzig formalism.

Although many results presented in this chapter can be extended to more general stochastic equations, we are going to focus on the relatively simpler case of Brownian dynamics:[129]

$$dx(t) = (\nabla D_{\mathrm{B}} - \beta D_{\mathrm{B}}(x)\nabla U(x))dt + R_{\mathrm{B}}(x)dW(t) \tag{7.60}$$

where $W(t)$ is a Wiener process (see p. 66 in Gardiner [1997]),[10] D_B is the diffusion tensor, $\beta^{-1} = k_{\mathrm{B}}T$, and R_{B} satisfies $R_{\mathrm{B}}(x)R_{\mathrm{B}}(x)^T = 2D_{\mathrm{B}}(x)$.

7.5.1 Eigenvectors and Eigenvalues of the Transition Matrix

The rate is typically computed by considering the eigenvalues of the transition matrix:

$$P_{ij}(\tau) = \mathbf{P}(\text{particle in cell } i \text{ at time 0 is in cell } j \text{ at time } \tau) \tag{7.61}$$

where we assume that the dynamics is given by Eq. (7.60). This matrix is basically used to construct a Markov state model of the system.

From the stochastic Eq. (7.60), we can define the conditional probability $\rho(x,t|x_0,0)$, which is the probability to be at x at time t if the system was at x_0 at time 0. This probability can be expanded in terms of the eigenfunctions $\rho_k(x)$ of the forward Fokker-Planck equation (for Eq. [7.60]), and the eigenfunctions $\psi_k(x_0)$ of the backward Fokker-Planck equation:[10]

$$\rho(x,t|x_0,0) = \sum_k \psi_k(x_0)\,\rho_k(x)e^{-\lambda_k t} \tag{7.62}$$

where λ_k are real and positive eigenvalues (the two sets of eigenfunctions are associated with the same eigenvalues).[10,36,94] Since

$$\int \psi_k(x)\rho_{k'}(x)dx = \delta_{k,k'} \tag{7.63}$$

$e^{-\lambda_k t}$ is an eigenvalue of the kernel $\rho(x,\, t|x_0,\, 0)$:

$$\int \rho(x,t|x_0,0)\rho_k(x_0)dx_0 = e^{-\lambda_k t}\rho_k(x) \tag{7.64}$$

We will denote:

$$\langle \psi_k \rangle_i = \frac{\int_{V_i} \psi_k(x_0)\rho(x_0)dx_0}{\int_{V_i} \rho(x_0)dx_0} \tag{7.65}$$

$$\rho_{jk} = \left[\int_{V_i} \rho k(x_0) dx_0 \right] \quad (7.66)$$

where $\rho(x)$ is the equilibrium distribution of the system.

We sort the λ_k in increasing order. For most systems, there is a single eigenvalue $\exp(-\tau\lambda_1)$ equal to 1 ($\lambda_1 = 0$) and the corresponding eigenvector is the stationary distribution $\rho(x)$. We are interested in estimating λ_2 by computing the eigenvalues μ_2 of the matrix P_{ij}, and using $\lambda_k \sim -\ln(\mu_k)/\tau$. In general, λ_2 and $-\ln(\mu_2)/\tau$ differ leading to inaccurate estimates. However, under certain assumptions, which will be discussed, $-\ln(\mu_2)/\tau$ provides an accurate estimate. In these circumstances, the Markov assumption made in building the model becomes accurate.

The term P_{ij} can then be written as:

$$P_{ij}(\tau) = \frac{\int_{x_0 \in V_i} \int_{x \in V_j} \rho(x, \tau | x_0, 0) \rho(x_0) dx dx_0}{\int_{V_i} \rho(x_0) dx_0} \quad (7.67)$$

$$= \sum_k e^{-\lambda_k \tau} \rho_{jk} \langle \psi_k \rangle_i \quad (7.68)$$

In the rest of this chapter we sometimes have to make a distinction between the two minimum energy regions A and B and a partitioning of the space Ω into two metastable regions $\bar{A}$ and $\bar{B}$ (see Figure 7.5). Typically A is defined as a small region around a stable conformation of interest (the reactant state), and similar for B (the product state). The regions $\bar{A}$ and $\bar{B}$ are defined as metastable regions, that is the rate of transition between these regions is the smallest among all other pairs of sets.

Properties of the eigenvectors. The second eigenvector $\rho_2(x)$ can be used to define a partition into two regions $\bar{A}$ and $\bar{B}$, with $A \subset \bar{A}$, $B \subset \bar{B}Z$. We will provide a more rigorous analysis later on but roughly speaking, the function ρ_2/ρ is nearly constant in two regions, which define $\bar{A}$ and $\bar{B}$. This function changes sign between these two regions. The narrow region where $\rho_2/\rho \sim 0$ defines the transition region between these two metastable regions. See, for example, Chapter 6,

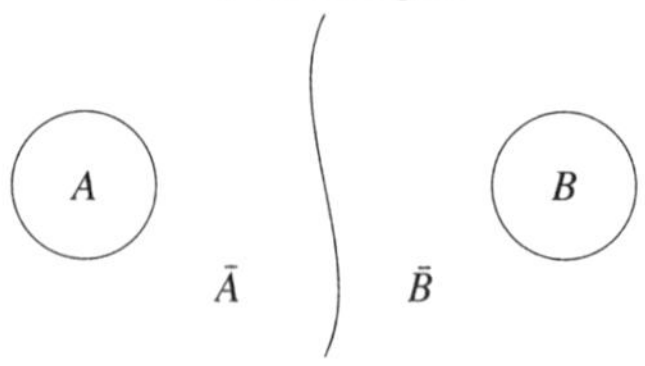

Figure 7.5 Definition of the different regions, A, B, $\bar{A}$, and $\bar{B}$.

p. 91–119 in Hill and Moore (2004).[130] The molecule is assumed to have a high probability of being in region A when in $\bar{A}$, and similarly for B.

We now discuss in more details the properties of the eigenvectors. If we assume that the relaxation time in basin $\bar{A}$ and $\bar{B}$ is short compared to the reaction rate, *i.e.* $0 < \lambda_2 \ll \lambda_k$ for $k>2$, then for $x_0 \in \bar{A}$, $x \in \bar{B}$, and $\lambda_k{}^{-1} \ll t \ll \lambda_2{}^{-1}$ we have

$$\rho(x, t|x_0, 0) \approx \rho(x) + \psi_2(x_0)\rho_2(x) \approx 0$$

from which we see that $\rho_2 \propto \rho$ in $\bar{B}$ (and similarly $\bar{A}$), and ψ_2 must be approximately constant in $\bar{A}$ (and similarly $\bar{B}$). Moreover, we have $\int \rho_2(x)dx = 0$ from which (with the appropriate normalization):

$$\rho_2(x) = \begin{cases} \sqrt{\rho(\bar{B})/\rho(\bar{A})}\,\rho(x), & \text{in } \bar{A} \\ -\sqrt{\rho(\bar{A})/\rho(\bar{B})}\,\rho(x), & \text{in } \bar{B} \end{cases} \tag{7.69}$$

Finally, we have the general relation

$$\rho_k(x) = \psi_k(x)\rho(x) \tag{7.70}$$

A short proof is provided in the appendix (see Proof 1). The eigenvectors ρ_2 and ψ_2 are depicted in Figure 7.6.

Near the transition region $\psi_2(x) \sim$ constant no longer holds, and in fact, the function ψ_2 varies rapidly near the transition region. The committor function $\pi(x)$ is defined as the probability to reach region B before reaching region A starting from x. By definition, at the transition region, $\pi(x) = 1/2$. The function ψ_2 has a natural interpretation in terms of a committor function[131] and

$$\pi(x) \sim \frac{\psi_2(x) - \psi_2(a)}{\psi_2(b) - \psi_2(a)} \tag{7.71}$$

See the appendix for a proof of this result (Proof 2).

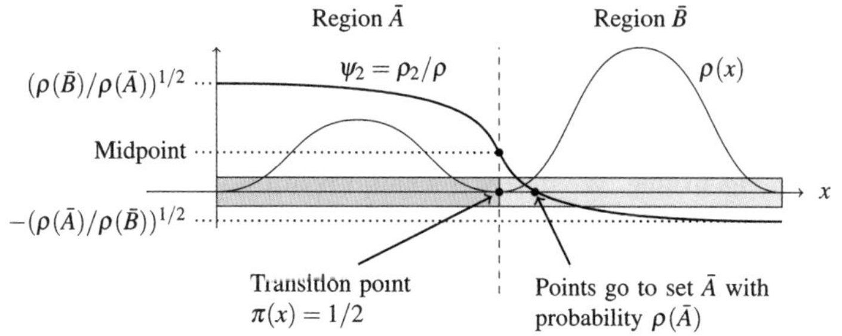

Figure 7.6 Definition of the metastable regions $\bar{A}$ and $\bar{B}$ and the second eigenvector ρ_2.

Therefore, from Eq. (7.71), the transition region corresponds to $\psi_2(x) = 1/2$ $(\psi_2(A) + \psi_2(B))$, where $\psi_2(A)$ denotes the constant value of ψ_2 in A (similarly for B). Therefore, at the transition point:

$$\psi_2(x) = \frac{1}{2}\left(\sqrt{\frac{\rho(\bar{B})}{\rho(\bar{A})}} - \sqrt{\frac{\rho(\bar{A})}{\rho(\bar{B})}}\right) \tag{7.72}$$

The transition point is shown on Figure 7.6. The hypersurface corresponding to $\rho_2(x) = 0$ is located nearby but corresponds to a slightly different situation. Points on this hypersurface are not at the transition region but rather are such that they move to basin $\bar{A}$ with probability $\rho(\bar{A})$ and to $\bar{B}$ with probability $\rho(\bar{B})$. (This is true only after a short time τ, with $\tau \gg \lambda_k^{-1}$, $k > 2$.) This is shown on Figure 7.6.

Forward and backward rates. Now that we have derived estimates for $\rho_2(x)$, we can clarify the relation between different rates. To calculate the forward rate, we construct a linear combination $\rho_f(x, 0)$ of ρ and ρ_2 such that:

$$\int \rho_f(x,0)dx = 1, \qquad \rho_f(x,0) \approx 0 \quad \text{for } x \in \bar{B}. \tag{7.73}$$

With these conditions we find that the unique solution is:

$$\rho_f(x,0) = \rho(x) + \sqrt{\frac{\rho(\bar{B})}{\rho(\bar{A})}}\rho_2(x) \tag{7.74}$$

Since ρ and ρ_2 are eigenvectors:

$$\frac{\partial \rho_f(x,t)}{\partial t} = -\lambda_2 \sqrt{\frac{\rho(\bar{B})}{\rho(\bar{A})}}\rho_2(x) - e^{\lambda_2 t}, \quad -\int_{\bar{B}} \frac{\partial \rho_f(x,t)}{\partial t} dx\Big|_{t=0} = \lambda_2 \rho(\bar{B}) \tag{7.75}$$

Therefore, the forward rate from A to B is equal to $\lambda_2\ \rho(\bar{B})$. Similarly the rate from B to A is equal to $\lambda_2\ \rho(\bar{A})$:

$$\text{rate}_{A\to B} = \lambda_2 \rho(\bar{B}), \qquad \text{rate}_{B\to A} = \lambda_2 \rho(\bar{A}). \tag{7.76}$$

7.5.2 Sensitivity of Eigenvalues and Systematic Errors

Sensitivity of eigenvalue. We derive a general result regarding the sensitivity of an eigenvalue to perturbations in the matrix entries. We consider a matrix P and assume that we have a full set of eigenvalues. The eigenvalue of interest is μ_2 and:

$$P - \mu_2 I = R(P)^T(\Lambda(P) - \mu_2 I)S(P), \qquad \text{where } S(P) = \left[R(P)^T\right]^{-1}, \tag{7.77}$$

and $\Lambda(P)$ is a diagonal matrix containing the eigenvalues. We denote $\bar{P}$ the exact matrix P and consider small deviations $\Delta P = P - \bar{P}$. We assume that in some neighborhood around $\bar{P}$ the matrices $R(P)$, $S(P)$, and $\Lambda(P)$ are differentiable. Calculate the derivative with respect to one of the entries P_{ij}:

$$\frac{\partial P}{\partial P_{ij}} = \frac{\partial R(P)^T}{\partial P_{ij}}(\Lambda(P) - \mu_2 I)S(P) + R(P)^T \frac{\partial \Lambda(P)}{\partial P_{ij}} S(P) + R(P)^T (\Lambda(P) - \mu_2 I)\frac{\partial S}{\partial P_{ij}} \tag{7.78}$$

We denote e_i a row vector such that $[e_i]_j = \delta_{ij}$, $r_2 = R_{2,:}$ (second row of R), $s_2 = S_{2,:}$. If we evaluate the partial derivative at $P = \bar{P}$, and multiply Eq. (7.78) to the left by s_2 and to the right by r_2^T we get:

$$s_2 e_i^T e_j r_2^T = [s_2]_i \, [r_2]_j = \frac{\partial \mu_2}{\partial P_{ij}} \tag{7.79}$$

because $s_2 R^T (\Lambda - \mu_2 I) = 0$ and $(\Lambda - \mu_2 I)S\, r_2^T = 0$. This matrix will be plotted later on, in Figure 7.9.

Systematic error due to the lag time τ. We will use this result regarding the sensitivity of μ_2 to study the systematic error or bias using the Markov state model. To simplify the discussion, we will define a fine matrix, which is assumed to accurately capture the continuous dynamics using the Brownian dynamics (7.60):

$$P_{ij}^f(\tau) = \frac{\int_{x_0 \in V_i^f} \int_{x \in V_j^f} \rho(x, \tau | x_0, 0)\rho(x_0)\, dx\, dx_0}{\int_{V_i^f} \rho(x_0)\, dx_0} \tag{7.80}$$

over some fine states. We consider that P^f gives the exact rate (this can be achieved using states that are fine enough). We could work with $\rho(x,t|x_0,0)$ directly but it is easier to discuss the results using P^f. The matrix P can be written interms of P^f:

$$P_{ij} = \sum_{k \in V_i} \sum_{l \in V_j} \frac{\rho_k}{\sum_{k' \in V_i} \rho_{k'}} P_{kl}^f = \left[\Pi\left(P^f\right)\right]_{kl} \tag{7.81}$$

From P, we can reconstruct an approximation P^c of P^f where an entry in P is mapped to a block in P^c:

$$P_{kl}^c = \frac{P_{ij}}{|V_j|} = [\mathscr{I}(P)]_{kl} \tag{7.82}$$

with $k \in V_i$, $l \in V_j$, and where $|V_j|$ is the number of fine states in V_j. We have $\prod \mathscr{I} = I$, the identity operator. We will use our result for the sensitivity of μ_2 with $\Delta P = P^c - P^f$.

We now show that every eigenvalue of P is an eigenvalue of P^c, which allows us to study the eigenvalues of P^c instead of P. Let us define $\tilde{s}_2$ the left eigenvector of P with eigenvalue $\tilde{\mu}_2$. Define $\tilde{s}_2^c$:

$$[\tilde{s}_2^c]_k = \frac{[\tilde{s}_2]_i}{|V_i|} \tag{7.83}$$

with $k \in V_i$. Then for $l \in V_j$:

$$\sum_k [\tilde{s}_2^c]_k P^c_{kl} = \sum_i \frac{[\tilde{s}_2]_i}{|V_i|} \sum_{k \in V_i} P^c_{kl} = \sum_i \frac{[\tilde{s}_2]_i}{|V_i|} |V_i| \frac{P_{ij}}{|V_j|} \tag{7.84}$$

$$= \tilde{\mu}_2 \frac{[\tilde{s}_2]_j}{|V_j|} = \tilde{\mu}_2 [\tilde{s}_2^c]_l \tag{7.85}$$

Therefore $\tilde{s}_2^c$ is a left eigenvector of P^c with eigenvalue $\tilde{\mu}_2$.

Using Eq. (7.79), the sensitivity of the second eigenvalue is:

$$\frac{\partial \mu_2}{\partial P^f_{ij}} = [s_2^f]_i [r_2^f]_j \tag{7.86}$$

where s_2^f is the left eigenvector, and r_2^f is the right eigenvector of P^f. We are going to use the following linear approximation:

$$\tilde{\mu}_2 - \mu_2 \approx \sum_{kl} [s_2^f]_k [r_2^f]_l (P^c_{kl} - P^f_{kl}) \tag{7.87}$$

We start by studying the systematic error at long lag times. Then:

$$P^f_{kl} = \rho(V_l^f) + e^{-\lambda_2 \tau} \rho^f_{l,2} \langle \psi_2 \rangle^f_k + O(e^{-\lambda_3 \tau}) \tag{7.88}$$

In the following, we will make the following approximations:

- $\rho(x)$ is negligibly small near the transition region.
- $\psi_2(x)$ is nearly constant away from the transition region.

In practice this is not true but the error due to these approximations is typically much smaller than $O(e^{-\lambda_3 \tau})$ and so we will simply ignore it.[§] As a consequence, from Eq. (7.88), we either have: P^f_{kl} negligible when l is near

[§]We will not pursue this point further but the analysis suggests that as $\tau \to \infty$ in fact $\tilde{\mu}_2$ does not converge exactly to μ_2 although as explained above this discrepancy is of no practical importance.

the transition region or $[r_2^f]_l$ is nearly constant. Denote $[r_2^f]_A$ the value in the left basin and $[r_2^f]_B$ in the right basin. Then:

$$\sum_l [r_2^f]_l P_{kl}^f = [r_2^f]_A \sum_{l\in\bar{A}} P_{kl}^f + [r_2^f]_B \sum_{l\in\bar{B}} P_{kl}^f + O(e^{-\lambda_3\tau}) \tag{7.89}$$

In the long lag time assumption, the choice of cells is not important. However, there is one property, which must be satisfied which is that no cell V_j can overlap significantly with both $\bar{A}$ and $\bar{B}$, formally: either $\int_{\bar{A}\cap V_j} \rho(x)dx$ or $\int_{\bar{B}\cap V_j} \rho(x)\,dx$ must be negligible. In that case we have the following three possibilities, assuming that k is away from the transition region:

$$\sum_{l\in V_j} [r_2^f]_l P_{kl}^c \approx \begin{cases} [r_2^f]_A \sum_{l\in V_j} P_{kl}^f & \text{if } V_j \text{ falls in } \bar{A}, \\ [r_2^f]_B \sum_{l\in V_j} P_{kl}^f & \text{if } V_j \text{ falls in } \bar{B}, \\ 0 & \text{if } V_j \text{ has support in a region where } \rho \text{ is} \\ & \text{negligible (transition region).} \end{cases} \tag{7.90}$$

with an error of order $O(e^{-\lambda_3\tau})0$. This results from the fact that by construction P_{kl}^c is constant inside $l\in V_j$ and that $\sum_{l\in V_j} P_{kl}^c = \sum_{l\in V_j} P_{kl}^f + O(e^{-\lambda_3\tau})$. Therefore:

$$\sum_l [s_2^f]_k [r_2^f]_l (P_{kl}^c - P_{kl}^f) = O(e^{-\lambda_3\tau}) \tag{7.91}$$

At long lag times the error in the eigenvalue is therefore:

$$\tilde{\mu}_2 - \mu_2 \approx O\left(e^{-\lambda_3\tau}\right) \tag{7.92}$$

At short lag times, we have a small systematic error provided the cells are chosen adequately. Let us assume that inside each cell $[r_2^f]_l$ is nearly constant.[¶] We will therefore denote $[r_2^f]_j$ the value of $[r_2^f]_l$ for $l\in V_j$ in the equation below. The fluctuations of $[r_2^f]_l$ in a given cell are assumed to be of order ε. We also have $[s_2^f]_k = [r_2^f]_k\, \rho(V_k^f)$ (see Eq. [(7.70)]), so that:

$$\sum_{kl} [s_2^f]_k [r_2^f]_l (P_{kl}^c - P_{kl}^f) = \sum_{ij} [r_2^f]_i [r_2^f]_j \sum_{k\in V_i} \sum_{l\in V_j} \rho(V_k^f)(P_{kl}^c - P_{kl}^f) + O(\varepsilon) \tag{7.93}$$

$$= O(\varepsilon) \tag{7.94}$$

from the definition of P^c and P.

¶Since r_2^f is an approximation of ψ_2 using the fine state discretization (which can be made arbitrarily fine), r_2^f is an approximation of the committor function.

This suggests the following choice of cell V_i that satisfies our assumption:

$$V_i = \left\{ k | i\varepsilon \leq [r_2^f]_k < (i+1)\varepsilon \right\} \tag{7.95}$$

The interpretation is therefore that the cells provide a fine partitioning based on the iso-surfaces of the eigenvector $\psi_2(x)$ or equivalently the committor function π, and we can equivalently write:

$$V_i = \{ k | i\varepsilon \leq \psi_2(x) < (i+1)\varepsilon \} \tag{7.96}$$

In that case:

$$\tilde{\mu}_2 - \mu_2 \approx O(\varepsilon) \tag{7.97}$$

This result is consistent with Eq. (7.35) in Prinz *et al.* (2011), which gives a bound on the error that depends on the eigenfunction approximation error when projecting onto the cells. This requirement is very important as constructing cells with small volume is prohibitive in high dimension. However, the definition (7.96) is much less restrictive in terms of computational cost since it requires refining only along a single direction, given by $\nabla\psi_2$. The dimensionality of the problem has little impact on the number of cells that are required for an accurate calculation.

The main caveat in this discussion is that computing or even approximating ψ_2 or the committor function π is very challenging, and consequently this approach remains difficult to use in practice.

7.5.3 Statistical Errors

Even though we have outlined methods to reduce the Markovian approximation error, *e.g.* by refining the spatial discretization, it remains the case in practice that cells need to have a relatively large volume and the committor function is difficult to approximate at best. As a result, a long lag time is more or less necessary for an accurate estimate. We will now discuss how statistical errors vary as a function of the lag time.

At small lag times, the statistics are typically expected to be reasonably accurate and the statistical error can be made satisfactorily small. At long lag times however the situation worsens. In this section, we will again use Eq. (7.79) that expresses the sensitivity of the eigenvalue μ_2 to perturbations in the matrix entries P_{ij}, this time due to statistical errors. For this analysis we will assume that the macro states are fine enough, or that τ is large enough, so that $P(2\tau) = P(\tau)^2$. This simplifies the analysis, but the results mostly carry over to the case where non-Markovian effects are important.

The sensitivity of μ_2 with respect to the matrix entries can be combined with the statistical errors in the entries P_{ij} $\{P_{i,1},\ldots,P_{i,n_{\text{cell}}}\}$ is a multinomial distribution,[38,128] to obtain an expression for the statistical error in the form:

$$\sigma^2(\mu_2) = \frac{1}{n+1}\sum_i \sum_{kl} \frac{\partial \mu_2}{\partial P_{ik}} \frac{\partial \mu_2}{\partial P_{il}} [P_{ik}\delta_{kl} - P_{ik}P_{il}] \tag{7.98}$$

$$= \frac{1}{n+1}\left[\sum_{ik} P_{ik}([s_2]_i[r_2]_k)^2 - \sum_i \left(\sum_k P_{ik}[s_2]_i[r_2]_k\right)^2\right] \tag{7.99}$$

$$= \frac{1}{n+1}\sum_i ([s_2]_i)^2 \left[\sum_k P_{ik}([r_2]_k)^2 - \left(\sum_k P_{ik}[r_2]_k\right)^2\right] \tag{7.100}$$

$$= \frac{1}{n+1}\sum_i ([s_2]_i)^2 \sigma_i^2(r_2) \tag{7.101}$$

where $\sigma_i(r_2)$ is the standard deviation of r_2 computed using the probabilities $P_{i,:}$ (row i of P).

The rate is given by $\lambda_2 = -\ln(\mu_2)/\tau$ and the relative error can be estimated using $\sigma(\lambda_2)/\lambda_2$:

$$\frac{\sigma(\lambda_2)}{\lambda_2} \approx \frac{\sigma(\mu_2)}{\mu_2 \tau \lambda_2} \tag{7.102}$$

We may be interested in minimizing the statistical error, given a computational cost. The cost is proportional to $n\tau/\Delta t$, the number of samples multiplied by the length of the trajectories with lag-time τ. We express the error in the form:

$$\frac{\sigma(\lambda_2)}{\lambda_2} \approx \frac{1}{\sqrt{n\tau/\Delta t}\sqrt{\lambda_2 \Delta t}} \frac{\sqrt{\sum_i ([s_2]_i)^2 \sigma_i^2(r_2)}}{e^{-\lambda_2 \tau}\sqrt{\lambda_2 \tau}} \tag{7.103}$$

With our assumption that the states are fine enough, we have that $[s_2]_i \approx \rho_{i,2}$ and $[r_2]_i \approx \langle \psi_2 \rangle_i$. Therefore, if we assume that $\lambda_2 \ll \lambda_3$, we have that r_2 is nearly constant. Hence $\sigma_i^2(r_2)$ is expected to be small and $\sigma(\lambda_2)/\lambda_2$ can remain bounded even as $\lambda_2 \to 0$.

Again, using the assumption that the states are fine enough, the eigenvectors s_2 and r_2 of P are independent of τ. At short times, the probability $P_{i,:}$ is concentrated around the diagonal (see Figure 7.9). In fact for $\tau = 0$ the matrix is equal to the identity and $\sigma_i(r_2) = 0$. As τ increases the spread of the entries in $P_{i,:}$ becomes larger and as a consequence $\sigma_i(r_2)$ must increase with τ. The factor $1/(e^{-\lambda_2 \tau}\sqrt{\tau})$ on the denominator in Eq (7.103) results in an initial increase of the error at small τ, then a plateau is reached, and as $\lambda_2 \tau \gg 1$ the error starts increasing again (τ is at this point large compared to the relaxation rate of the system). See Figure 7.9.

An interpretation is that as $\tau \gg \lambda_3^{-1}$, the system has time to relax within basin $\bar{A}$ or $\bar{B}$. Therefore, independent of where the system starts from, the states that were started in basin $\bar{A}$ will be distributed as $\sim \rho(V_i)/\rho(\bar{A})$ while the states in $\bar{B}$ are distributed as $\sim \rho(V_i)/\rho(\bar{B})$. Therefore, in that regime, the method degenerates to a direct calculation of the rate where trajectories are initiated in basin $\bar{A}$ (resp. $\bar{B}$) and we observe how many transitions to basin $\bar{B}$ (resp. $\bar{A}$) occur. For this type of calculation, statistical errors are large whenever $\lambda_2\tau$ is very small.

More numerical results will be shown later on, but to illustrate the point above we present a simple example. Figure 7.7 shows a 1D system with $0 \le x \le 1$. A random walker is moving between discrete states. The probability to attempt a move to the left is 0.25 (same for the right). A Metropolis criterion is used to accept or reject this move using:[132]

$$U(x) = 400\left(0.98(x-0.2)^4 + (x-0.8)^4 - 1.5(x-0.5)^2\right) \quad (\beta = 1) \quad (7.104)$$

The eigenvalues of the matrix are shown on Figure 7.8, along with$_{100}$ the decay of $\exp(-\lambda_k\tau)$ as a function of τ for $k=2$, 3, 4. In Figure 7.9, the matrix P is shown along with $-\log 10(P^{100})$. This shows how the matrix P^τ progressively goes from a tri-diagonal form to a rank-2 matrix given by the first two eigenvectors. Recall the assumptions regarding $\rho(x)$ and $\psi_2(x)$. In Figure 7.9, this can be seen from the fact that in the top and bottom portions of the matrix the entries in the columns are nearly constant. In the center of the matrix, entries in the column vary rapidly ($\langle\psi_2\rangle_k^f$ is changing sign) but this is also the region where ρ is small. This can be seen by observing that the columns in the center have small values ($\sim 10^{-4}$). See also Equation (7.88). The same behavior can be observed in Figure 7.7 where for $x < 0.4$, ψ_2 is constant; for $0.4 \le x \le 0.6$, ρ is small; for $x > 0.6$, ψ_2 becomes constant again.

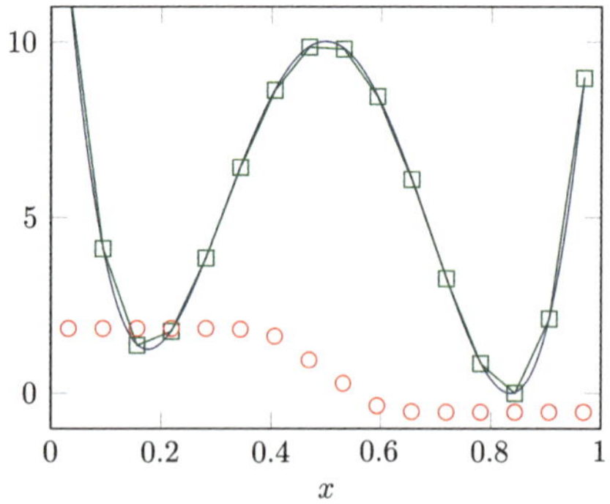

Figure 7.7 Solid blue line: energy $U(x)$; green line and green squares: fine states used to define the transition matrix P_{ij}; red circles: ψ_2. The probability density is defined as $\rho = \exp(-U)$.

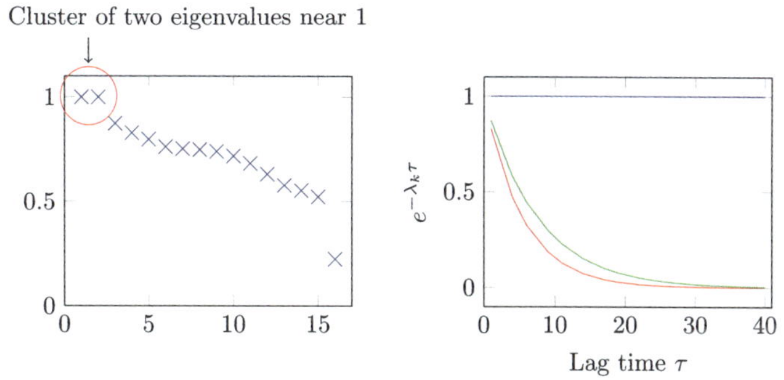

Figure 7.8 Eigenvalues of the transition matrix P. Left panel: Cluster of eigenvalues near 1. Right panel: Decay of $\exp(-\lambda_k \tau)$ vs τ for $k=2, 3, 4$.

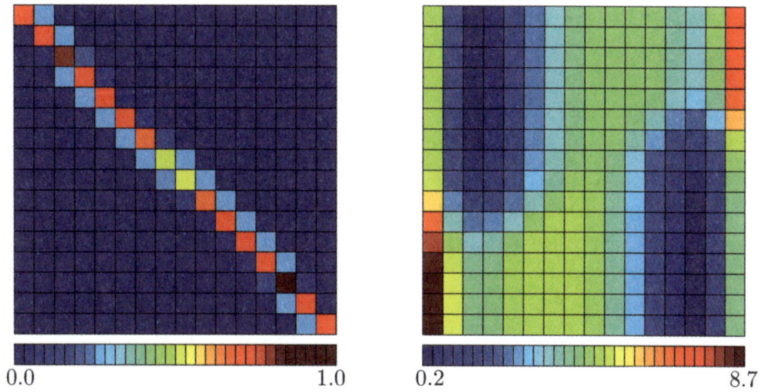

Figure 7.9 Left panel: matrix P; right panel: matrix $-\log 10(P^{100})$. The color represents the magnitude of the entries. A color bar is shown at the bottom for reference.

Figure 7.10 shows the relative statistical error calculated as $2\sigma(\lambda_2)/\lambda_2$. As τ increases the statistical error increases because the system has more time to relax in each basin, thereby reducing the computational benefit of using coarse states.

The overall behavior is therefore as follows. At small lag times, we have non-Markovian effects leading to systematic errors. As τ increases, the terms $e^{-\lambda_k \tau}$, $k>2$, become small when $\lambda_k \tau \gg 1$, such that systematic non-Markovian effects progressively disappear. As this happens, the statistical errors increase because the system has more time to relax in each basin (see Figure 7.10). When we reach $\tau \gg \lambda_3^{-1}$ (around $\tau \sim 40$), we see a plateau. Systematic errors are now negligible but the statistical errors are large.

The implication is that it may be difficult in practice to apply these methods accurately. The reasoning above shows that if the cells V_i are not fine enough (see Eq. [7.96]), we are caught between systematic errors at small lag times and

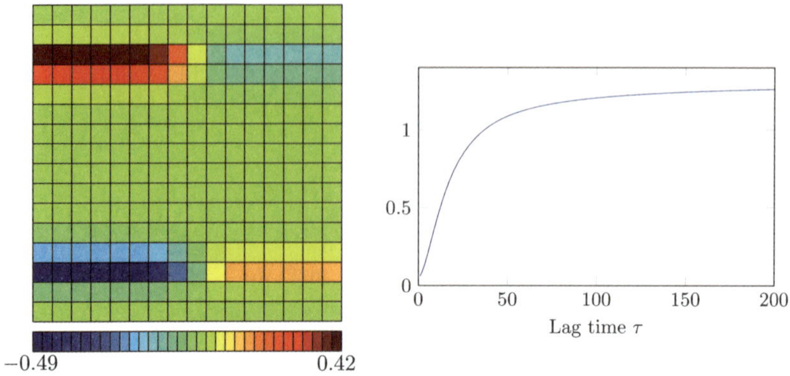

Figure 7.10 Left panel: sensitivity matrix $\partial\mu_2/\partial P_{ij} = [s_2]_i\ [r_2]_j$. Right panel: relative statistical error for the rate estimate as a function of τ computed using Eq. (7.103). The statistical error was computed as two standard deviations (95% confidence interval). The number of sample points used in each state is $10^5/\tau$ ($n\tau/\Delta t = 10^5$ in Eq. [7.103]).

large statistical errors at large lag times, and a trade-off must be found between these two extremes to maximize the efficiency of the calculation. This issue can be mitigated by choosing cells V_i such that $\psi_2(x) \sim$ constant in each cell, although this can lead to a large computational cost if the number of cells becomes too large or may be intractable if the committor function cannot be well approximated.

The advantage of RTS is that none of these issues are present. Systematic errors are absent and convergence is easy to monitor. There is no lag time t that needs to be adjusted to control the accuracy and computational cost. One benefit of MSM though is that it requires only independent sampling inside each macro-state, making the calculation embarrassingly parallel. In contrast, RTS requires a "global" convergence of the macro-state weights. This is a necessary trade-off however resulting from the fact that the Markov approximation is not used and therefore transition pathways, in effect, must be computed.

7.6 Some Numerical Benchmarks for Markov State Models and Reactive Trajectory Sampling

7.6.1 Set up of Numerical Simulations

To illustrate the performance and numerical behavior of Markov state models (MSM) and the reactive trajectory sampling method (RTS), we present results from some numerical simulations on simple problems in 1D and 2D. They are meant to illustrate the concepts introduced in the previous sections. These results can not really be used as benchmarks of the various methods presented, as conclusions using actual models from bio-chemistry will inevitably be different. Nonetheless, they illustrate the key concepts and confirm to some extent the validity of the analysis.

We will consider two types of dynamics. In 1D, we will consider Brownian Dynamics as introduced in Equation (7.60). Using a time step Δt, the numerical integrator is given by Ermak and McCammon (1978):[129]

$$x_{n+1} = x_n + \nabla D_{\mathrm{B}} \nabla t - \beta D_{\mathrm{B}} \nabla U \nabla t + \sqrt{2\Delta t D_{\mathrm{B}}}\, W \tag{7.105}$$

where D_{B} is the diffusion matrix, W is a vector of independent standard Gaussian random variables (mean $\mu = 0$, standard deviation $\sigma = 1$). A special difficulty of this discretization is that depending on the choice of time step, the walker x_{n+1} may (infrequently) jump over many macro states. In fact there is a finite chance to jump in one step directly from A to B. This is an artifact of the discretization. This case is a little artificial but does highlight a limitation of the method, which is that the statistical error typically increases when such "long-range" jumps are allowed. In practice however, in the context of molecular systems, this is not an issue as the system typically diffuses slowly and can only cross to neighboring cells during a single time step.

In 2D, the walkers are moving on a 2D Cartesian grid and we consider the Metropolis algorithm.[132] The scheme is outlined in algorithm 1.

The exact rate is obtained by computing the eigenvalues of

$$Q_{ij} = \mathbf{P}(\text{walker at } j \text{ the next step} \mid \text{walker currently at } i) \approx \rho\left(x_j, \Delta t \mid x_i, 0\right)$$

In 2D the system already has discrete states so Q is well-defined. In 1D, we discretize the interval of interest in order to define the fine states i.

The coarse transition probability matrix P represents the transition probability between the coarse cells or macro states. The definition is simply as follows:

$$P_{ij} = \mathbf{P}\left(\text{walker in } V_j \text{ the next step} \mid \text{walker currently in } V_i\right) \tag{7.106}$$

Algorithm 1: `Metropolis Algorithm`

/* `Walkers are moving on a Cartesian grid in dimension` d. */

while more samples are required **do**

  Move walker to an adjacent position x_{new} with uniform probability of $1/2^d$ (d-dimensional random walk.)

  Accept this move with probability

$$p = \min\left\{1, \frac{e^{-\beta U(x_{\text{new}})}}{e^{-\beta U(x_{\text{old}})}}\right\}$$

  /* `Note that if` $U(x_{\text{new}}) < U(x_{\text{old}})$ `then` $p = 1$. */

  If move is rejected, stay at x_{old}, otherwise move to x_{new}.

This matrix can be computed analytically in our examples because everything is low dimensional. For example, we can use:

$$P_{ij} = \sum_{y \in V_j} \sum_{x \in V_i} \mathbf{P}(y|x)\mathbf{P}(x|V_i) \tag{7.107}$$

where $\mathbf{P}(y|x)$ is obtained from Q and $\mathbf{P}(x|V_i)$ is known exactly. Now given a measurement of P, one can compute the second eigenvalue μ_2 and let $-\ln(\mu_2)/\tau$ be an approximation to λ_2.

The RTS algorithm proceeds as follows. In our examples we only have two basins and therefore only use two colors, red and blue. Each time a red particle enters B, its color changes to blue, and vice versa. We use Algorithm **resample** to maintain in each cell or macro state a constant number of walkers of each color. At each step n, we calculate

$$J_n^{\mathrm{b}\to\mathrm{r}} = \Delta t^{-1} \frac{\text{Sum of the weights of all the blue particles who turn red during step } n}{\text{Sum of the weights of all the blue particles at step } n} \tag{7.108}$$

The rate from A to B is then given by:

$$\mathrm{rate}_{A\to B} = \lim_{n\to\infty} \frac{1}{n} \sum_{i=1}^{n} J_i^{\mathrm{b}\to\mathrm{r}} \tag{7.109}$$

The reverse rate, $\mathrm{rate}_{B\to A}$, is obtained similarly.

7.6.2 Numerical Benchmarks

7.6.2.1 1D Brownian Dynamics Setup

The 1D simulation takes place in the domain $\Omega = [-10, 10]$. The system parameters are $\beta^{-1} = 0.2$, $D = 0.06$ (diffusivity), $\Delta x = 0.03$ (fine state discretization), $\Delta t = 0.03$, $n_{\text{walkers per cell}} = 10$, and $n_{\text{cell}} = 32$ (coarse states). We define: $\bar{A} = [-10, 0]$, $A = [-7, -5]$, $\bar{B} = [0, 10]$, and $B = [5, 7]$. The potential $U(x)$ is

$$U(x) = \frac{(x+5)^2(x-5)^2}{1000} + 3e^{-\frac{x^2}{10}} - \frac{x}{10}$$

The exact rates computed using the fine states (with spacing Δx and time step Δt) are $\mathrm{rate}_{A\to B} = 1.59 \times 10^{-1}$ and $\mathrm{rate}_{B\to A} = 6.70 \times 10^{-11}$. The overall setup is illustrated in Figure 7.11.

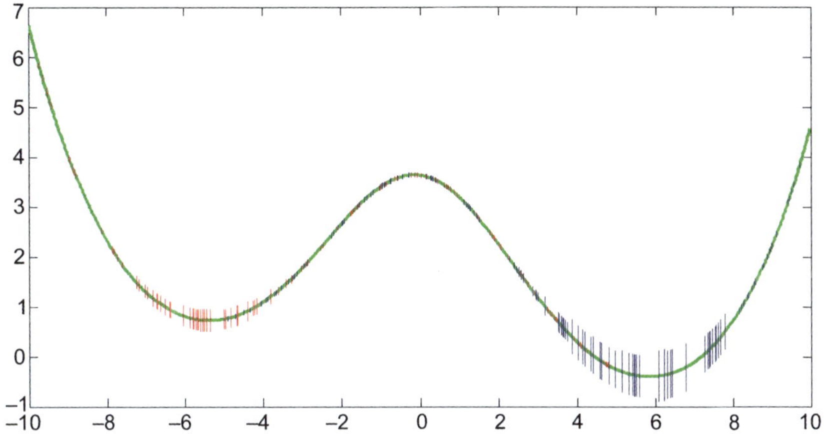

Figure 7.11 A snapshot of the 1D simulation setup. The green line shows the potential $U(x)$ and the red and blue lines show the positions and weights of the walkers. The weight has large variations and as a result we used a non-linear scaling to determine the length of the short vertical lines with $l=\beta(\log(\alpha w+1)+1)$, where l is the plotted length, w is the weight of the walker, and α, β are constants. (A simple log-scale is not suitable since it would lead to negative vertical line lengths.)

The initialization of walkers can impact the convergence time of the rates. A simple choice is to first scatter red walkers uniformly in region A and blue walkers in region B and then assign weight to the walkers proportional to $\rho(x)$. The resulting distribution of red and blue walkers is not the steady-state distribution as the steady-state distribution has red walkers in region B with small weights. However, this is a good approximation of this steady-state distribution. The computed rate from the algorithm will not be correct until the distribution of the walkers has converged to steady-state.

7.6.2.2 Discussion

Figure 7.14(a) shows results using the coarse grained matrix and estimating the rate using the second eigenvalue $\mu_2=e^{-\lambda_2\tau}$. As predicted by the theoretical analysis of the previous sections we observe two trends. At small lag times τ, the rate is over-predicted because of non-Markovian effects. At longer lag times, the statistical error increases. A good value of the lag time in this example is probably around $\tau\approx 300$, for which the systematic and statistical errors are both reasonably small.

The rates obtained using the reactive trajectory sampling method (RTS) is shown on Figure 7.12. This case is actually relatively difficult as the system is allowed to jump across multiple cells in one time step. As a result there might be some infrequent jump of particles with large weights from basin A to B (or *vice versa*). This results in larger statistical errors, which leads to a large sampling to reach an acceptable accuracy. The initial bias at small times (underestimation

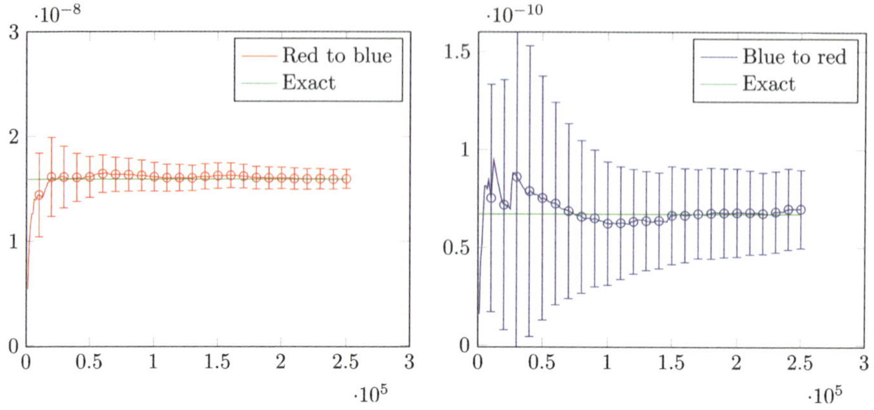

Figure 7.12 RTS Rates for the 1D simulation.

of the rate) is caused by the initial distribution of particles, which needs to equilibrate before correct statistics can be obtained. In practice a more judicious choice of initial distribution for the particles can reduce this initial equilibration.

7.6.2.3 2D Metropolis Method Setup

The 2D simulation takes place in the domain $\Omega = [-1, 1]^2$. The system parameters are $\beta^{-1} = 0.1$, $\Delta x = 0.01$, $n_{\text{walkers per cell}} = 100$, and $n_{\text{cell}} = 20$. We define the basins as: $\bar{A} = [-1, 0] \times [-1, 1]$, $A = \{(x, y) : \sqrt{(x+1)^2 + y^2} \leq 0.4\}$, $\bar{B} = [0, 1] \times [-1, 1]$, $B = \{(x, y) : \sqrt{(x+1)^2 + y^2} \leq 0.4\}$, (see Figure 7.13b). The potential in use is $U(x, y) = e^{-x^2} + y^2$ and is illustrated in Figure 7.13(a).

The domain Ω is divided into n_{cell} cells where the boundaries are equally spaced lines with a slant angle of θ. Figure 7.13b is a snapshot of the simulation with a value of $\theta = 20°$ where the boundary of the cells are depicted with the black slanted lines. The time stepping was done using the metropolis algorithm.

If basin A was defined as a vertical slab, the iso-surfaces of the committor function would be vertical lines because of the special form of U in which x and y are basically decoupled. Since A and B are defined as half circles, the choice of $\theta = 0°$ makes the cell boundaries close to the iso-surfaces, and thus should be the optimal choice. We will see that as the angle θ increases the statistical error in the prediction increases.

The exact rates are $\text{rate}_{A \to B} = \text{rate}_{B \to A} = 5.9 \times 10^{-7}$.

7.6.2.4 Discussion

Results obtained using the coarse grained matrix are shown in Figures 7.14b, 7.14c, and 7.14d. As before, the non-Markovian effects are visible. From the

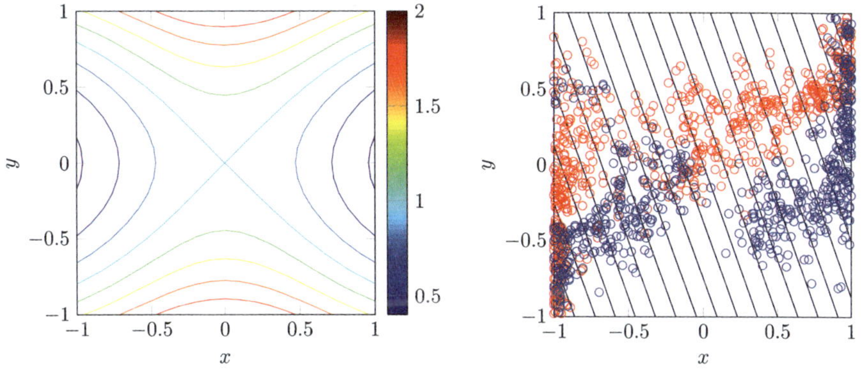

(a) **(b)**

Figure 7.13 Illustration of the 2D simulation setup. (a) Potential function $U(x, y) = e^{-x^2} + y^2$. (b) A snapshot of the simulation with $\theta = 40°$. We can see the two half circular regions A and B the absence of blue and red walkers near $(-1, 0)$, $(1, 0)$, respectively.

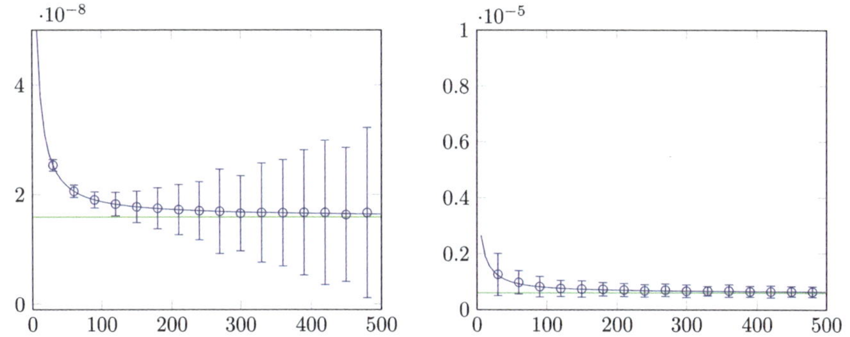

(a) 1D Example. $n_{\text{walker per cell}} = 5 \times 10^5$, $n_{\text{trial}} = 1000$. **(b)** 2D Example. $n_{\text{walker per cell}} = 500$, $n_{\text{trial}} = 100$, and $\theta = 0$.

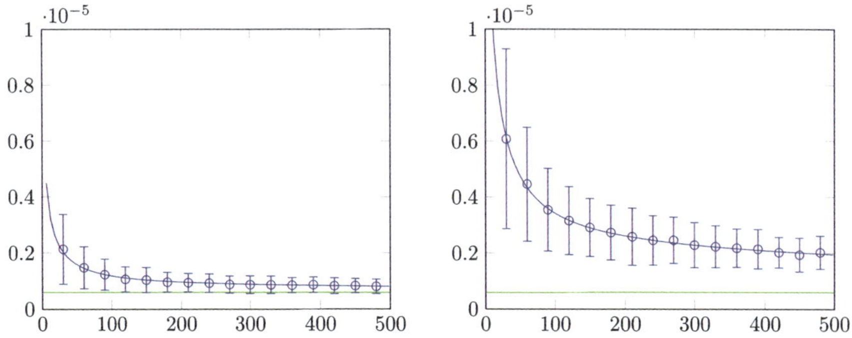

(c) 2D example with same parameters as above except $\theta = 20$. **(d)** 2D example with $\theta = 40$.

Figure 7.14 Rates computed with the transition matrix for the coarse grained model. The green line represents the exact rate. The error bars correspond to 2 standard deviations, estimated using n_{trial} samples. The x axis is the lag time τ.

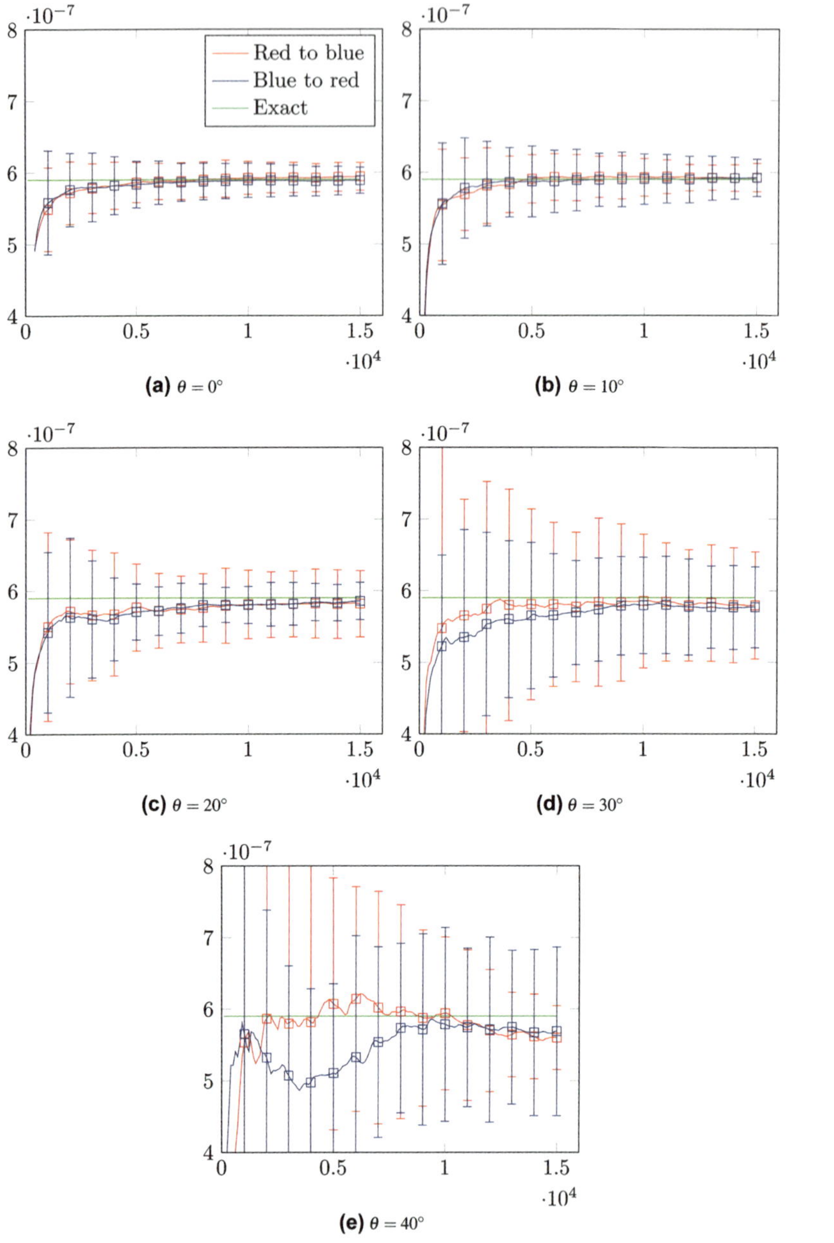

Figure 7.15 Rates computed using RTS for different angles θ for the cells. As θ increases the method shows no systematic bias but statistical errors increase. The initial relaxation to the exact rate is a result of the initial distribution of particles. The legend is the same for all plots and is only shown for the first plot.

theoretical analysis, we established that non-Markovian effects are reduced when the cells boundary approximate the iso-surface of the committor function and the cells are narrow in the direction $\nabla\pi$. Here the width of the cells is kept fixed but we vary the angle of the cells boundaries, which are straight lines. At $\theta = 0$, the lines are relatively good approximation of the iso-surfaces and therefore memory effects are small. As θ increases, the deviation from the iso-surfaces increases and we observe longer lag times. Note that the cell width is the same so that the degradation does not correspond to a coarsening of the cell but rather a poor choice of their geometry. In complex examples from biochemistry we expect that the choice of cells is far from ideal since it is difficult to guess the iso-committor surfaces and therefore the situation in Figures 7.14c and 7.14d is somewhat representative. We also note that in addition to larger memory effects, the statistical errors increase with θ.

The rates computed using RTS are shown in Figure 7.15a–7.15e. We see that, as expected, the rate always converges towards the correct value. As θ increases the statistical errors become larger and at $\theta = 40$ the method is no longer very efficient. Even though it is shown, the choice $\theta = 90$ would basically ruin the method since the cells would no longer provide any sampling enhancement. In that case, all the cells would have horizontal boundaries, parallel to x.

Figure 7.13b illustrates the point made in Section 4.3. The minimum energy path in this case is the centerline $y = 0$. As the cells become tilted particles tend to move away from the center line. In this case, on the left side, particles tend to move up and on the right side particles tend to move down. Roughly speaking, particles accumulate near the point where ∇A is orthogonal to the cell boundary. As θ increases, this point moves away from $y = 0$. As a result fewer particles are found near the center lines, even though these are the particles that make the largest contributions to the rate. This translates into infrequent events in which a particle with large weight changes color. At $\theta = 20$ this effect is small but it increases with θ. At $\theta = 90$, we would see very few particles changing color but with a very large weight. The rate would therefore be still correctly calculated, however, with a large standard deviation.

7.7 Conclusion

This chapter has discussed a number of approaches to calculate reaction rates. We reviewed the reactive flux approach, transition path sampling, transition interface sampling, forward flux sampling, conformation dynamics, Markov state models, non-equilibrium umbrella sampling, and an extension of weighted ensemble Brownian dynamics (renamed reactive trajectory sampling in this chapter). These methods differ in their assumptions and computational cost.

The reactive flux approach is probably the least computationally expensive but requires a fair amount of knowledge about the system (transition pathways, transition region, location of saddle point, *etc.*). Transition interface sampling

(TIS) and forward flux sampling (FFS) both rely on a reaction coordinate ξ or an order parameter (variable that increases monotonically with ξ). In cases where the mechanism is unknown or multiple pathways are present such methods may be more difficult to apply.

Markov state models (MSM) attempt to alleviate some of these problems by considering a general partitioning of space in macro-states. They can therefore more easily accommodate multiple pathways and complex reaction mechanisms. MSM depends on a lag time τ, equal to the length of trajectories that need to be run. At short times τ, significant non-Markovian effects are present, resulting in a systematic bias, while at long times, statistical errors increase. MSM are "relatively" easy to set up and calculate as they involve running a large number of independent short trajectories. It is therefore embarrassingly parallel and can be run efficiently even on loosely connected computers, for example, with cloud computing.

Some methods attempt to improve on transition path sampling methods (TIS, FFS) and MSM. They include the non-equilibrium umbrella sampling and weighted ensemble Brownian dynamics, which we called reactive trajectory sampling (RTS) when extended to the case of general macro-states. In this case the assumption of Markovian dynamics can be relaxed. This comes at the cost of a global convergence of statistics across macro-states. For example, in RTS and non-equilibrium umbrella sampling, the weights of the macro-states need to be converged, typically using a fixed-point iteration scheme. Although this is typically fast, this does lead in general to a computational cost that is larger than with MSM. In addition, some amount of global communication is required at regular intervals (*e.g.* for the `resample` algorithm in RTS), although the information that needs to be communicated is minimal. It includes the weights of walkers w_i and macro-state index. Depending on the setup, particle coordinates (x) may need to be communicated in some cases.

Techniques exist to construct optimal macro-states or improve an initial guess. They rely, for example, on computing minimum energy pathways (MFEP). Although there have been many successful implementations, this remains a challenging problem. For example, computing all the MFEPs that make significant contributions to the reaction rate and building macro-states from this data remains a challenge.

Nevertheless, these methods offer promising avenues to calculating reaction rates, transition states and reaction mechanisms. They make excellent use of modern parallel computers as most of these methods involve running a large number of fairly independent trajectories (perhaps with a small amount of global communication required). This puts less pressure on developing software that can run long trajectories using many processors. The efficiency of these methods is independent of the degree of metastability of the system (Perron cluster) and their convergence is often dictated by the shorter mixing time scales inside each metastable basin (at least in an optimal set up of the method).

7.8 Technical Proofs

These proofs can be found in classical textbooks such as Gardiner (1997). We provide them here as they can be helpful to understand some of the results and are also sufficiently simple to be succinctly explained.

Proof 1. We prove that

$$\rho_k(x) = \psi_k(x)\rho(x) \tag{7.110}$$

Consider simply Eq. (7.60) in one dimension (although the proof can be extended to the multi-dimensional case):

$$dx(t) = A_{\mathrm{B}}(x)dt + \sqrt{B_{\mathrm{B}}(x)}\,dW(t) \tag{7.111}$$

The function $\rho\ (x,t|x_0,\ 0)$ satisfies the forward Chapman-Kolmogorov differential equation:

$$\frac{\partial\rho(x,t|x_0,0)}{\partial t} = -\frac{d}{dx}(A_{\mathrm{B}}(x)\rho(x,t|x_0,0)) + \frac{1}{2}\frac{d^2}{dx^2}(B_{\mathrm{B}}(x)\rho(x,t|x_0,0)) \tag{7.112}$$

From the eigenfunction expansion we therefore have:

$$-\lambda_k\rho_k(x) = -\frac{d}{dx}(A_{\mathrm{B}}(x)\rho_k(x)) + \frac{1}{2}\frac{d^2}{dx^2}(B_{\mathrm{B}}(x)\rho_k(x)) \tag{7.113}$$

Similarly the backward equation is satisfied. In this case we write:

$$\rho(x,t|x_0,0) = \rho(x,0|x_0,-t) \tag{7.114}$$

and

$$\begin{aligned}\frac{\partial\rho(x,t|x_0,0)}{\partial t} &= -\frac{\partial\rho(x,0|x_0,-t)}{\partial(-t)}\\ &= A_{\mathrm{B}}(x_0)\frac{d}{dx_0}\rho(x,t|x_0,0) + \frac{1}{2}B_{\mathrm{B}}(x_0)\frac{d^2}{dx_0^2}\rho(x,t|x_0,0)\end{aligned} \tag{7.115}$$

This leads to:

$$-\lambda_k\psi_k(x_0) = A_{\mathrm{B}}(x_0)\frac{d}{dx_0}\psi_k(x_0) + \frac{1}{2}B_{\mathrm{B}}(x_0)\frac{d^2}{dx_0^2}\psi_k(x_0) \tag{7.116}$$

For the equilibrium density ρ (which is equal to ρ_1):

$$0 = -\frac{d}{dx}(A_{\mathrm{B}}(x)\rho(x)) + \frac{1}{2}\frac{d^2}{dx^2}(B_{\mathrm{B}}(x)\rho(x)) \quad \Rightarrow \quad A_{\mathrm{B}}\rho = \frac{1}{2}\frac{d}{dx}(B_{\mathrm{B}}\rho) \tag{7.117}$$

Consider now $\psi_k(x)\rho(x)$. From Eqs (7.113), (7.116) and (7.117), we can prove that:

$$-\lambda_k \psi_k \rho = -\frac{d}{dx}(A_{\mathrm{B}} \psi_k \rho_k) + \frac{1}{2}\frac{d^2}{dx^2}(B_{\mathrm{B}} \psi_k \rho) \tag{7.118}$$

so that:

$$\rho_k(x) = \psi_k(x)\rho(x) \tag{7.119}$$

□

Proof 2. We prove that

$$\pi(x) \sim \frac{\psi_2(x) - \psi_2(a)}{\psi_2(b) - \psi_2(a)} \tag{7.120}$$

We again consider simply a one-dimensional system. Then, the forward Chapman-Kolmogorov equation can be re-written as:

$$\begin{aligned} \frac{\partial \rho(x,t|x_0,0)}{\partial t} &= \frac{d}{dx}(-A_{\mathrm{B}}(x)\rho(x,t|x_0,0) + \frac{1}{2}\frac{d}{dx}(B_{\mathrm{B}}(x)\rho(x,t|x_0,0)) \\ &= -\frac{d}{dx}J(x,t|x_0,0) \end{aligned} \tag{7.121}$$

where J is interpreted as a probability flux. Assuming absorbing boundary conditions at A and B, the probability to reach B before A is obtained by integrating the flux from 0 to ∞:

$$\pi(x) = \int_0^\infty J(b,t|x,0)\, dt \tag{7.122}$$

assuming that $B = [b, \infty)$. From the definition of the flux:

$$\pi(x) = \int_0^\infty \left(A_{\mathrm{B}}(b)\rho(b,t|x,0) - \frac{1}{2}\frac{d}{db}(B_{\mathrm{B}}(b)\rho(b,t|x,0)) \right) dt \tag{7.123}$$

This time, we use the backward Chapman-Kolmogorov equation in order to obtain derivatives with respect to x:

$$A_{\mathrm{B}}\frac{d\pi}{dx} + \frac{1}{2}B_{\mathrm{B}}\frac{d^2\pi}{dx^2} = \int_0^\infty \frac{\partial J(b,t|x,0)}{\partial t}\, dt = -J(b,0|x,0) \tag{7.124}$$

Since for $x \neq b$, $J(b, 0|x, 0) = 0$, we have:

$$A_B \frac{d\pi}{dx} + \frac{1}{2} B_B \frac{d^2\pi}{dx^2} = 0 \tag{7.125}$$

This is the differential equation satisfied by the committor function. This expression can be extended to the multidi-mensional setting. The boundary conditions are $\pi(a) = 0$ and $\pi(b) = 1$. The function ψ_2 satisfies a slightly different equation:

$$A_B \frac{d\psi_2}{dx} + \frac{1}{2} B_B \frac{d^2\psi_2}{dx^2} = -\lambda_2 \psi_2 \tag{7.126}$$

However, with λ_2 very small we can conclude that ψ_2 satisfies nearly the same differential equation as π. From the boundary conditions, we conclude that:

$$\pi(x) \sim \frac{\psi_2(x) - \psi_2(a)}{\psi_2(b) - \psi_2(a)} \tag{7.127}$$

□

Acknowledgements

The authors gratefully acknowledge the work of Jesús A. Izaguirre and Haoyun "Michelle" Feng (University of Notre-Dame) who made several suggestions to improve the manuscript.

References

1. A. Warmflash, P. Bhimalapuram and A. R. Dinner, Umbrella sampling for nonequilibrium processes, *J. Chem. Phys.*, 2007, **127**(15), 154112.
2. A. Dickson, A. Warmflash and A. R. Dinner, Nonequilibrium umbrella sampling in spaces of many order parameters, *J. Chem. Phys.*, 2009, **130**(7), 074104.
3. G. A. Huber and S. Kim, Weighted-ensemble Brownian dynamics simulations for protein association reactions., *Biophys. J.*, Jan. 1996, **70**(1), 97–110.
4. R. Marcelin, Contribution à l'étude de la cinétique physico-chimique, *Ann. Phys*, 1915, **3**, 120–231.
5. H. Eyring and M. Polanyi, Ubcr einfache gasreaktionen, *Z. Physik. Chem*, 1931, **B12**, 279–311.
6. H. Eyring, The activated complex in chemical reactions, *J. Chem. Phys.*, 1935, **3**(2), 107.

7. I. Horiuti *On the statistical mechanical treatment of the absolute rate of chemical reaction*, volume 13. Bulletin of the Chemical Society of Japan, 1938.
8. E. Wigner, The transition state method. *Trans. Faraday Soc.*, 1938, pp. 29–41.
9. D. Chandler, Statistical mechanics of isomerization dynamics in liquids and the transition state approximation, *J. Chem. Phys*, 1978, **68**(6), 2959.
10. C. W. Gardiner, *Handbook of Stochastic Methods for Physics, Chemistry and the Natural Sciences*. Springer, second edition, 1997.
11. P. Hänggi and M. Borkovec, Reaction-rate theory: fifty years after Kramers., *Rev. Mod. Phys.*, Apr. 1990, **62**(2), 251–341.
12. P. G. Bolhuis, D. Chandler, C. Dellago and P. L. Geissler, Transition path sampling: throwing ropes over rough mountain passes, in the dark, *Annu. Rev. Phys. Chem.*, 2002, **53**, 291–318.
13. C. Dellago, P. G. Bolhuis, F. S. Csajka and D. Chandler, Transition path sampling and the calculation of rate constants, *J. Chem. Phys.*, 1998, **108**(5), 1964–1977.
14. C. Dellago, P. G. Bolhuis and P. L. Geissler, Transition Path Sampling., *Adv. Chem. Phys.*, 2002, **123**, 1–78.
15. E. Vanden-Eijnden, Transition-path theory and path-finding algorithms for the study of rare events, *Annu. Rev. Phys. Chem*, 2010, **61**, 391–420.
16. T. S. van Erp, D. Moroni and P. G. Bolhuis, A novel path sampling method for the calculation of rate constants, *J. Chem. Phys.*, 2003, **118**(17), 7762.
17. R. J. Allen, P. B. Warren and P. R. ten Wolde, Sampling rare switching events in biochemical networks., *Phys. Rev. Lett.*, Jan. 2005, **94**(1), 018104.
18. R. J. Allen, D. Frenkel and P. R. ten Wolde, Forward flux sampling-type schemes for simulating rare events: Efficiency analysis, *J. Chem. Phys.*, 2006, **124**(19), 194111.
19. R. J. Allen, D. Frenkel and P. R. ten Wolde, Simulating rare events in equilibrium or nonequilibrium stochastic systems, *J. Chem. Phys.*, 2006, **124**(2), 024102.
20. A. K. Faradjian and R. Elber, Computing time scales from reaction coordinates by milestoning, *J. Chem. Phys.*, 2004, **120**(23), 10880.
21. E. Vanden-Eijnden, M. Venturoli, G. Ciccotti and R. Elber, On the assumptions underlying milestoning, *J. Chem. Phys.*, 2008, **129**(17), 174102.
22. P. Májek and R. Elber, Milestoning without a reaction coordinate, *J. Chem. Theory Comput.*, June 2010, **6**(6), 1805–1817.
23. P. Deuflhard, M. Dellnitz, O. Junge, and C. Schütte. *Computation of Essential Molecular Dynamics by Subdivision Techniques I: Basic Concept*, volume SC 96-45. Konrad-Zuse-Zentrum für Informationstechnik Berlin, 1996.
24. C. Schütte and W. Huisinga. Biomolecular conformations can be identified as metastable sets of molecular dynamics. In *Handbook of*

numerical analysis, Vol. X, North-Holland, Amsterdam, 2003, pp. 699–744.

25. N. Singhal, C. D. Snow and V. S. Pande, Using path sampling to build better Markovian state models: predicting the folding rate and mechanism of a tryptophan zipper beta hairpin, *J. Chem. Phys*, 2004, **121**(1), 415.
26. W. C. Swope, J. W. Pitera and F. Suits, Describing protein folding kinetics by molecular dynamics simulations. 1. Theory., *J. Phys. Chem. B*, May 2004a, **108**(21), 6571–6581.
27. W. C. Swope, J. W. Pitera, F. Suits, M. Pitman, M. Eleftheriou, B. G. Fitch, R. S. Germain, A. Rayshubski, T. J. C. Ward, Y. Zhestkov and R. Zhou, Describing protein folding kinetics by molecular dynamics simulations. 2. Example applications to alanine dipeptide and a β-hairpin peptide., *J. Phys. Chem. B*, May 2004b, **108**(21), 6582–6594.
28. J. D. Chodera, W. C. Swope, J. W. Pitera and K. A. Dill, Long-time protein folding dynamics from short-time molecular dynamics simulations, *Multiscale Model. Simul.*, 2006, **5**(4), 1214.
29. J. D. Chodera, N. Singhal, V. S. Pande, K. A. Dill and W. C. Swope, Automatic discovery of metastable states for the construction of Markov models of macromolecular conformational dynamics, *J. Chem. Phys.*, 2007, **126**, 155101.
30. F. Noé, I. Horenko, C. Schütte and J. C. Smith, Hierarchical analysis of conformational dynamics in biomolecules: Transition networks of metastable states, *J. Chem. Phys.*, 2007, **126**(15).
31. C. Shalloway, Macrostates of classical stochastic systems, *J. Chem. Phys.*, 1996, **105**(22), 9986.
32. B. W. Zhang, D. Jasnow and D. M. Zuckerman, The "weighted ensemble" path sampling method is statistically exact for a broad class of stochastic processes and binning procedures, *J. Chem. Phys*, 2010, **132**(5), 054107.
33. D. Bhatt and D. M. Zuckerman, Heterogeneous Path Ensembles for Conformational Transitions in Semiatomistic Models of Adenylate Kinase., *J. Chem. Theory Comput.*, Nov. 2010, **6**(11), 3527–3539.
34. C. Schütte, F. Noé, J. Lu, M. Sarich and E. Vanden-Eijnden, Markov state models based on milestoning., *J. Chem. Phys.*, May 2011, **134**(2), 4105.
35. M. Sarich, F. Noé and C. Schütte, On the approximation quality of Markov state models, *Multiscale Model. Simul.*, 2010, **8**(4), 1154–1177.
36. J.-H. Prinz, H. Wu, M. Sarich, B. Keller, M. Senne, M. Held, J. D. Chodera, C. Schütte and F. Noé, Markov models of molecular kinetics: generation and validation, *J. Chem. Phys.*, 2011, **134**(17), 174105.
37. N. S. Hinrichs, *Algorithms for Building Models of Molecular Motion from Simulations*. PhD thesis, Stanford University, Stanford University, 2007.
38. N. Singhal and V. S. Pande, Error analysis and efficient sampling in Markovian state models for molecular dynamics, *J. Chem. Phys*, 2005, **123**(20), 204909.

39. P. Metzner, F. Noé and C. Schütte, Estimating the sampling error: distribution of transition matrices and functions of transition matrices for given trajectory data., *Phys. Rev. E*, Aug. 2009a, **80**(2), 21106.
40. T. Lelièvre, G. Stoltz, and M. Rousset, *Free Energy Computations*, A Mathematical Perspective, World Scientific, Oct. 2010.
41. C. Chipot and A. Pohorille, editors, *Free Energy Calculations: Theory and Applications in Chemistry and Biology*, Springer, 2007, 2.
42. E. Darve and A. Pohorille, Calculating free energies using average force, *J. Chem. Phys.*, Nov. 2001, **115**(2), 9169–9183.
43. J. Hénin and C. Chipot, Overcoming free energy barriers using unconstrained molecular dynamics simulations, *J. Chem. Phys*, 2004, **121**(7), 2904–2914.
44. D. Rodríguez-Gómez, E. Darve and A. Pohorille, Assessing the efficiency of free energy calculation methods, *J. Chem. Phys.*, 2004, **120**(8), 3563–3578.
45. T. Lelièvre, M. Rousset and G. Stoltz, Computation of free energy differences through nonequilibrium stochastic dynamics: the reaction coordinate case, *J. Comput. Phys.*, Mar. 2007, **222**(2), 624–643.
46. E. Darve, D. Rodríguez-Gómez and A. Pohorille, Adaptive biasing force method for scalar and vector free energy calculations, *J. Chem. Phys.*, Apr. 2008, **128**(14), 144120.
47. E. Darve, J. Solomon and A. Kia, Computing generalized Langevin equations and generalized Fokker-Planck equations, *P. Natl. Acad. Sci. USA*, July 2009, **106**(27), 10884–10889.
48. D. Truhlar. Variational transition state theory. *Annu. Rev. Phys. Chem.*, 1984, pp. 159–189.
49. S. C. Tucker. Variational transition state theory in condensed phases. In P. Talkner and P. Hänggi, editors, *New Trends in Kramers' Reaction Rate Theory*, Springer, 1995, p. 251.
50. C. H. Bennett. Molecular dynamics and transition state theory: the simulation of infrequent events. In *Algo. for Chem. Comp.*, 1977, pp. 63–97.
51. C. Dellago and P. G. Bolhuis. Transition path sampling and other advanced simulation techniques for rare events. *Adv. Polym. Sci.*, 2009, pp. 167–233.
52. G. M. Torrie and J. P. Valleau, Nonphysical sampling distributions in Monte Carlo free-energy estimation: umbrella sampling, *J. Comput. Phys.*, 1977, **23**, 187–199.
53. P. Metzner, C. Schütte and E. Vanden-Eijnden, Transition path theory for Markov jump processes, *Multiscale Model. Simul.*, 2009b, **7**(3), 1192–1219.
54. W. E and E. Vanden-Eijnden, Towards a Theory of Transition Paths, *J. Stat. Phys.*, May 2006, **123**(3), 503–523.
55. P. Metzner, C. Schütte and E. Vanden-Eijnden, Illustration of transition path theory on a collection of simple examples, *J. Chem. Phys.*, 2006, **125**(8), 1–17.

56. C. Schütte, A. Fischer, W. Huisinga and P. Deuflhard, A direct approach to conformational dynamics based on hybrid Monte Carlo, *J. Comput. Phys.*, 1999, **151**(1), 146–168.
57. S. C. Harvey and H. A. Gabb, Conformational transitions using molecular dynamics with minimum biasing, *Biopolymers*, Aug. 1993, **33**(8), 1167–1172.
58. C. Valeriani, R. J. Allen, M. J. Morelli, D. Frenkel and P. Rein ten Wolde, Computing stationary distributions in equilibrium and nonequilibrium systems with forward flux sampling, *J. Chem. Phys.*, 2007, **127**(11), 114109.
59. E. E. Borrero and F. A. Escobedo, Reaction coordinates and transition pathways of rare events via forward flux sampling, *J. Chem. Phys.*, 2007, **127**(16), 164101.
60. M. Weber, Meshless methods in conformation dynamics, *Multiscale Model. Simul.*, 2007, **6**(2), 396–416.
61. R. Elber, Long-timescale simulation methods., *Curr. Opin. Struct. Biol.*, Apr. 2005, **15**(2), 151–156.
62. R. Elber, A milestoning study of the kinetics of an allosteric transition: atomically detailed simulations of deoxy Scapharca hemoglobin, *Biophys. J.*, May 2007, **92**(9), L85–L87.
63. E. Vanden-Eijnden and M. Venturoli, Markovian milestoning with Voronoi tessellations, *J. Chem. Phys.*, May 2009, **130**(19), 194101.
64. A. Amadei and A. Linssen, Essential dynamics of proteins, *Proteins: Struct., Funct., Genet*, 1993, **17**, 412–425.
65. H. Grubmüller and P. Tavan, Molecular-dynamics of conformational substates for a simplified protein model, *J. Chem. Phys*, 1994, **101**(6), 5047–5057.
66. P. Deuflhard. From molecular dynamics to conformation dynamics in drug design. In M. Kirkilionis, S. Krömker, R. Rannacher, and F. Tomi, editors, *Trends in Nonlinear Analysis*, Springer Verlag, 2003, p. 269.
67. C. Schütte, W. Huisinga, and P. Deuflhard, Transfer operator approach to conformational dynamics in biomolecular systems. In B. Fiedler, editor, *Ergodic Theory, Analysis, and Efficient Simulation of Dynamical Systems*, Springer, Berlin, 2001, pp. 191–223.
68. W. Huisinga, *Metastability of Markovian systems*. PhD thesis, Freien Universität Berlin, 2001.
69. W. Huisinga and B. Schmidt, Metastability and dominant eigenvalues of transfer operators. *New Algorithms for Macromolecular Simulation*, 2006, pp. 167–182.
70. P. Deuflhard, W. Huisinga and A. Fischer, Identification of almost invariant aggregates in reversible nearly uncoupled Markov chains., *Lin. Alg. App.*, 2000, **315**, 39–59.
71. C. Schütte and W. Huisinga, Biomolecular conformations as metastable sets of Markov chains, *Proc. 38 Annu. Allerton Conf. on Comm., Cont., and Comput.*, 2000b, pp. 1106–1115.

72. P. Deuflhard and M. Weber, Robust Perron cluster analysis in conformation dynamics, *Lin. Alg. App*, 2005, **398**, 161–184.
73. M. Weber and W. Rungsarityotin, *Perron Cluster Analysis and its Connection to Graph Partitioning for Noisy Data*. Konrad-Zuse-Zentrum für Informationstechnik Berlin, 2004.
74. D. Fritzsche, V. Mehrmann, D. B. Szyld and E. Virnik, An SVD approach to identifying metastable states of Markov chains, *Elec. Trans. Num. Anal.*, 2007, **29**, 46–69.
75. S. Kube and M. Weber, Conformation Kinetics as a Reduced Model for Transition Pathways, Technical report, Konrad-Zuse-Zentrum für Informationstechnik Berlin, 2005.
76. S. Kube and M. Weber, A coarse graining method for the identification of transition rates between molecular conformations, *J. Chem. Phys*, 2007, **126**, 024103.
77. C. Schütte and W. Huisinga. On conformational dynamics induced by Langevin processes, In *International Conference on Differential Equations, Vol. 1, 2 (Berlin, 1999)*, World Sci. Publ., River Edge, NJ, 2000a, pp. 1247–1262.
78. F. Cordes, M. Weber, and J. Schmidt-Ehrenberg, *Metastable Conformations via Successive Perron-cluster Cluster Analysis of Dihedrals*, volume 02-40. Konrad-Zuse-Zentrum für Informationstechnik Berlin, 2002.
79. T. Galliat, P. Deuflhard, R. Roitzsch, and F. Cordes, *Automatic Identification of Metastable Conformations via Self-organized Neural Networks*, volume 00-51. Konrad-Zuse-Zentrum für Informationstechnik Berlin, 2000.
80. V. Schultheis, T. Hirschberger, H. Carstens and P. Tavan, Extracting Markov models of peptide conformational dynamics from simulation data, *J. Chem. Theory Comput.*, 2005, **1**(4), 515–526.
81. M. Kloppenburg and P. Tavan, Deterministic annealing for density estimation by multivariate normal mixtures, *Phys. Rev. E*, Mar. 1997, **55**(3), R2089–R2092.
82. E. Meerbach, C. Schütte and A. Fischer, Eigenvalue bounds on restrictions of reversible nearly uncoupled Markov chains, *Lin. Alg. App.*, 2005, **398**, 141–160.
83. I. Horenko, E. Dittmer, A. Fischer and C. Schütte, Automated model reduction for complex systems exhibiting metastability, *Multiscale Model. Simul.*, 2006, **5**(3), 802–827.
84. I. Horenko and C. Schütte. Likelihood-based estimation of multidimensional Langevin models and its application to biomolecular dynamics. *Multiscale Model. Simul.*, 2008, pp. 1–39.
85. A. Fischer, S. Waldhausen, I. Horenko, E. Meerbach and C. Schütte, Identification of biomolecular conformations from incomplete torsion angle observations by hidden Markov models, *J. Comput. Chem.*, 2007, **28**(15), 2453–2464.
86. R. R. Coifman, I. G. Kevrekidis, S. Lafon, M. Maggioni and B. Nadler, Diffusion maps, reduction coordinates, and low dimensional

representation of stochastic systems, *Multiscale Model. Simul.*, 2008, **7**(2), 842–864.
87. M. Oreššič and D. Shalloway, Hierarchical characterization of energy landscapes using Gaussian packet states, *J. Chem. Phys.*, 1994, **101**(11), 9844.
88. B. W. Church, M. Orešič, and D. Shalloway. Tracking metastable states to free-energy global minima. In P. M. Pardalos, D. Shalloway, and G. Xue, editors, *DIMACS Series in Discrete Mathematics and Theoretical Computer Science, Global Minimization of Nonconvex Energy Functions: Molecular Conformation and Protein Folding*, American Mathematical Society, Providence, RI, 1996, pp. 41–64.
89. A. Ulitsky and D. Shalloway, Variational calculation of macrostate transition rates, *J. Chem. Phys.*, 1998, **109**(5), 1670.
90. D. Korenblum and D. Shalloway, Macrostate data clustering., *Phys. Rev. E*, May 2003, **67**(5), 056704.
91. B. W. Church, A. Ulitsky, and D. Shalloway. Macrostate dissection of thermodynamic Monte-Carlo integrals, In I. Prigogine and S. A. Rice, editors, *Advances in Chemical Physics*, John Wiley & Sons, Inc., Hoboken, NJ, USA, Jan. 1999, pp. 273–310.
92. G. Fleming and P. Hänggi, *Activated Barrier Crossing: Applications in Physics, Chemistry and Biology*. World Scientific, 1993.
93. V. S. Pande, K. A. Beauchamp and G. R. Bowman, Everything you wanted to know about Markov state models but were afraid to ask, *Methods*, 2010, **52**(1), 99–105.
94. C. Schütte. Conformational dynamics: modelling, theory, algorithm, and application to biomolecules. *Konrad-Zuse-Zentrum für Informationstechnik Berlin*, Preprint SC 99-18: 1999, pp. 1–139.
95. N. I. Fisher, *Statistical Analysis of Circular Data*, Cambridge University Press, Cambridge, 1993.
96. N. I. Fisher and A. J. Lee, A correlation coefficient for circular data, *Biometrika*, 1983, **70**(2), 327–332.
97. A. C. Pan and B. Roux, Building Markov state models along pathways to determine free energies and rates of transitions, *J. Chem. Phys*, 2008, **129**(6), 064107.
98. S. Park and V. S. Pande, Validation of Markov state models using Shannon's entropy, *J. Chem. Phys.*, 2006, **124**(5), 054118.
99. D. Nerukh, C. H. Jensen and R. C. Glen, Identifying and correcting non-Markov states in peptide conformational dynamics, *J. Chem. Phys.*, 2010, **132**(8), 084104.
100. J. P. Crutchfield, The calculi of emergence: computation, dynamics and induction., *Physica D*, Aug. 1994, **75**(1–3), 11–54.
101. J. P. Crutchfield and K. Young, Inferring statistical complexity, *Phys. Rev. Lett.*, July 1989, **63**(2), 105–108.
102. C. R. Shalizi and J. P. Crutchfield, Computational mechanics: pattern and prediction, structure and simplicity, *J. Stat. Phys.*, 2001, **104**(3/4), 817–879.

103. X. Huang, G. R. Bowman, S. Bacallado and V. S. Pande, Rapid equilibrium sampling initiated from nonequilibrium data, *P. Natl. Acad. Sci. USA*, 2009, **106**(47), 19765–19769.
104. U. Hansmann and Y. Okamoto, New Monte Carlo algorithms for protein folding, *Curr. Opin. Struct. Biol.*, 1999, **9**(2), 177–183.
105. Y. Sugita, Replica-exchange molecular dynamics method for protein folding, *Chem. Phys. Lett.*, Nov. 1999, **314**(1–2), 141–151.
106. A. P. Lyubartsev, A. A. Martsinovski, S. V. Shevkunov and P. N. Vorontsov-Velyaminov, New approach to Monte Carlo calculation of the free energy: method of expanded ensembles, *J. Chem. Phys.*, 1992, **96**(3), 1776.
107. E. Marinari, Simulated tempering: a new Monte Carlo scheme, *Europhys. Lett.*, 1992, **19**(6), 451–458.
108. G. R. Bowman, Using generalized ensemble simulations and Markov state models to identify conformational states, *Methods*, 2009, **49**, 197–201.
109. T. H. Chiang, D. Hsu and J. C. Latombe, Markov dynamic models for long-timescale protein motion., *Bioinformatics*, June 2010, **26**(12), i269–i277.
110. S. P. Elmer, S. Park and V. S. Pande, Foldamer dynamics expressed via Markov state models. II. State space decomposition, *J. Chem. Phys.*, 2005, **123**(11), 114903.
111. S. P. Elmer, S. Park and V. S. Pande, Foldamer dynamics expressed via Markov state models, I. Explicit solvent molecular-dynamics simulations in acetonitrile, chloroform, methanol, and water, *J. Chem. Phys.*, 2005, **123**(11), 114902.
112. P. M. Kasson, N. W. Kelley, N. Singhal, M. Vrljic, A. T. Brunger and V. S. Pande, Ensemble molecular dynamics yields submillisecond kinetics and intermediates of membrane fusion, *P. Natl. Acad. Sci. USA*, 2006, **103**(32), 11916–11921.
113. G. Jayachandran, V. Vishal and V. S. Pande, Using massively parallel simulation and Markovian models to study protein folding: examining the dynamics of the villin headpiece, *J. Chem. Phys*, 2006, **124**(16), 164902.
114. G. R. Bowman, K. A. Beauchamp, G. Boxer and V. S. Pande, Progress and challenges in the automated construction of Markov state models for full protein systems, *J. Chem. Phys.*, 2009, **131**(12), 1–11.
115. F. Noé and S. Fischer, Transition networks for modeling the kinetics of conformational change in macromolecules, *Curr. Opin. Struct. Biol.*, Apr. 2008, **18**(2), 154–162.
116. F. Noé, C. Schütte, E. Vanden-Eijnden, L. Reich and T. R. Weikl, Constructing the equilibrium ensemble of folding pathways from short off-equilibrium simulations, *P. Natl. Acad. Sci. USA*, Nov. 2009, **106**(45), 19011–19016.
117. F. Morcos, S. Chatterjee, C. L. McClendon, P. R. Brenner, R. López-Rendón, J. Zintsmaster, M. Ercsey-Ravasz, C. R. Sweet, M. P. Jacobson, J. W. Peng and J. A. Izaguirre, Modeling conformational ensembles of

slow functional motions in Pin1-WW, *PLoS Comput. Biol.*, 2010, **6**(12), e1001015.
118. V. A. Voelz, G. R. Bowman, K. A. Beauchamp and V. S. Pande, Molecular simulation of ab initio protein folding for a millisecond folder NTL9(1-39), *J. Am. Chem. Soc.*, 2010, **132**(5), 1526–1528.
119. E. Vanden-Eijnden and M. Venturoli, Exact rate calculations by trajectory parallelization and tilting, *J. Chem. Phys.*, 2009, **131**(4), 044120.
120. A. Dickson, A. Warmflash and A. R. Dinner, Separating forward and backward pathways in nonequilibrium umbrella sampling, *J. Chem. Phys.*, 2009, **131**(15), 154104.
121. A. Dickson and A. R. Dinner, Enhanced Sampling of Nonequilibrium Steady States, *Annu. Rev. Phys. Chem.*, Mar. 2010, **61**(1), 441–459.
122. A. Dickson, M. Maienschein-Cline, A. Tovo-Dwyer, J. R. Hammond, and A. R. Dinner, Flow-dependent unfolding and refolding of an RNA by nonequilibrium umbrella sampling. *arXiv*, cond-mat.stat-mech, Apr. 2011, pp. 1–24.
123. B. W. Zhang, D. Jasnow and D. M. Zuckerman, Efficient and verified simulation of a path ensemble for conformational change in a united-residue model of calmodulin, *P. Natl. Acad. Sci. USA*, 2007, **104**(46), 18043–18048.
124. B. W. Zhang, D. Jasnow and D. M. Zuckerman. *Weighted Ensemble Path Sampling for Multiple Reaction Channels. arXiv, physics.bio-ph.* Feb. 2009, pp. 1–14.
125. G. A. Huber and J. A. McCammon, Weighted-ensemble simulated annealing: faster optimization on hierarchical energy surfaces, *Phys. Rev. E*, Apr. 1997, **55**(4), 4822–4825.
126. L. Maragliano, A. Fischer, E. Vanden-Eijnden and G. Ciccotti, String method in collective variables: minimum free energy paths and iso-committor surfaces., *J. Chem. Phys.*, July 2006, **125**(2), 24106.
127. D. Bhatt, B. W. Zhang and D. M. Zuckerman, Steady-state simulations using weighted ensemble path sampling, *J. Chem. Phys.*, 2010, **133**(1), 014110.
128. N. S. Hinrichs and V. S. Pande, Calculation of the distribution of eigenvalues and eigenvectors in Markovian state models for molecular dynamics, *J. Chem. Phys*, 2007, **126**(24), 244101.
129. D. L. Ermak and J. A. McCammon, Brownian dynamics with hydrodynamic interactions., *J. Chem. Phys.*, Aug. 1978, **69**(4), 1352–1360.
130. J. Hill and R. Moore. *Applied Mathematics Entering the 21st Century: Invited Talks from the ICIAM 2003 Congress*. 116. Society for Industrial Mathematics, 2004.
131. W. Ren, E. Vanden-Eijnden, P. Maragakis and W. E., Transition pathways in complex systems: application of the finite-temperature string method to the alanine dipeptide, *J. Chem. Phys.*, Oct. 2005, **123**(1), 4109.

132. D. Frenkel and B. Smit, *Understanding Molecular Simulation: from Algorithms to Applications*. Academic Press, Inc. Orlando, FL, USA, 1996.
133. D. Chandler, *Introduction to Modern Statistical Mechanics*, volume 1, Oxford University Press, Sept. 1987.
134. A. M. A. West, R. Elber and D. Shalloway, Extending molecular dynamics time scales with milestoning: example of complex kinetics in a solvated peptide, *J. Chem. Phys.*, 2007, **126**(14), 145104.

CHAPTER 8

Challenges in Applying Monte Carlo Sampling to Biomolecular Systems

M. MEZEI

Mount Sinai School of Medicine, Department of Structural and Chemical Biology, One Gustave L. Levy Pl., New York, NY 10029, USA
Email: mihaly.mezei@mssm.edu

Dedicated to the memory of Professor Edwin S. Campbell

8.1 Introduction

The era of atomic-level simulations was ushered in by the 1952 paper of Metropolis and coworkers.[1] Molecular dynamics was introduced much later, by Rahman and Stillinger,[2] but Monte Carlo was the preferred choice of simulation technique for quite a while. Currently, however, many simulations of biomolecular systems use molecular dynamics.

On a purely philosophical level, however, the Monte Carlo approach should have the edge since the problem of accurately solving the equation of motion for a very large number of degrees of freedom ($O(10^6)$) appears to be a much more exacting task than the generation of a sample of conformation that 'just' follows a certain distribution (*i.e.* the Boltzmann distribution corresponding to the ensemble in which the simulation was done). Indeed, the Monte Carlo approach has been successfully applied in diverse areas.[3] This success, however, did not extend to the field of simulating macromolecular assemblies.

The aim of this chapter is – instead of reviewing the considerable progress made so far – to discuss the obstacles that prevent the wider use of the Monte

RSC Biomolecular Sciences No. 23
Innovations in Biomolecular Modeling and Simulations: Volume 1
Edited by Tamar Schlick

Published by the Royal Society of Chemistry, www.rsc.org

Carlo method for macromolecular simulations. Successful adoption of the Monte Carlo method for conformational sampling of macromolecular assemblies requires solution(s) to the following problems: (1) convince investigators that it is worth it; (2) devise move sets that generate large enough correlated changes that can be accepted with reasonable probability; (3) develop efficient treatment of non-pairwise additive potentials; (4) develop efficient treatment of long-range contributions to the system's energy; and (5) the efficient parallelization of the algorithm. In the remainder of this chapter these issues will be treated one by one. For many fundamental details see refs. 4–6 and for some recent applications see refs. 7 or 8. Richey provided a historical account of the development of the Markov-chain Monte Carlo method and the widening of the scope of its application.[34]

Note that there are two distinct approaches for the enhancement of conformational sampling: (a) enhancing the algorithm generating successive conformations during the simulation and (b) manipulating the treatment of the energy surface governing the simulation. Typical examples for the second approach are umbrella sampling[9] or replica exchange;[10] but they are all equally applicable to molecular dynamics and Monte Carlo and thus will not be discussed in this chapter – it is for the algorithm generating successive conformations where the choice between Monte Carlo and molecular dynamics arises.

8.2 Basic Ideas of Monte Carlo Sampling

The Metropolis method[1] obtains a Boltzmann-averaged sample of configurations by generating a Markov chain. Markov chains have the basic property that each member of the chain (series of configurations in the case of molecular systems) is obtained solely based on the member preceding it. For each probability distribution defining a new member (called transition probability) there is a corresponding distribution describing the members of the Markov chain defined by the transition probability. The most general relation between transition probabilities has been described by Hastings.[11]

The general pattern for selecting the next member of the Markov chain, usually called a Monte Carlo move, is by making a random change in the current configuration and accepting it with a probabilistic filter – the next member of the chain will be either the configuration just generated (if accepted) or the repeat of the previous configuration (if rejected). The classic case is the move of one atom by selecting $\Delta \boldsymbol{x}_i$ with uniform probability within a cube of fixed size around $\boldsymbol{x}_i$, and accepting the new configuration with probability $\min\{1,\exp(-\Delta E/kT)\}$ where ΔE is the energy difference between the two configurations. In other words, if the change lowers the energy, accept for sure, if not, then accept it with exponentially diminishing probability as the difference increases. The beauty of this method lies in the fact that it leads to Boltzmann-averaged distribution without having to calculate the partition function that is the normalizing factor in calculating the Boltzmann factor. Further, it requires only the *change* in the energy, leading to simplifications in the energy calculations in certain cases – this issue will be discussed in Sections 5 and 6 in more detail.

The design of a successful Monte Carlo sampling algorithm consists of the selection of the trial moves (move set) and the distribution these moves are sampled from. First requirement is that the move set should provide for the sampling of all the degrees of freedom in the system. This obviously includes the change in the atomic coordinates. For thermodynamic ensembles other than the canonical, further moves can include the change in the volume (in the isobaric-isothermal ensemble) or even the number of particles (in the grand-canonical ensemble), requiring the creation or annihilation (attempts) of part of the system.

The guiding principle in selecting the type of moves and the distribution(s) they are sampled from is to simultaneously maximize the change in the configuration and minimize the increase in the energy since the computational effort in calculating ΔE is (in most cases) independent of the magnitude of the change. In the basic example above, this involves optimizing the edge of the cube within which the random change is made. The selection probability can also be different from the uniform distribution – for example, moves can be biased in the direction of force acting on the atom moved[12] – with a concomitant modification in the acceptance probability – the so-called force-biased sampling. This idea can be applied to the volume changes[13] and the insertion/deletion steps[14] in the isothermal-isobaric and in the grand-canonical ensembles, respectively. The important thing in the use of such biased sampling is that the probability distribution of the bias is well defined since it is needed in the modified acceptance expression.

While molecules can, in principle, be considered 'just' a collection of atoms and sampled accordingly, the resulting simulation would be very inefficient since the intramolecular energy changes very steeply with the change in bondlength. Sampling of molecules involves the sampling of orientation and conformation as well. This is usually achieved by some random rotation of the molecule and by some random changes in the intramolecular coordinates, usually torsion or bond angles. This is an area rich in possibilities whose discussion is outside the scope of this chapter. Some examples will be discussed in Section 8.3.

8.3 Why 'Bother' with Monte Carlo?

The spectacular success of molecular dynamics raises the question of the worthiness of the effort in trying to apply the Monte Carlo method for such problems. However, success begets success, and the effort expended in improving both the theoretical and computational aspects of molecular dynamics has far exceeded the effort spent on developing Monte Carlo. Reversing this disparity in the efforts expended thus can lead to development of efficient Monte Carlo applications. One example of an issue that has never been thoroughly examined is the fact that while the most general form of generating a Metropolis move has been described by Hastings;[11] most applications use a more limited form and no systematic study has been performed to determine the optimal choice.

Besides the philosophical argument cited in the Introduction it has also been frequently remarked that the very fact that is the source of molecular dynamics' success – the strict reliance on the time evolution of the system – is also an intrinsic limitation since each step the simulation can make is very small. Monte Carlo, not having this tie to time, is free, at least in principle, both to make much larger steps and to take shortcuts in the configurational space. The problem is that when one takes shortcuts, it is easy to get lost. It is also important to note that with the Monte Carlo approach only indirect kinetic information (*via* analysis of activation barriers) can be obtained.

The superiority of current molecular dynamic applications over Monte Carlo is, however, not absolute. In a comparison of Monte Carlo and molecular dynamics on liquid hexane[15] Monte Carlo outperformed molecular dynamics. Also, simulation of aqueous systems in the grand-canonical ensemble[14] is an order of magnitude more efficient with Monte Carlo than with molecular dynamics.[16] However, these successes have not (yet) carried to simulations of biomacromolecules.

8.4 Correlated Moves

Another significant difference between molecular dynamics and Monte Carlo is that while devising the best numerical method for the calculation of the next conformation is a strictly scientific enterprise, there is an art to the design of Monte Carlo moves – followed, of course, by the scientific analysis and implementation. As discussed in Section 8.2, a successful Monte Carlo move makes a large change in the conformation that only changes the energy by an amount that is commensurable to kT so that the acceptance probability is high enough. This generally requires the selection a 'soft' degree of freedom (*e.g.* changing torsion angles instead of translating atoms). An additional limitation on the change is that the ratio of the probability of selecting this change and its reverse be known (or be computable).

The major question in the design is the selection of the set of atoms to be moved. First, it is clear that if the change in energy when moving two sets of atoms is the sum of energy changes for moving the two sets independently (and, concomitantly, the computational efforts are also additive or nearly so) then moving them separately is more advantageous, since collision caused by move in one of the sets would cause the rejection of the combined move while it would still accept the move of the rest of the set if the moves are performed separately. This would suggest that generally it is advantageous to select as small a set of atoms to move as possible.

Biomolecular systems, however, are generally dense, interconnected, and heterogeneous. This makes the design of efficient moves rather difficult since the additivity of the energy changes assumed in the argument above is unlikely to hold. Instead, the chance for finding larger moves with smaller energy change lies in finding correlated moves of selected atoms that end up avoiding clashes while making significant change in the conformation. Changing torsion angles on a side chain or performing a local backbone moves[17] are examples of such correlated moves using a small set of atoms.

The opposite to moving a single atom or a small subset is the moving all atoms in each step. This was shown to be feasible[18] if the change in the conformation is along one of the eigenvectors of the systems Hessian. Unfortunately, the complexity of the calculation of the eigenvectors is $O(n^3)$, making this technique difficult to extend to large systems. Furthermore, its efficiency is lost when applied to systems in explicit waters.[19] It remains to be seen if using judiciously chosen Hessian blocks of limited size that could reduce the complexity to $O(n^2)$, (as suggested for the simulation of lipid bilayers)[6] can be found to be implemented efficiently.

To illustrate the frustration that can result in attempts of 'clever' Monte Carlo moves, Figure 8.1 shows the average displacement and orientational correlation of lipid molecules in two simulations. In one, simulations were performed using a move set that includes whole molecule translations, rotations and torsion angle changes on a bilayer of DMPC molecules[20] using the program MMC.[21] In the other, selected rotations of one lipid around the bilayer normal were accompanied by a similar rotation of the lipid nearest to the first lipid, but in the opposite direction. It was expected that such correlated rotations would act like two cogwheels and reduce the clashes between lipids, resulting in higher acceptance rates and thereby improved sampling. However, no significant change was observed either in the diffusion rate or in the decay of orientational correlations, indicating that different approaches are needed for accelerated the sampling.

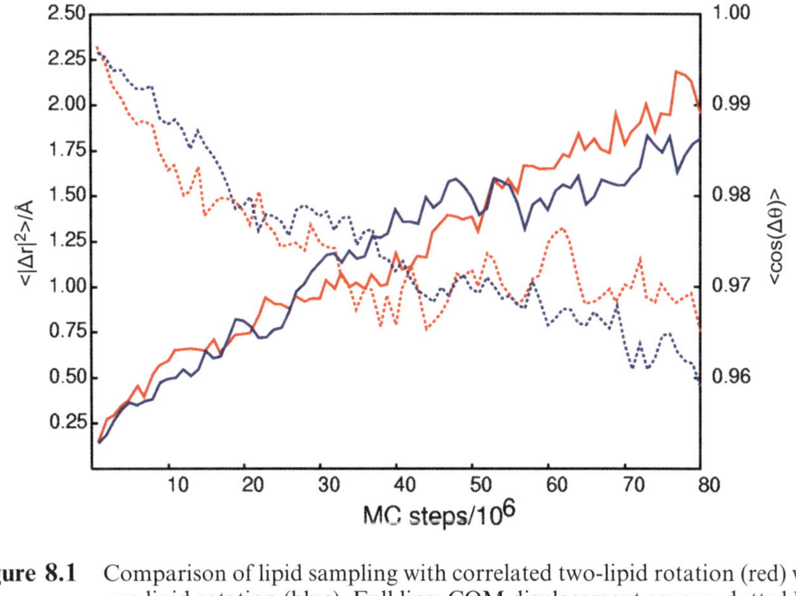

Figure 8.1 Comparison of lipid sampling with correlated two-lipid rotation (red) with one lipid rotation (blue). Full line: COM displacement square; dotted line: decay of the orientational correlation around the membrane normal.

8.5 Cooperative Potentials

Most of the successful Monte Carlo techniques change a small fraction of the atoms at each single step. This means that the efficiency of the method degrades when the energy update involves atom pairs that were not changed – which is just the case for most Monte Carlo moves.

Cooperativity is generally introduced either by the use of three-body, four-body, *etc.*, terms[22] or by the introduction of polarization – either by induced dipoles (or, possibly, higher order multipoles)[23,24] or by induced changes in the partial charges or molecular geometry,[25] essentially mimicking induced multipoles. Since the multibody terms are generally short ranged, their use with Monte Carlo sampling will not change the computational complexity of the calculations. However, cooperative potentials using some form of polarizability will essentially raise the complexity of the energy calculation when the attempted move involves a small number of atoms only.

While Monte Carlo calculations have been performed with polarizable potentials on relatively small systems despite the significant additional computational cost, this is clearly an unsatisfactory solution since for larger systems the additional expense will become prohibitive. Possible solutions to this problem is either (a) limit the update of the polarization state to every n-th step assuming that the change in the polarization state in a single step is (can be considered) negligible or (b) run the simulation with the pairwise additive part of the potential and correct the Boltzmann probability of the conformations selected for averaging with the cooperative part during post-processing. However, the first solution clearly introduces an error while the second solution can exacerbate round-off errors if the cooperative terms fluctuate more than a few kT (due to the exponential in the Boltzmann factor). Other approximations to the calculation of the polarization contribution have been developed but tested for liquid water only.[26] So far no efficient and procedure with sufficient numerical precision has been developed for the use of cooperative potentials with local Monte Carlo moves with proteins or nucleic acids where the presence of fully charged atoms would make approximate solutions significantly less accurate than for water. Thus it may well be that use of cooperative potentials will be restricted to Monte Carlo moves where the whole (or, at least, a large part of the) system is changed at each step.

8.6 Long-range Energy Contributions

The standard technique for dealing with the long-range contributions to the electrostatic energy is the use of Ewald sums. This technique, however, raises the same problem for Monte Carlo methods where a small part of the system is changed in each step as the calculation of the energy with cooperative potentials, due to the need for the summation in the reciprocal space.

The root of the difficulty of treating long-range electrostatics is the slow *and* conditional convergence of the dipolar lattice sum in three dimensions. Without using an explicit treatment of the full extent of the long-range contributions, for acceptable accuracy, rather long cutoffs distances are needed that, incurs a large additional computational burden. There are, however, two possible options that may be amenable to efficient Monte Carlo implementation. First, R. Sperb[27] published a formalism that uses fast-converging series, without resorting to the reciprocal space. This can thus be implemented efficiently since these series would involve only the changed part of the system. Second, E. Campbell[28,29] has developed a formalism for calculating the Ewald sums (for multipoles of any order, not just for dipoles) from two components: one, called crystal constants, that are functions only of the periodic system (and include the contributions from the direct and reciprocal space) and an other component whose terms depend on the actual configuration. Change in a conformation thus affects the part not involved in lattice sums and thus could also be amenable to efficient Monte Carlo implementation.

8.7 Parallelization

Any algorithm that is designed for simulating large molecular assemblies has to be amenable to massive parallelization. Monte Carlo methods are usually considered amenable to 'embarrassingly parallel' treatment where the same calculation is repeated on different processors with different random-number seeds and the result is averaged at the end. The problem is more difficult if fine-grained parallelization is required. There are two obstacles to efficient implementation of massive fine-grained parallelization.

The tasks involved in a Monte Carlo move include some additional calculations beyond that of the change in the energy of interaction between the atoms changed in the trial move and the rest – the part that is amenable in general to massive parallelization. Other tasks include the generation of the trial change in the configuration and (in most cases) the change in the interaction energy between the moved atoms. While these usually take only a small fraction of the computational effort, if massive fine grained parallelization is planned then this fraction will increase proportionally to the number of processors, and ultimately limiting the parallel efficiency. Since these smaller tasks are rarely amenable to massive parallelization the only possibility is to perform them simultaneously on different parts of the system; but that is only possible if the system is large enough and the moves are local enough that they can be generated in such a way that all changes are independent of each other. For example, torsion angle changes on different side chains can be generated on different protein side chains and the energy change involved can be calculated in parallel and then tested for acceptance sequentially using the massively parallelized energy calculation. Such a decomposition has been discussed by Heffelfinger and Lewwitt.[30] The idea of 'pre-fetching' (*i.e.* calculate multiple likelihoods ahead of time and only use the ones that are needed) were introduced and tested by Brockwell.[33]

The biggest obstacle to massive parallelization on distributed memory systems is the need of communication at the Metropolis decision stage, since the acceptance decision depends on the energy of the *whole* system. While shared memory systems can initiate communication with little or no latency, this is not the case for distributed memory system – the ones that are likely to be used for massively parallel applications, given their much lower price. Similar problems would arise for GPU implementations, since communication with current GPUs have relatively large latencies.[31]

The effect of latency on the parallel efficiency of the fine-grained parallelization using MPI was tested on two systems of TIP3P[32] water molecules under periodic boundary conditions using the minimum-image convention. Calculations were performed on two systems containing 3000 and 30 000 molecules, respectively, on computers employing either distributed memory (marked as DM) or shared memory architecture (marked as SM). The distributed memory runs were on our G5 cluster using gigabit Ethernet connections and the shared memory runs were on an 8-CPU SGI Origin system using R12K processors and the simulations were run with the program MMC.[21] Figure 8.2 shows the results for up to 8 processors, plotting the efficiency factor $E = T(N_{cpu})*N_{cpu}/T(N_{cpu} = 1)$ as a function of the number of processors, N_{cpu} (note the logarithmic scale on the horizontal axis). Here $T(N_{cpu})$ and $T(N_{cpu} = 1)$ are the run times for simulations using N_{cpu} and 1 processors, respectively. Ideal parallel efficiency would result in $E = 1$.

The results on the parallelization of the Monte Carlo steps also show nearly ideal efficiency on the shared memory system indicating that the parallelization is distributing the workload efficiently among the processors. However, due to

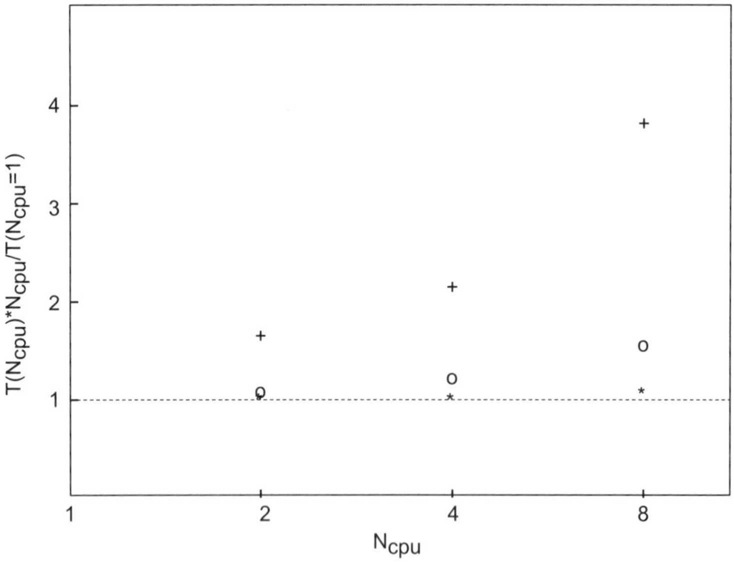

Figure 8.2 Parallel efficiencies of water simulations. +: 3000 waters, distributed memory; ○: 30 000 waters, distributed memory; *: 3000 waters, shared memory; broken line: ideal speedup.

the latency on the communication on the distributed memory system the efficiency is disappointingly low. As expected, there is an improvement with the increased system size.

8.8 Conclusion

This chapter discussed the open problems that need to be resolved for the Monte Carlo algorithms to be used for simulating macromolecular systems. One of the problems facing an efficient Monte Carlo implementation is partly the result of the extensive freedom one has in the choice of sampling algorithms since definitive comparison of such choices require large-scale simulations on a variety of systems. Other problems arise when the energy change is needed to be calculated for cooperative potentials, for the use of the well-tested Ewald summation to calculate the long-range electrostatics contributions and for massive parallelization on distributed memory computer systems. Possible solutions were suggested for the calculation of long-range electrostatics and for some aspects of parallelization. It is hoped that further developments, both in algorithm design and test as well as in hardware/software development will solve most of these problems.

References

1. N. A. Metropolis, A. W. Rosenbluth, M. N. Rosenbluth, A. H. Teller and E. Teller, *J. Chem. Phys.*, 1953, **21**, 1087.
2. A. Rahman and F.H. Stillinger, *J. Chem. Phys.*, 1971, **55**, 3336.
3. P. Diaconis, *Bull. Amer. Math. Soc.*, 2009, **46**, 179.
4. W. W. Wood, in *Physics of Simple Liquids*, ed. H. N. V. Temperley, J. S. Rowlinson and G. S. Rushbrooke, 1968, North Holland, Amsterdam, pp. 115–229.
5. M. O. Allen and D. J. Tildesley, *Computer Simulation of Liquids*, 1987, Oxford University Press.
6. M. Mezei, in *Third International Workshop for Methods for Macromolecular Modeling Conference Proceedings*, ed. H. H. Gan and T. Schlick, 2002, Springer, New York, pp. 177–198.
7. J. P. Ulmschneider and W. L. Jorgensen, *J. Am. Chem. Soc.*, 2004, **126**, 1849.
8. M. Cui, M. Mezei and R. Osman, *Protein Eng. Des. Sel.*, 2008, **21**, 729.
9. G. M. Torrie and J. P. Valleau, *J. Comp. Phys.*, 1977, **23**, 187.
10. D. D. Franz, D. L. Freeman and J. D. Doll, *J. Chem. Phys.*, 1979, **93**, 2769.
11. W. K. Hastings, *Biometrika*, 1970, **57**, 97.
12. M. Rao, C. S. Pangali and B. Berne, *Molec. Phys.*, 1979, **71**, 3366.
13. P. Jedlovszky and M. Mezei, *Mol. Phys.*, 1999, **96**, 293–296.
14. M. Mezei, *Molec. Phys.*, 1987, **61**, 565; Erratum, 1989, **67**, 1207.
15. W. L. Jorgensen and J. Tirado-Rives, *J. Phys. Chem.*, 1996, **100**, 14508.
16. M. Mezei, *J. Chem. Phys.*, 2000, **112**, 1059.

17. L. R. Dodd, T. D. Boone and D. N. Theodoru, *Molec. Phys.*, 1993, **78**, 961.
18. T. Noguti and N. Go, *Biopolymers*, 1985, **24**, 527.
19. N. Go, *Personal Communication*, 1999.
20. P. Jedlovszky and M. Mezei, *J. Chem. Phys.*, 1999, **111**, 10770.
21. M. Mezei, URL: http://inka.mssm.edu/~mezei/mmc
22. J. Detrich, G. Corongiu and E. Clementi, *Int. J. Quant. Chem: Quant. Chem Symp.*, 1984, **18**, 701.
23. F. H. Stillinger and C.W. David, *J. Chem. Phys.*, 1978, **69**, 1473.
24. E. S. Campbell and M. Mezei, *J. Chem. Phys.*, 1977, **67**, 2338.
25. H. J. C. Berendsen and G. A. van der Velde, *CECAM Workshop Report*, 1972, 63.
26. M. W. Mahoney and W. L. Jorgensen, *J. Chem. Phys.*, 2001, **114**, 9337.
27. R. Sperb, *Molecular Simulation*, 1999, **22**, 199.
28. E. S. Campbell, *J. Phys. Chem. Solids*, 1965, **26**, 1395.
29. E. S. Campbell and M. Mezei, *J. Comp. Phys.*, 1976, **21**, 114.
30. G. S. Heffelfinger and M. E. Lewitt, *J. Comp. Chem.*, 1996, **2**, 250.
31. J. D. Owens, M. Houston, D. Luebke, S. Green, J. E. Stone, and J. C. Phillips, *Proceedings of the IEEE*, 2008, **96**, 879.
32. W. L. Jorgensen, J. Chandrasekhar, J. D. Madura, R. W. Impey and M. L. Klein, *J. Chem. Phys.*, 1983, **79**, 926.
33. A. E. Brockwell, *J. Comp. Graph. Stat.*, 2006, **15**, 246.
34. M. Richey, *The Am. Math Monthly*, 2010, **117**, 383.

Coarse Graining and Multiscale Models

CHAPTER 9
Coarse-grain Protein Models

N. CERES AND R. LAVERY*

Bases Moléculaires et Structurales des Systèmes Infectieux, Université Lyon1/ CNRS, UMR 5086, IBCP, 7 Passage du Vercors, 69367, Lyon, France
*Email: richard.lavery@ibcp.fr

9.1 Introduction

Coarse-grain models of macromolecules appeared at the very beginning of the molecular modeling era. By replacing groups of atoms with single pseudoatoms, and often by simplifying interaction potentials, they fulfill multiple goals: accelerating energy and force calculations; reducing the dimensionality of conformational space; smoothing the corresponding energy hypersurface and reducing the number of local minima. Despite the growing power of computers, coarse-grain models are still attractive, as demonstrated by the dramatic increase in publications in this area over the last 20 years.[1] Coarse-grain models continue to offer the possibility of treating larger, or more numerous, molecular systems and of tackling slower, and more complex, conformational changes. In the protein field, many new coarse-grain models are being developed and applied to an increasingly wide range of problems ranging from the properties of individual proteins (mechanics, folding pathways, structure prediction, *etc.*) to aggregation and assembly processes, and even to models of the intra-cellular environment. In this chapter, we will give an overview of how these models have developed over the last few decades in terms of protein representation, conformational or interaction energy formulation and areas of application. Our selection of references is necessarily partial, and we have also chosen to concentrate on the modeling of soluble proteins and their assemblies, excluding a significant number of studies of membrane

RSC Biomolecular Sciences No. 23
Innovations in Biomolecular Modeling and Simulations: Volume 1
Edited by Tamar Schlick

Published by the Royal Society of Chemistry, www.rsc.org

proteins (see recent reviews)[2,3] and an increasing interest in protein-nucleic acid interactions and complexes such as the nucleosome[4–6] or the ribosome.[7]

9.2 Coarse-grain Representations

9.2.1 All-atom and United-atom Models

Although protein structures represented at the atomic level are not coarse-grained models, all-atom representations can be used with energy functions that are characteristic of coarse-grain models, the most well-known example being Tirion's all-atom elastic network model,[8] or Clementi's all-atom Gō model. In the latter case, heavy atoms are clustered to generate a reduced alphabet composed of binary polar/hydrophobic atom types.[9] Rigid envelopes derived from all-atom models are also commonly used in protein docking and multi-protein assembly procedures as discussed later.

The first step away from all-atom representations involves the creation of united-atom models of the type available in standard molecular dynamics (MD) programs for many years. This modest type of coarse-graining typically eliminates non-polar hydrogens by replacing groups such as CH, CH_2 and CH_3 with single "pseudoatoms", but can also be used to model large "pseudoions", which integrate their first hydration shell. Pushing the idea of united-atom models further leads to the more general set of so-called bead models.

9.2.2 Bead Models

These models aim to smooth out the details of an all-atom representation by replacing groups of atoms with single pseudoatoms or "beads". Protein bead models can be roughly grouped into three categories, following the N/P ratio, where N is the number of residues in a protein and P is the number of pseudoatoms.

$N/P = 1$ (*i.e.* one bead per residue) models generally place the beads at the Cα position (Figure 9.1B), although Cβ can also be used,[10] and some models place the bead at off-atom positions, such as side chain centroids.[11,12] Note that $N/P = 1$ models imply that the peptide backbone will be composed of effective valence angles and torsions that have more complex angular distributions, and stronger coupling between these variables, than the underlying valence angles and torsions of the all-atom peptide backbone. This difficulty needs to be overcome when treating backbone flexibility. Clearly, one-bead models also make it difficult to fully integrate the range of properties of each amino acid side chain, adding to the complexity linked to the simplified backbone representation.[13]

$N/P < 1$ (*i.e.* several beads per residue). These models generally use a combination of backbone and side chain beads. For example, Cα can be combined with a single side chain bead,[14–16] one or two beads, following the size of the side chain,[14,17] or up to four side chain beads, allowing more conformational freedom.[18] Scheraga's UNRES (UNited-RESidue) representation introduces side chain structure in another way by using an ellipsoidal side chain bead

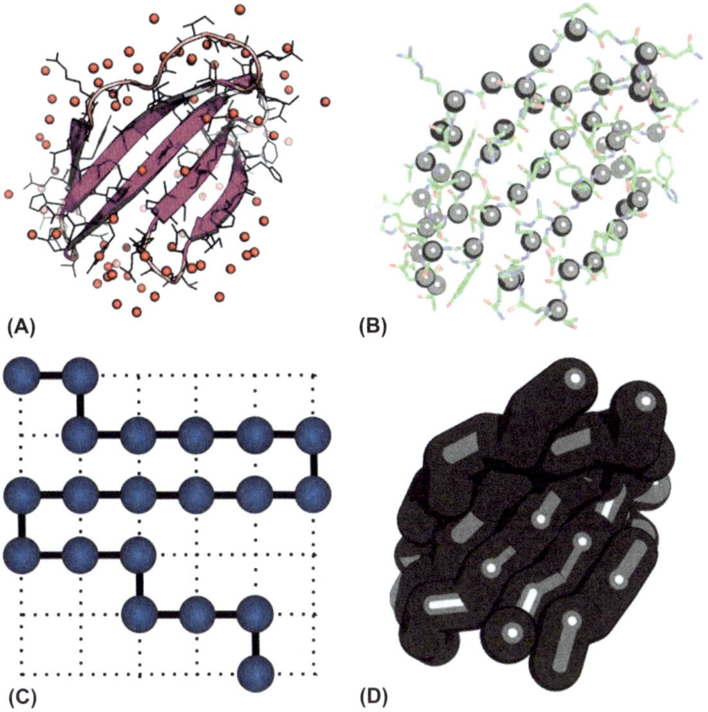

Figure 9.1 Globular proteins at different levels of resolution. (A) Three-dimensional all-atom model (line drawing, without hydrogens). Secondary structures are shown in cartoon representation and red spheres show experimentally resolved water molecules. (B) Bead representation with N/P = 1 ratio: each residue is represented by a single bead (grey sphere) at the Cα position. The corresponding all-atom structure is shown by the translucent sticks. (C) Square lattice model with one bead per residue forming a 2D self-avoiding chain. (D) Polymer model. A fixed-radius tube representing a protein with an all-β conformation.

having its principal axis aligned with a vector from Cα through the side chain center of mass.[19] Side chain beads can also be combined with various backbone representations, for example, with a backbone centroid bead plus an attached mobile bead to account for the dipolar interactions.[20] Attempts to account for peptide dipole interactions were indeed made in the earliest bead models such as that of Levitt by incorporating "N" and "O" beads at the center of the peptide groups.[14] Other models have followed similar routes, with beads at C', N and Cα sites,[21] or at O, H(N) and Cα.[22]

N/P > 1 (*i.e.* several residues per bead). These low-resolution models are used for representing very large systems (in extreme cases, an entire protein may be represented by a single bead).[23] In such models, it is particularly important to decide how to place the beads. This can be done intuitively, as, for example, in an F-actin filament model where each G-actin monomer is represented by four beads, each corresponding to a protein subdomain (constituted of amino acids that may come from many segments of the primary sequence).[24] A more formal approach uses a topological rigidity analysis of the constituent proteins in order to guide bead positioning.[25] Alternative methods define domains on the basis of the collective movements revealed by an essential dynamics analysis of all-atom simulations,[26] or by the normal mode analysis of an elastic network model.[27] It is also possible to use an atom-clustering approach to maintain not only the shape, but also the moment of inertia of the macromolecular system,[28] or Markov chain propagation theory, that allows a succession of increasingly coarse-grained representations to be constructed.[29]

9.2.3 Lattice Models

Simplifying still further, we arrive at lattice models, equivalent to $N/P = 1$ representations, but with beads placed on the nodes of a pre-defined 2D[30,31] (Figure 9.1C) or 3D lattice,[32,33] rather than freely positioned in 3D space. The beads are linked into a linear polymer and only one bead can occupy a given node (thus forming a self-avoiding walk). The main advantage of these models is that a complete conformational search becomes possible in the discretized space. Since simple cubic lattices are poorly adapted to representing secondary structures, other geometries, including body-centered, face-centered and diamond lattices have been exploited,[34–36] as has the idea of using subsets of cubic lattice vertices, as in the "chess-knight" model.[37] Lattice models have also been extended in an attempt to improve the representation of specific amino acids by combining on-lattice Cα beads with off-lattice side chain beads.[38,39]

9.2.4 Polymer and Material Models

Finally, we arrive at models derived from polymer or material physics that no longer attempt to distinguish individual residues, or even residue groups. An example of this is the flexible tube model,[40] whose folded forms, taking into account excluded volume, can mimic polypeptide secondary structures (Figure 9.1D). Methods related to tensegrity analysis, which replace the molecular structure with sets of rigid links and then analyze residual degrees of freedom,[41] or the origins of structural stability,[42] can also be placed in this category.

9.3 Formulating the Energy

9.3.1 Gō Potentials

Early lattice modeling of proteins by the group of Gō put forward the idea of using interaction potentials that directly reflect the native structure of a protein: attractive interactions between residues in contact in the native structure and repulsive interactions between other residues.[32,43] This choice ensures that the native structure will be energetically favored over partially folded or misfolded states, but at the cost of eliminating the frustration that is a feature of the conformational energy landscape of proteins.

Residue contacts are generally defined using a Cα-Cα cutoff distance,[44,45] typically 7–8 Å, with optional constraints on the proximity, or the interactions, of the associated amino acid side chains.[46,47] Alternatively, "chemically favorable" interactions can be used to define native contacts.[48]

Most simply, Gō potentials can be attractive or repulsive square wells,[49] but distance dependence can be included using potentials of a Lennard-Jones form,[44] optionally incorporating an energy barrier to mimic a desolvation penalty.[47,50] Repulsive potentials commonly impose a single minimal approach distance (≈4–7 Å),[48,51] but can also incorporate amino acid dependent radii.[52] In basic Gō models all attractive interactions are set equal,[45,46,48,51,53] but it is again possible to account for different types of interaction, for example, by dividing residues into polar and non-polar families,[9] or by using amino acid dependent parameters.[47] Off-lattice Gō models must also naturally include bonded interactions, typically with equilibrium values derived from the native conformation and including backbone dihedral angle terms that favor the native secondary structures[48,51] or maintain chirality.[52,54]

A number of attempts have been made to refine Gō models,[55] for example, by introducing additional attractive potentials between residues not in contact in the native conformation, enabling some elements of frustration to be introduced;[52] by introducing hydrogen bonds to improve cooperative folding behavior;[49,52,56] or by adding sequence-dependent hydrophobic potentials.[57] Some studies have even eliminated the notion of predefined non-native contacts, choosing rather to optimize the nature (attractive or repulsive) and the strength of interactions in the native conformation in order to ensure an energy gap with other unfolded forms.[58]

9.3.2 Elastic Network Models

In common with basic Gō potentials, elastic network models automatically define the native structure as the global energy minimum and, in addition, assume that the energy surface around this minimum is quadratic (Figure 9.2A). The first work in this area comes from Levy *et al.*, who used quadratic potentials for the bond lengths, bond angles and dihedrals of a Cα virtual bond model with force constants derived from an all-atom MD simulation.[59] Tirion carried this simplification further by representing the molecule as an ensemble of Hookean springs between all atoms pairs closer than a chosen cutoff distance. Equilibrium distances were derived from the native conformation,[8] avoiding the need for energy minimization that could lead to conformational drift.[60] The Tirion approach, the so-called "elastic network model" (ENM), represents the protein as an irregularly-shaped elastic solid with variable density, rather than directly considering any mechanics due to a particular backbone or side chain geometry. This model has been adapted to a variety of coarse-grain representations[61,62] and is widely used for studying protein conformational fluctuations. It should be noted that while standard elastic network

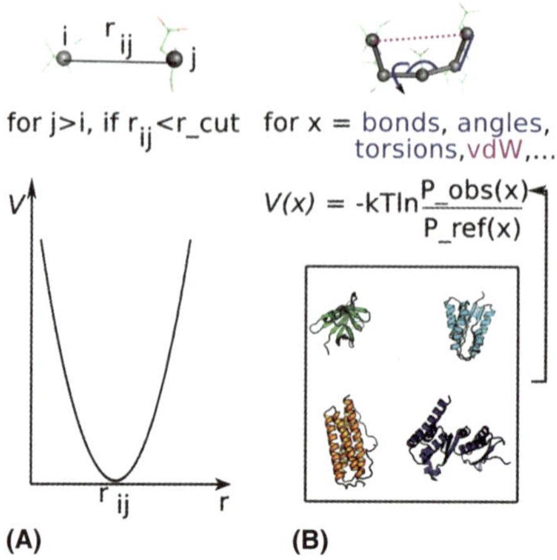

Figure 9.2 Examples of energy functions. (A) In typical elastic network models, any two beads (gray spheres), here at the Cα positions of two amino acids i and j in the native structure, interact with a harmonic function if their separation r_{ij} is below a given cutoff (r_cut). It follows that the native coarse-grain structure occupies the global energy minimum of a quadratic energy surface. (B) Statistical potentials derive from the analysis of a database of known three-dimensional structures. The dataset is used to compute the observed probability (P_obs) of bonded and/or non-bonded interactions for the coarse-grain model of interest, here a Cα-based one. These probabilities are converted in (free) energy by Boltzmann inversion, once the reference state chosen.

models calculate vibrational modes using Cartesian coordinates, early work based on harmonic fluctuations around internal coordinates[59] showed that larger movements could be modeled before the harmonic approximation of the energy hypersurface breaks down.[63]

Only a single spring force constant and a single cutoff distance (in addition to the native protein conformation) are necessary for constructing the network. The force constant can be chosen to match all-atom normal mode frequencies or molecular dynamics fluctuations,[8,64] or to fit crystallographic mean-square atomic fluctuations (Debye-Waller, or so-called B-factors).[61,62,65] It is also possible to use other experimental data, such as hydrogen exchange rates.[66] Alternatively multiple force constants can be used to reflect residue interaction classes,[67] spring length intervals,[68] or inter- and intra-domain interactions.[69,70] Cutoff distances typically vary between 6 Å and 10 Å, with longer distances being used for more coarse-grain representations. All-atom distances, rather than pseudoatom distances, can also be used in determining cutoffs.[67] Alternatively, a cutoff can be avoided altogether by making the spring constants a function of the inter-node distances.[71,72]

Although basic elastic network models are only valid for fluctuations around the native conformation, some attempts have been made to extend them to larger conformational changes, including transitions between conformational states using harmonic double-well potentials,[73] and even to complete unfolding, via the successive elimination of the springs that show the largest fluctuations.[74]

9.3.3 Statistical Potentials

Statistical potentials (SP), also termed knowledge-based potentials, or database-derived mean force potentials, are based on a statistical analysis of a set of protein structures (Figure 9.2B).[75,76] It is most commonly assumed that the observed probability p of finding a given interaction of type i within these structures is related to an independent Boltzmann ensemble, allowing the free energy E_i of a given interaction to be written as

$$p_i = \frac{1}{Z} exp^{-\frac{E_i}{kT}} \quad (9.1)$$

where k is the Boltzmann constant, T the absolute temperature and Z the appropriate partition function. Taking the logarithm of this expression, introducing a reference state P_R, and ignoring the constant term involving the partition function, we can write the potential of mean force for the interaction as:

$$\varepsilon_i = -kTln\left(\frac{p_i}{P_R}\right) \quad (9.2)$$

Despite its obvious applications, this approach can be criticized on a number of grounds.[77,78] First, it is not clear that each interaction constitutes a thermodynamic ensemble at equilibrium (and at a given temperature), although this may be true for certain cases (such as proline cis/trans distributions, or hydrophobic/hydrophilic residue distributions with respect to the protein surface).[77] Multiple errors can arise from excluded volume and hydrophobic effects that make statistical potentials dependent on the size and composition of the structural database analyzed. These effects were demonstrated using lattice protein models, but other authors have questioned whether these results also apply to databases of real protein structures.[12,79] Another problem arises because it is not clear that inter-residue potentials deduced in this way are pairwise additive. Indeed, neighboring residues will generally contribute to the observed distance distributions between any given residue pair.[77] One way to overcome this problem is to limit the analysis to distances where neighboring residues do not intervene along the residue interaction vector.[80] Alternatively, cooperative non-bonded interactions involving 3- or 4-body terms could be considered, as suggested by Betancourt and Thirumalai (1999).[81] Finally, other errors can come from chain connectivity effects, which can lead to a bias in the distance distributions between residues lying close in the primary sequence of a protein.[82]

The simplest forms of statistical potentials are contact potentials, which only generate energy contributions for residues directly in contact with one another.[11,83] Later approaches extended this idea to include distance-dependent inter-residue interactions[84] and also took into account topology-dependent criteria and dealt with errors due to sparse data sets. The notion of anisotropic interactions has also been added to both contact potentials[85] and distance-dependent potentials,[86,87] including a detailed treatment of hydrogen bonds.[88] SPs may also be obtained for internal coordinates, either using conventional bond, valence angles and torsions, or effective torsion angles along the Cα backbone.[13,20] In the latter case, as mentioned above, both valence angles and torsions can exhibit multimodal distributions that reflect the secondary structure preferences of a polypeptide chain and need to be fitted with appropriate mathematical functions.

Whatever their form, statistical potentials require the definition of a reference state. The three principle approaches are the so-called random mixing, ideal gas and random coil methods. In the random mixing reference state, it is assumed that every amino acid has an equal chance of occurring at any position within a folded protein structure. Problems associated with chain connectivity are reduced by averaging over a database,[11] leading to a collection of effectively disconnected residues. In this case, it is only necessary to correct for the probability of occurrence of each amino acid type by mole fraction,[83] or by pair contact fraction.[11] In the ideal gas reference state,[84] the radial distribution function for a residue pair is averaged over all possible residue combinations at each distance. This basic approach can be corrected for the finite size of proteins[89] and for the composition dependence as a function of size.[90] Lastly, the random coil reference state[91] is based on comparisons with a Gaussian random polymer. It can thus take into account chain connectivity explicitly. It can also be extended to take into account for excluded volume effects and limited sequence-dependency.[92] It should be added that rescaling Miyazawa-Jernigan potentials by replacing an implicit solvent-based reference state with a "mean" residue, optimally threonine, leads not only to better scaled residue interactions, but also to better correlations with potentials derived from the random-coil reference state.[81]

Unfortunately the ideal reference state is not experimentally accessible. However, in an insightful study, Godzik et al. have noted that this may not be a major problem.[93] By comparing early CG approaches that mainly used contact, rather than distance-dependent pair potentials, these authors showed that apparently very different choices of reference state can in fact result in surprisingly similar parameter sets. Different reference states lead to amino acid transfer energies (a mainly 1-body term) being included to different extents in the 2-body interactions between pseudoatoms. These differences become visible when comparing the interaction potentials for amino acid types i and j with the mean interaction $(ii+jj)/2$, which, in an analogy with liquids, is equivalent to deviations from ideal mixing. This comparison further showed that potentials obtained by Boltzmann inversion, using existing protein structures, may give very similar results to apparently unrelated approaches that aim at differentiating the energy of native and unfolded protein conformations.

A promising way to overcome both correlation effects and the need to define the reference state in statistical potentials derived by Boltzmann inversion is offered by iterative techniques.[94,95] The first of these was based on a succession of Monte Carlo simulations in order to iteratively derive a set of pair potentials that could reproduce experimental radial distribution functions for ions in solution.[96] A number of variants of this idea have been developed and used to solve a range of protein-related problems.[97–102] Applied to the extraction of pairwise potentials from observed pairwise probability distributions, the following equation is used,

$$e_i^n = e_i^{n-1} - kTln\left(\frac{p_i^{obs}}{p_i^{n-1}}\right) \tag{9.3}$$

Successive iterations n enable the potentials e_i to be adjusted until the simulated probability distributions (from the preceding iteration) p_i^{n-1} match the experimental distributions p_i^{obs}. As highlighted in the work of Thomas and Dill,[97] this procedure can implicitly integrate chain connectivity, sequence dependency, and conformational heterogeneity in the reference state without the need to determine it analytically. More generally, iteration avoids defining an *a priori* reference state for terms such as the hydrogen-bond interaction in folded proteins,[103] or desolvation in protein-ligand interactions.[101] Although these approaches cannot overcome basic assumptions, such as representing conformational energies using pairwise additive terms, they have two important advantages: (1) they can be applied to any subset of the statistical potential parameters and (2) they can be used to effectively combine the statistical potentials with physics-based terms (for example, Coulomb electrostatics) by iteratively removing the corresponding part of the conformational energy from the statistical potentials. Note that good results have also been obtained with the Monte Carlo iterative procedure applied to lattice models, where, in contrast to using a quasi-chemical reference state, the procedure was able to extract pre-defined potentials from strongly interacting folded states.[104]

Finally, we remark that it is possible to avoid Boltzmann inversion completely by using learning algorithms to optimize potentials that distinguish native protein structures from non-native or misfolded ones. A possible goal for such approaches is to maximize the energy gap between these two sets of structures using inter-residue contact potentials,[105,106] or distance-dependent potentials.[107] Optimization can also be carried out using a perceptron,[108] for example, targeting the destabilization of artificially generated local conformational energy minima with respect to the native conformation,[109] or with methods based on spin-glass theory.[110]

Optimization procedures naturally lead to questions concerning the convergence, transferability and quality of the resulting potentials. Some answers have been provided, notably, one study has shown that convergence to a unique solution is possible both for on-lattice and off-lattice models using relatively few structures in the training set, although the composition of this set will influence the transferability of the resulting potential.[97,105] The impact of the number of parameters on convergence and performance has also been analyzed.[97,106]

9.3.4 Physics-based Potentials

Physics-based (PB) potentials are a "bottom-up" attempt to subdivide the potential energy of a system onto a number of terms representing bonded and non-bonded interactions, each having a clear physical basis that could, in principle, be obtained as the components of an *ab initio* quantum chemical study of the system (although, in practice, experimental data is always necessary to obtain usable parameters). This approach is the basis of the conventional force fields used in all-atom simulations.[111–113]

At the coarse-grain level, PB potentials between beads can be derived by integrating out neglected degrees of freedom to generate potentials of mean force (PMF). A pioneering model from Levitt and Warshel[14,114] used Boltzmann-weighted averaging of the non-bonded energies from an all-atom dipeptide in vacuum to generate effective pseudo-torsion angle potentials. A similar approach was used to generate interaction potentials between side chain beads. The united-residue model (UNRES) from the Scheraga group[19,115,116] similarly employs effective PMFs obtained by averaging over the degrees of freedom eliminated from an all-atom system. This group took particular care to clearly separate different contributions to the potential and introduced multi-body potential terms, including correlations, almost all of which could be obtained from *ab initio* quantum mechanical PMF calculations for appropriate all-atom model systems.[117] Distance and orientation dependent side chain interaction potentials for hydrophobic[118] and charged[119,120] residues have recently been introduced using PMFs from umbrella sampling simulations of all-atom models in water.

Other groups have also adopted a physics-based approach to study protein-protein interactions with CG van der Waals interactions extracted as multi-body PMFs from *in vacuo* molecular dynamic simulations by numerical integration of the mean force between uncharged amino acid pairs, and electrostatic interactions fitted to the vacuum electrostatic potentials generated by AMBER atomic partial charges (leaving the option of adding any desired solvent model).[121]

Although being able to fit physics-based potentials term by term is nominally a significant advantage, it is also possible to take a more holistic approach and derive complete interaction potentials in a single step from all-atom simulations. This is the method adopted in the so-called force matching approach.[122] Here, snapshots from a simulation are divided into blocks and the forces acting on the atoms constituting a bead in the coarse-grain model are summed and used to solve a system of linear equations, separating the forces into long- and short-range components. These solutions are then averaged over all simulation blocks. No assumption is made concerning the functional form of these interactions. Obviously, these potentials apply to the specific system simulated and nominally to the range of conformations sampled during the simulation. However, recent extensions of the method target transferable potentials that could be used for arbitrary amino acid sequences.[18]

Another approach to obtaining good overall behavior for protein folding simulations is to directly target the free energy that not only favors folded conformations, but also reproduces the correct ordering of conformations ranging from fully folded to fully unfolded states. Liwo *et al.* have used a procedure they term "hierarchical optimization"[123] to achieve this goal with the help of multiplexing replica exchange MD simulations to extensively sample the conformational space of small proteins. Simulations are carried out for proteins with different folds at their folding temperature, and also above and below this value, in order to adjust the parameters and obtain correct folded-unfolded energy gaps at each temperature. Explicit temperature dependence is introduced into the free energy function via prefactors multiplying the various terms.[124]

9.3.5 Hybrid Potentials

Today, the boundaries between statistical/knowledge-based and physics-based potentials are fuzzy. Statistical potentials often involve terms with functional forms that are derived from physics-based force fields, while physics-based potentials can include at least some terms parameterized on the basis of existing structures. The popular MARTINI model[125] is an example of this, since the non-bonded interactions are physics-based, being calibrated against the oil/water partition coefficients of amino acid side chain analogues using all-atom MD simulations of the component functional groups. In contrast, bonded potentials are derived from a statistical analysis of protein structures (although they were subsequently optimized with respect to data obtained from simulations) and notably include dihedral terms that specifically favor the native secondary structure of the protein being modeled. Other approaches have made similar choices. In the work of Ha-Duong,[20] for example, molecular dynamics derived non-bonded interactions are coupled with bonded terms obtained by Boltzmann inversion from a protein structure database. Similarly, Betancourt[126] has developed a CG non-bonded potential, incorporating both distance and orientation dependency, using equilibrium ensembles of residue pairs from all-atom MD simulations in water, coupled with knowledge-based bonded interactions.

9.3.6 Solvent

Water is an important component of all living systems and it is crucial to take into account its effects when modeling biomolecules and their interactions. The computational cost of an explicit, all-atom treatment of the solvent can be overcome in various ways. A first level of approximation is again based on bead models, which still allow an explicit description of the solvent, but at a reduced computational cost. Each bead can represent one or several water molecules (presently up to five). In one-to-one mapping, the bead can be located at the water oxygen,[127] the geometrical center of the water molecule[122] or at its center of mass.[128] Different approaches, including iterative Boltzmann inversion, force-matching and reverse Monte Carlo, have been used to derive an effective pairwise potential that implicitly incorporates electrostatic interactions. Explicit electrostatics, and even solvent polarization, can be treated in lower resolution bead models, where each bead replaces more than one water molecule. Polarization can be introduced via a bead dipole interacting with the solute electrostatic field, but not with one another,[129] or by allowing oppositely charged interacting centers inside each bead to move.[130]

In order to further reduce the computational effort, and to remove the risks of incomplete conformational sampling of the solvent, it is possible to replace discrete particles with solvation free energy functions and/or continuum solvent models. Solvation free energy functions describe the transfer free energy of the solute from vacuum to the appropriate solvent environment. These functions can be approximated in terms of solute exposed surface area, intra-solute contacts, the volume of the first hydration shell accessible, or by using group contributions to the solvation free energy, taking into account solvent exclusion.[131]

Continuum solvent models deal with the impact of the solvent (and optionally salt) on the electrostatic interactions within the solute by solving the Poisson-Boltzmann equation, or simplified versions such as the generalized Born approach. Hydrophobic effects (cavity creation within the solvent) can be added *via* accessible surface area calculations. Lastly, hydrodynamic solvent drag can be introduced *via* stochastic dynamics, using an approximate diffusion tensor and random forces to mimic friction and the collision of water molecules with the solute.

9.4 Applications: From Folding to Function

9.4.1 Protein Folding

Coarse-grain models have significantly contributed to our current view of how protein folding occurs with natural peptide sequences that have been carefully selected by evolution (as opposed to the behavior of random polypeptides). The folding of natural sequences is favored both kinetically and thermodynamically, leading to folding timescales (typically milliseconds to seconds) that are much faster than the random conformational search envisaged by Levinthal.[132] In line with the Anfinsen hypothesis,[133] the native state is generally the Gibbs free energy minimum for the polypeptide in its solvent environment at ambient temperature, but the passage from the unfolded to the folded state (each involving dynamic conformational ensembles), is characterized by a funnel-like free energy landscape with saddle points and local non-native free energy minima corresponding to partially folded structures.[134–136] In contrast, random polypeptides exhibit a much more rugged energy landscape and lack the significant free energy gap that separates the folded state ensemble from all other conformations with natural protein sequences.

Early lattice-based, coarse-grain, polypeptide models played a fundamental role in building this picture of folding. By studying linear polymers constructed from a very limited variety of bead types (of the $N/P = 1$ family) it was possible to completely scan the corresponding conformational space. The first models were limited to binary hydrophobic/polar bead models[31] with interaction energies being either randomly assigned,[33] or based on statistical potentials.[11] More refined approaches considered explicit side chains,[137] or moved to other lattice types which could model secondary structures better.[36] Multi-body contact potentials have also been considered.[138] Conformational sampling is generally carried out with Monte Carlo techniques using appropriately designed local polymer moves.

Lattice Gō models,[32,43] even if not ensuring exhaustive conformational sampling,[139] and also ignoring the problem of frustration, laid the foundations for using reduced models to understand folding.[33] Besides their computational advantages, their success can be attributed to the fact that folding follows the principle of minimal frustration, with cooperative interactions along the polypeptide chain helping to lower energy barriers and guide the system towards the folded state.[140]

Today, off-lattice CG models are continuing to play an important role in analyzing protein folding. Their use in studying such complex, multi-dimensional, conformational problems is justified both by their computational speed and by the fact that it seems possible to successfully describe the average behavior of a polypeptide with a limited number of parameters and without the complications of an all-atom representation.[141] Off-lattice Gō-like models combined with stochastic dynamics,[46,50,53] discrete molecular dynamics [142,143] or Newtonian dynamics,[48,144,145] and eventually with enhanced sampling methods[50] have been widely used to probe both the kinetics and thermodynamics of folding for specific proteins.

For *ab initio* folding studies, where no prior information on the folded state is used, physics-based models have shown promise since the very early work of Levitt.[114] Over the years, many refinements have been incorporated into both the representation of the polypeptide chain, the interaction potentials and the search procedures. A good example is provided by the UNRES model from the Scheraga group that now uses a hierarchical optimization procedure,[123,124] stochastic dynamics, generalized ensemble Newtonian dynamics and principal component analysis to study the energy landscapes and the folding kinetics of small proteins.[117] Derreumaux and collaborators have also developed a coarse-grained folding approach, termed OPEP (Optimized Potential for Efficient peptide-structure Prediction).[146,147] Originally developed for "*de novo*" structure prediction, it has emerged as a good model for exploring folding pathways,[148] free energy surfaces[149] and peptide aggregation processes.[150,151] It can use both MD with replica exchange, or the activation-relaxation technique (ART) that enhances sampling by moving over saddle points from one local minimum to the next. The force-matching approach, developed by the Voth group,[122] has also been used to study free energy landscapes[22] and the folding of small peptides with both conventional dynamics and replica exchange approaches.[18] Lastly, we mention an interesting study of the role of macromolecular crowding on protein folding, using Langevin dynamics and coarse-grain protein models that incorporate information on predicted secondary structure regions *via* inter-residue springs. The results of this work suggest that going beyond known cellular crowding levels can strongly hinder folding.[152]

9.4.2 Structure Prediction

We distinguish protein structure prediction from folding methods since the aim here is to obtain reliable native structures, without attempting to understand the folding process. Although a difficult goal, steady progress is being made, thanks in part to the CASP experiment.[153] Two basic prediction strategies have emerged, depending on whether or not a homologous protein is available to use as a structural template.

Template-based methods attempt to fit (''thread'') the target sequence into the known structure of one or more homologous proteins. The fit is evaluated using a score that can include single-residue properties (amino acid preferences for given structural environments)[154] or inter-residue interactions.[155] The latter approach generally begins with CG protein representations and uses statistical potentials to compare different fits of the target sequence into the template structure. Various potential types using distance-dependence,[84] residue contact,[156] residue accessibility[157] and local secondary-structure preference[158] have been tried. The quality assessment of the threading procedures generally involves comparison with all-atom or CG "decoy" structures.[15,80,87,159,160] It has been found that statistical threading potentials generally outperform standard physics-based force fields. Although it has been shown that recent all-atom force field implementations can distinguish decoys from native structures, they cannot generally be used to refine predicted structures,[161] particularly in the absence of explicit solvent.[162]

When no reasonable structural templates are available, it is possible to use fragment-based methods. These methods attempt to construct a model for the target sequence using many oligopeptide fragments derived from a database analysis of existing protein structures.[163] This nominally "*de novo*" approach is the basis of the successful Rosetta program, where an evolutionary algorithm guides the optimization of the assembled target structure (using a CG model and restraints on Cβ contacts), prior to Monte Carlo refinement using a physics-based potential with carefully weighted terms.[164]

Other approaches based on fragment libraries are FRAGFOLD[165,166] and TASSER.[167,168] The latter approach is distinguished by the use of both lattice and real space CG models. Both FRAGFOLD and TASSER use multi-term statistical potentials and conformational sampling with Monte Carlo, or, in the case of FRAGFOLD, simulated annealing and an optional genetic algorithm.

The importance of a correct combination of local features (defined by the fragment library) and global features (such as the packing of the protein core) in the scoring functions has been discussed and a framework for properly balancing their contributions has been proposed in the context of statistical potentials and iterative optimization procedures.[103]

It is also possible to generate models without using fragments. This is the case with the on/off-lattice CABS model, which again uses statistical potentials and a Monte Carlo search algorithm,[39] or the UNRES algorithm, where models for small proteins are initially generated at low resolution, using conformational searches with a genetic algorithm, before building and refining all-atom models with energy minimization.[169] Replica-exchange MD using CG models and potentials derived from the force-matching procedure have also been tested for their ability to fold small proteins. Despite smoothing the energy landscape and speeding up folding, this approach does not correctly model inter-residue repulsions in non-native conformations and cannot clearly distinguish native-like conformations.[18]

Lastly, we come to the sub-domain of protein design. This can involve attempts to re-engineer existing protein functions, to design protein sequences that are compatible with existing folds, or the even more adventurous goal of developing new protein folds. A major challenge in this field is finding the conformation/sequence combination corresponding to the lowest free energy minimum. The sequence search problem is often tackled as a local minimization problem, on the basis of the observation that a given fold is often compatible with very different sequences. However, it is also important to find the lowest energy basin in conformational space, in order to generate a structure that is both stable and specific (with no other conformations having the same sequence and similar stabilities).[170]

Sequence and conformational searches use algorithms already mentioned, including Monte Carlo approaches, genetic algorithms and simulated annealing. Energy evaluation and conformational optimization use potentials that can include both physics-based and knowledge-based terms.

The RosettaDesign algorithm[171] can identify amino acid sequences that are energetically well-adapted to a given fixed backbone conformation on the basis of a stochastic sampling algorithm using Monte Carlo optimization with simulated annealing. In order to reduce computational costs, amino acid side chains only sample a discrete set of favorable conformations. The potential used in the energy evaluations takes into account packing, desolvation and angle-dependent hydrogen bonds. Search steps in sequence space and discrete side chain moves are performed alternately.

Including backbone optimization significantly increases the conformational space that needs to be treated, but can naturally improve the quality of the protein design results. Coupling search steps in both backbone and side-chain conformational space to sequence design has in fact been used to successfully design proteins with novel fold topologies.[172] Alternatively, a static conformational ensemble can be generated from a single crystallographic structure and used in subsequent sequence design and side chain conformational optimization. The ensemble can be generated by normal mode analysis,[173] or by considering "backrub motion" that involves a rotation around a $C\alpha_{i-1} - C\alpha_{i+1}$ axis.[174] This approach has recently been implemented in the RosettaBackrub webserver.[175]

In the future, structure prediction may become a routine part of protein design procedures. In a recent work,[176] new sequences fitting a given fold are generated by Monte Carlo mutations and clustering of optimal sequences, followed by structure prediction to validate the selected sequence and conformation.

As always, the final validation of *in silico* protein design requires the experimental characterization of the new protein in terms of structure and function, but this field is certainly opening the route for many exciting applications.[177]

9.4.3 Mechanical and Dynamical Properties

Proteins have dynamic structures and, even in their native state, sample a variety of conformations separated by a range of energy barriers. The resulting mechanical and dynamical properties can make significant contributions to their functional behavior.[178] Several experimental methods can provide information on protein dynamics, notably including NMR spectroscopy and hydrogen exchange kinetics,[179] while studies of mechanics have developed largely thanks to single molecule approaches.[180,181] In parallel, computational methods offer the chance to investigate spontaneous, or controlled, conformational fluctuations in detail.[182] When sufficient computer resources are available, all-atom dynamics coupled with enhanced sampling techniques such as metadynamics[183] or replica exchange[184] can be very powerful, but useful information can also be obtained much faster, particularly for large systems, at the coarse-grain level.

CG approaches have typically used normal mode analysis (NMA) to analyze the harmonic fluctuations around a native conformation. Using elastic network protein representations means that the native state automatically becomes the global energy minimum and allows fluctuations to be obtained rapidly in the anharmonic (ANM) or Gaussian (GNM) formalisms, by respectively diagonalizing either the Hessian[62] or Kirchhoff[61] matrices. In either case, the lowest frequency modes characterize the most collective and, generally, the most functionally relevant motions. In the case of enzymes, numerous studies have shown that a limited combination of the lowest frequency modes can effectively model the global protein movements linked to both enzymatic catalysis and protein deformation upon binding[70,185] (although, because of the construction of elastic networks, the passage from "open" to "closed" states is easier than the reverse).[186] Many proteins, and protein complexes, have been explored in this way including HIV reverse transcriptase,[187] tryptophan synthase,[188] tubulin,[189] myosin[190] and even, by exploiting symmetry constraints, viral capsids.[191]

In order to investigate single residue properties with an elastic network it is possible to perturb a given residue with a random force,[192] to modify the force constants of the associated springs,[193] or to calculate effective force constants for moving a residue within the overall network.[194] Despite the simplistic nature of elastic networks, local properties can vary rapidly from one residue to the next[194,195] and, in $N/P<1$ models, can be significantly affected by removing a single side-chain node (corresponding to a point mutation of the protein sequence).[196] The influence of one protein domain on another has also been studied using elastic networks both by calculating force constants for displacements within a single domain (without modifying the overall elastic network)[197] or by testing the impact of domain interface restraints on the mechanics of a given domain.[198]

In order to deal with large conformational changes, various extensions of the elastic network approach have been developed. When the starting and ending conformations of a pathway are known, it is possible to use double-well potentials[73] or, in the so-called plastic network model,[199] to follow the lowest elastic energy pathway between the basins representing the two conformational energy minima. Alternatively, the network can be modified as the transition occurs, on the basis of the instantaneous residue contact topology, as in a study of the GroEL/GroES complex containing some 8000 amino acids.[200] Gō potentials can also be used to analyze transitions by combining the native interactions from two distinct conformations.[201]

Studying more general dynamic equilibria is still a challenging problem. As an example, Tozzini *et al.* tackled flap opening in HIV-1 protease using a CG model and sampling conformations using Newtonian and Langevin dynamics.[13,202] In this case, the conformational energy was composed of statistical potentials derived from Boltzmann inversion, but local, non-bonded, interactions were biased to favor secondary structure formation. The same group has attempted to avoid such biasing using more complex functions for local geometry based on $C\alpha$ pseudo-angles and pseudo-dihedrals.[203] Introducing dipoles to represent hydrogen bonding, combined with appropriate conformation-dependent angle/dihedral weighting factors, enabled a helix to sheet transition to be modeled in a generic polypeptide.[204]

9.4.4 Protein-protein Interactions

Many proteins function *via* interactions. Consequently, tracing protein-protein interactions (PPI) is a key step to understanding cell function. A number of high-throughput methods have now been developed for analyzing cellular interactomes, but they remain labor intensive and suffer from noise, yielding significant numbers of false positives and false negatives.[205–207] In addition, high-throughput methods do not provide structural information on the protein complexes formed, information that can be crucial for understanding functional mechanisms. Structural information on complexes will also determine whether PPIs can be classified into a limited number of families in the same way as protein folds.[208]

Computational methods should be able to help in solving these problems, although the scale of typical interactomes is daunting; the 6300 proteins in a yeast cell could form almost 20 million different binary complexes.[209] Until recently, most computational studies have concentrated on the simpler problem of protein docking, namely predicting structures for binary complexes formed by proteins that are experimentally known to interact.[210]

Docking algorithms can be separated into those representing proteins as rigid bodies and those treating flexibility. Coarse-grain models have played a major role in both approaches, since they can smooth energy hypersurfaces, especially if combined with "soft" Lennard-Jones potentials. Encouragingly, it has been shown that even relatively imprecise models can still predict the gross features of complexes,[211] suggesting that successively more refined, and more costly, methods can be chained together to achieve accurate predictions.[212]

The most static approach to docking takes advantage of the fact that homologous proteins not only have similar structures, but also tend to form similar interactions,[213] although there are always exceptions to the rule.[214] This led to approaches where interactions are predicted from the structure of known complexes by "threading" homologous peptide sequences into each of the partners.[215,216] The resulting structures are typically scored using residue contact potentials. The success of such approaches is naturally linked to the quality and coverage of the database of known binary complexes. Redesigning protein-protein interfaces can be seen as a derivative of the threading approach to docking. It involves adapting one or both partners to improve affinity and/or specificity[217,218] and generally requires more refined interaction potentials.[219,220]

Returning to docking in the strict sense, the fastest algorithms digitize rigid protein structures onto Cartesian grids. Fast Fourier transform (FFT) methods are then used to find orientations leading to good steric complementarity.[221] Whereas most FFT approaches have been applied to the translational degrees of freedom,[222] the Hex algorithm exploits the rotational variables[223] and, particularly on GPU systems, leads to very fast docking.[224] Note that FFT methods can be supplemented by adding an electrostatic term based on exact or simplified Poisson-Boltzmann calculations.[225,226]

Moving from grids to real-space optimization, it is possible to use energy minimization, as long as a sufficient number of starting conformations are tested. The ATTRACT model from the Zacharias group is a good example of this approach, using 2- or 3-bead amino acid models combined with a soft Lennard-Jones term and damped Coulomb electrostatics,[227] or a composite non-bonded potential.[228] Other optimization methods such as genetic algorithms[229] or Monte Carlo[230] can be used in similar ways. At this level of precision, it is often desirable to include side chain flexibility. This can constitute a refinement step and can be carried out in torsional space,[230] or with a library of rotamer conformations.[231] Multicopy algorithms have also been used successfully to overcome the rugged energy hypersurfaces created by side chain interactions.[17]

Dealing with backbone flexibility during docking still poses considerable problems.[232,233] Multicopy approaches have been extended to peptide loops,[234] but more global changes are generally treated by docking independent, predetermined, protein conformations.[235] Both normal mode calculations, using elastic network models, and snapshots drawn from MD simulations can be used to generate these conformations.[233]

All these methods can be combined with statistical, physics-based or hybrid potentials and iterative approaches have also been used to overcome problems with reference states.[99] Note that the importance of treating orientation-dependent hydrogen bonds,[88] the role of interfacial waters[236–238] and the quality of interface packing[239] have all been stressed.

Docking has progressed considerably in recent years due to the availability of good benchmarks for known complexes (involving varying degrees of conformational rearrangement),[240] decoy datasets for testing interaction potentials[241] and, not least, the CAPRI blind docking challenges.[242] However, it is still difficult to predict binding affinities and thus to determine which proteins will form stable complexes.[243,244] Using bound protein conformations[245] or information on the interacting surfaces of each protein[243,246] can help, but better interaction potentials clearly need to be developed. The availability of a protein-protein binding affinity benchmark[247] is a useful tool for achieving this goal, but more quantitative information on non-interacting or weakly interacting proteins is also required since present databases listing proteins that don't interact are generally based on the absence of information to the contrary, or on the cellular localization of proteins,[248,249] with the associated risks.[250]

9.4.5 Large Protein Assemblies

Moving beyond binary protein complexes leads to the larger challenge of constructing the multi-component molecular machines that play so many vital roles in the cell.[251] This problem could be approached in a hierarchical way, starting with the prediction of binary interactions and then combining these fragments into larger and larger subassemblies. This approach suffers from the fact that binary interactions within a complex may not be stable, or optimal, in isolation and also that it can lead to a combinatorial explosion of the interactions that need to be considered.

Present approaches to building large complexes rely on a variety of experimental data. Virtually any experimental information can be converted into effective energy functions (or restraints), whose importance in determining the final structure can be adjusted to reflect the quality of the input data.[252,253] Electron tomography plays a notable role in providing low resolution structures for a variety of complexes,[254] but many other techniques including small-angle X-ray scattering, mass spectrometry, proteomics, FRET, cross-linking, gene expression data, and symmetry can also contribute. This type of approach has led to very impressive results on the nuclear pore complex that contains more than 450 proteins.[252,255,256]

In these methods, protein representations can range from simple spheres or ellipsoids to all-atom models, depending on the amount of experimental data available. Structure optimization can be achieved using multiple Monte Carlo runs, supplemented by energy minimizations, but initial searches can also use FFT techniques. To deal with structural conflicts, proteins can be divided into a number of linked rigid bodies, or multiple rigid conformations can be included in the search.[251]

A subdomain of macromolecular assembly concerns systems with fewer components and high symmetry. Fibers are one such case. As an example actin filaments have been studied using coarse-grain models derived from all-atom MD, with normal mode analysis to derive macroscopic properties such as persistence length.[24] A second example involves viral capsids, that have been the subject of studies looking at assembly mechanisms using coarse-grain protein models and Gō-like interaction potentials that directly reflect the interactions observed in the complete capsid.[257] Assembly has been achieved using discontinuous molecular dynamics,[257] but also, in probably the most coarse-grained protein model ever attempted, by shaking a flask containing plastic shapes with embedded magnets to represent the interaction potentials.[258] Coarse-grain representations combined with elastic network models,[191] with molecular dynamics,[28] or with both,[259] have also been used to investigate the dynamic properties of viral capsids.

Lastly, we come to models of the cellular cytoplasm itself. At least for bacteria, we have a good idea of the identity and the concentration of all the major components of the cytoplasm.[260] We also have the structures of many of the important proteins or macromolecular assemblies. Consequently, given enough computer power, it is possible to begin simulating the behavior of at least a zone within the cytoplasm, bringing David Goodsell's magnificent images to life. Such simulations combine rigid-body protein models with stochastic dynamics[23,261,262] and provide important information on physical properties at the molecular level, notably diffusion and crowding, which both play significant roles in protein folding[263] and protein complex assembly.[264]

9.5 Conclusions

Despite faster and faster computers, and the consequently increasing scope of all-atom models, coarse-grain approaches continue to play a major role in a wide variety of applications. Faster computation equally benefits coarse-grain models, allowing more detailed representations and interaction potentials. In recent years, iterative Boltzmann refinement techniques have proven particularly useful in improving the quality of both statistical and mixed physical/statistical potentials, largely overcoming the difficulties of defining reference states.

Coarse-grain protein models are becoming increasingly powerful in predicting and designing protein structure and in understanding both folding and aggregation pathways. Concerning protein-protein interactions, despite the success of docking methods, work still remains to be done to achieve reliable predictions of binding affinity. Such data is necessary not only for predicting which specific interactions will occur, but also for quantifying weak, non-specific interactions. The latter are a crucial factor in determining how processes take place within the dense and heterogeneous medium of the cytoplasm and they will be essential for making quantitative *in cellulo* models. There is however already a growing role for coarse-grain models as a means for refining low-resolution experimental data and building structures for multi-macromolecular assemblies. This will undoubtedly be an important area of future growth, notably given the striking progress in electron tomography.

Coarse-grain models are still in their infancy. They continue to develop rapidly and are likely to follow the same path as all-atom representations and force fields, where the effort of the developers combined with growing feedback from the user community has led to increased predictive power and to considerable focusing of initially divergent approaches. This is clearly a necessary step for powering both sequential and concurrent multi-scale modeling, which seems indispensable for understanding protein behavior within a realistic biological context.

Acknowledgements

The authors wish to thank the CNRS and the Rhône-Alpes project CIBLE for funding and Marco Pasi for helpful suggestions on this manuscript.

References

1. T. Schlick, R. Collepardo-Guevara, L. A. Halvorsen, S. Jung and X. Xiao, *Q. Rev. Biophys.*, 2011, **44**(2), 191.
2. E. Lindahl and M. S. Sansom, *Curr. Opin. Struct. Biol.*, 2008, **18**(4), 425.
3. G. S. Ayton and G. A. Voth, *Curr. Opin. Struct. Biol.*, 2009, **19**(2), 138.
4. G. Arya, Q. Zhang and T. Schlick, *Biophys. J.*, 2006, **91**(1), 133.
5. S. Sharma, F. Ding and N. V. Dokholyan, *Biophys. J.*, 2007, **92**(5), 1457.
6. K. Voltz, J. Trylska, V. Tozzini, V. Kurkal-Siebert, J. Langowski and J. Smith, *J. Comput. Chem.*, 2008, **29**(9), 1429.
7. J. Trylska, *J. Phys.: Condens. Matter*, 2010, **22**, 453101.
8. M. M. Tirion, *Phys. Rev. Lett.*, 1996, **77**(9), 1905.
9. C. Clementi, A. E. García and J. N. Onuchic, *J. Mol. Biol.*, 2003, **326**(3), 933.
10. M. Wiederstein and M. J. Sippl, *J. Mol. Biol.*, 2005, **345**(5), 1199.
11. S. Miyazawa and R. L. Jernigan, *Macromolecules (Washington, DC, U. S.)*, 1985, **18**(3), 534.
12. S. Miyazawa and R. L. Jernigan, *J. Mol. Biol.*, 1996, **256**(3), 623.
13. V. Tozzini, *Curr. Opin. Struct. Biol.*, 2005, **15**(2), 144.
14. M. Levitt, *J. Mol. Biol.*, 1976, **104**(1), 59.
15. F. Fogolari, L. Pieri, A. Dovier, L. Bortolussi, G. Giugliarelli, A. Corazza, G. Esposito and P. Viglino, *BMC Struct. Biol.*, 2007, **7**, 15.
16. C. Micheletti, P. Carloni and A. Maritan, *Proteins*, 2004, **55**(3), 635.
17. M. Zacharias, *Protein Sci.*, 2003, **12**(6), 1271.
18. R. D. Hills, L. Lu and G. A. Voth, *PLoS Comput. Biol.*, 2010, **6**(6), e1000827.
19. A. Liwo, S. Ołdziej, M. R. Pincus, R. J. Wawak, S. Rackovsky and H. A. Scheraga, *J. Comput. Chem.*, 1997, **18**(7), 849.
20. T. Ha-Duong, *J. Chem. Theory Comput.*, 2010, **6**, 761.
21. T. Bereau and M. Deserno, *J. Chem. Phys.*, 2009, **130**, 235106.
22. I. F. Thorpe, J. Zhou and G. A. Voth, *J. Phys. Chem. B*, 2008, **112**(41), 13079.
23. T. Ando and J. Skolnick, *Proc. Natl. Acad. Sci. U. S. A.*, 2010, **107**(43), 18457.
24. J. W. Chu and G. A. Voth, *Biophys. J.*, 2006, **90**(5), 1572.

25. H. Gohlke and M. F. Thorpe, *Biophys. J.*, 2006, **91**(6), 2115.
26. Z. Zhang, L. Lu, W. G. Noid, V. Krishna, J. Pfaendtner and G. A. Voth, *Biophys. J.*, 2008, **95**(11), 5073.
27. Z. Zhang, J. Pfaendtner, A. Grafmüller and G. A. Voth, *Biophys. J.*, 2009, **97**(8), 2327.
28. A. Arkhipov, P. L. Freddolino and K. Schulten, *Structure*, 2006, **14**(12), 1767.
29. C. Chennubhotla and I. Bahar, *Mol. Syst. Biol.*, 2006, **2**, 36.
30. H. Abe and N. Gō, *Biopolymers*, 1981, **20**(5), 1013.
31. K. F. Lau and K. A. Dill, *Macromolecules (Washington, DC, U. S.)*, 1989, **22**(10), 3986.
32. Y. Ueda, H. Taketomi and N. Gō, *Biopolymers*, 1978, **17**(6), 1531.
33. A. Sali, E. Shakhnovich and M. Karplus, *J. Mol. Biol.*, 1994, **235**(5), 1614.
34. A. Koliński, J. Skolnick and R. Yaris, *J. Chem. Phys.*, 1986, **85**, 3585.
35. A. Koliński, D. Gront, P. Pokarowski and J. Skolnick, *Biopolymers*, 2003, **69**(3), 399.
36. P. Pokarowski, A. Koliński and J. Skolnick, *Biophys. J.*, 2003, **84**(3), 1518.
37. A. Koliński, M. Milik and J. Skolnick, *J. Chem. Phys.*, 1991, **94**, 3978.
38. A. Koliński 1993, A. Godzik and J. Skolnick, *J. Chem. Phys.*, 1993, **98**, 7420.
39. A. Koliński, *Acta Biochim. Pol.*, 2004, **51**(2), 349.
40. J. R. Banavar, T. X. Hoang, J. H. Maddocks, A. Maritan, C. Poletto, A. Stasiak and A. Trovato, *Proc. Natl. Acad. Sci. U. S. A.*, 2007, **104**(44), 17283.
41. D. J. Jacobs, A. J. Rader, L. A. Kuhn and M. F. Thorpe, *Proteins*, 2001, **44**(2), 150.
42. Y. Luo, X. Xu, T. Lele, S. Kumar and D. E. Ingber, *J. Biomech.*, 2008, **41**(11), 2379.
43. H. Taketomi, Y. Ueda and N. Gō, *Int. J. Pept. Protein Res.*, 1975, **7**(6), 445.
44. C. Clementi, M. Vendruscolo, A. Maritan and E. Domany, *Proteins*, 1999, **37**(4), 544.
45. M. Cieplak, T. X. Hoang and M. O. Robbins, *Proteins*, 2002, **49**(1), 114.
46. J. E. Shea, J. N. Onuchic and C. L. Brooks, *Proc. Natl. Acad. Sci. U. S. A.*, 1999, **96**(22), 12512.
47. J. Karanicolas and C. L. Brooks III, *Protein Sci.*, 2002, **11**(10), 2351.
48. C. Clementi, H. Nymeyer and J. N. Onuchic, *J. Mol. Biol.*, 2000, **298**(5), 937.
49. A. Kleiner and E. Shakhnovich, *Biophys. J.*, 2007, **92**(6), 2054.
50. J. Karanicolas and C. L. Brooks III, *J. Mol. Biol.*, 2003, **334**(2), 309.
51. P. Das, C. J. Wilson, G. Fossati, P. Wittung-Stafshede, K. S. Matthews and C. Clementi, *Proc. Natl. Acad. Sci. U. S. A.*, 2005, **102**(41), 14569.
52. M. S. Cheung, J. M. Finke, B. Callahan and J. N. Onuchic, *J. Phys. Chem. B*, 2003, **107**(40), 11193.
53. B. Erman, *Biophys. J.*, 2001, **81**(6), 3534.

54. J. I. Sułkowska and M. Cieplak, *J. Phys.: Condens. Matter*, 2007, **19**, 283201.
55. J. I. Sułkowska and M. Cieplak, *Biophys. J.*, 2008, **95**(7), 3174.
56. F. Ding, N. V. Dokholyan, S. V. Buldyrev, H. E. Stanley and E. I. Shakhnovich, *J. Mol. Biol.*, 2002, **324**(4), 851.
57. A. Zarrine-Afsar, S. Wallin, A. M. Neculai, P. Neudecker, P. L. Howell, A. R. Davidson and H. S. Chan, *Proc. Natl. Acad. Sci. U. S. A.*, 2008, **105**(29), 9999.
58. P. Das, S. Matysiak and C. Clementi, *Proc. Natl. Acad. Sci. U. S. A.*, 2005, **102**(29), 10141.
59. R. M. Levy, A. R. Srinivasan, W. K. Olson and J. A. McCammon, *Biopolymers*, 1984, **23**(6), 1099.
60. B. Brooks and M. Karplus, *Proc. Natl. Acad. Sci. U. S. A.*, 1983, **80**(21), 6571.
61. I. Bahar, A. R. Atilgan and B. Erman, *Folding Des.*, 1997, **2**(3), 173.
62. A. R. Atilgan, S. R. Durell, R. L. Jernigan, M. C. Demirel, O. Keskin and I. Bahar, *Biophys. J.*, 2001, **80**(1), 505.
63. S. Sunada and N. Go, *J. Comput. Chem.*, 1995, **16**(3), 328.
64. E. Lyman, J. Pfaendtner and G. A. Voth, *Biophys. J.*, 2008, **95**(9), 4183.
65. T. Haliloglu, I. Bahar and B. Erman, *Phys. Rev. Lett.*, 1997, **79**(16), 3090.
66. I. Bahar, A. Wallqvist, D. G. Covell and R. L. Jernigan, *Biochemistry*, 1998, **37**(4), 1067.
67. D. A. Kondrashov, Q. Cui and G. N. Phillips, *Biophys. J.*, 2006, **91**(8), 2760.
68. D. Ming and M. E. Wall, *Phys. Rev. Lett.*, 2005, **95**(19), 198103.
69. G. Song and R. L. Jernigan, *Proteins*, 2006, **63**(1), 197.
70. L. Yang, G. Song and R. L. Jernigan, *Biophys. J.*, 2007, **93**(3), 920.
71. K. Hinsen, *Proteins*, 1998, **33**(3), 417.
72. L. Yang, G. Song and R. L. Jernigan, *Proc. Natl. Acad. Sci. U. S. A.*, 2009, **106**(30), 12347.
73. J. W. Chu and G. A. Voth, *Biophys. J.*, 2007, **93**(11), 3860.
74. J. G. Su, C. H. Li, R. Hao, W. Z. Chen and C. X. Wang, *Biophys. J.*, 2008, **94**(12), 4586.
75. M. J. Sippl, *J. Comput.-Aided Mol. Des.*, 1993, **7**(4), 473.
76. M. J. Sippl, *Curr. Opin. Struct. Biol.*, 1995, **5**(2), 229.
77. P. D. Thomas and K. A. Dill, *J. Mol. Biol.*, 1996, **257**(2), 457.
78. A. Ben-Naim, *J. Chem. Phys.*, 1997, **107**, 3698.
79. I. Bahar and R. L. Jernigan, *J. Mol. Biol.*, 1997, **266**(1), 195.
80. E. Ferrada and F. Melo, *Protein Sci.*, 2009, **18**(7), 1469.
81. M. R. Betancourt and D. Thirumalai, *Protein Sci.*, 1999, **8**(2), 361.
82. Y. Feng, A. Kloczkowski and R. L. Jernigan, *Proteins*, 2007, **68**(1), 57.
83. S. Tanaka and H. A. Scheraga, *Macromolecules (Washington, DC, U. S.)*, 1976, **9**(6), 945.
84. M. J. Sippl, *J. Mol. Biol.*, 1990, **213**(4), 859.
85. S. Miyazawa and R. L. Jernigan, *J. Chem. Phys.*, 2005, **122**(2), 024901.

86. N. V. Buchete, J. E. Straub and D. Thirumalai, *Protein Sci.*, 2004, **13**(4), 862.
87. Y. Wu, M. Lu, M. Chen, J. Li and J. Ma, *Protein Sci.*, 2007, **16**(7), 1449.
88. T. Kortemme, A. V. Morozov and D. Baker, *J. Mol. Biol.*, 2003, **326**(4), 1239.
89. H. Zhou and Y. Zhou, *Protein Sci.*, 2002, **11**(11), 2714.
90. D. Rykunov and A. Fiser, *Proteins*, 2007, **67**(3), 559.
91. J. Skolnick, L. Jaroszewski, A. Koliński and A. Godzik, *Protein Sci.*, 1997, **6**(3), 676.
92. J. Cheng, J. Pei and L. Lai, *Biophys. J.*, 2007, **92**(11), 3868.
93. A. Godzik, A. Koliński and J. Skolnick, *Protein Sci.*, 1995, **4**(10), 2107.
94. A. K. Soper, *Chem. Phys.*, 1996, **202**(2–3), 295.
95. T. Murtola, A. Bunker, I. Vattulainen, M. Deserno and M. Karttunen, *Phys. Chem. Chem. Phys.*, 2009, **11**(12), 1869.
96. A. P. Lyubartsev and A. Laaksonen, *Phys. Rev. E: Stat. Phys., Plasmas, Fluids, Relat. Interdiscip. Top.*, 1995, **52**(4), 3730.
97. P. D. Thomas and K. A. Dill, *Proc. Natl. Acad. Sci. U. S. A.*, 1996, **93**(21), 11628.
98. D. Reith, M. Pütz and F. Müller-Plathe, *J. Comput. Chem.*, 2003, **24**(13), 1624.
99. S. Y. Huang and X. Zou, *Proteins*, 2008, **72**(2), 557.
100. P. Májek and R. Elber, *Proteins*, 2009, **76**(4), 822.
101. S. Y. Huang and X. Zou, *J. Chem. Inf. Model.*, 2010, **50**(2), 262.
102. Y. Song, M. Tyka, A. Leaver-Fay, J. Thompson and D. Baker, *Proteins*, 2011, **79**(6), 1898.
103. T. Hamelryck, M. Borg, M. Paluszewski, J. Paulsen, J. Frellsen, C. Andreetta, W. Boomsma, S. Bottaro and J. Ferkinghoff-Borg, *PLoS ONE*, 2010, **5**(11), e13714.
104. M. R. Betancourt, *Proteins*, 2009, **76**(1), 72.
105. V. N. Maiorov and G. M. Crippen, *J. Mol. Biol.*, 1992, **227**(3), 876.
106. L. A. Mirny and E. I. Shakhnovich, *J. Mol. Biol.*, 1996, **264**(5), 1164.
107. D. Tobi and R. Elber, *Proteins*, 2000, **41**(1), 40.
108. M. Vendruscolo and E. Domany, *J. Chem. Phys.*, 1998, **109**, 11101.
109. C. Micheletti, F. Seno, J. R. Banavar and A. Maritan, *Proteins*, 2001, **42**(3), 422.
110. R. A. Goldstein, Z. A. Luthey-Schulten and P. G. Wolynes, *Proc. Natl. Acad. Sci. U. S. A.*, 1992, **89**(19), 9029.
111. W. L. Jorgensen and J. Tirado-Rives, *J. Am. Chem. Soc*, 1988, **110**(6), 1657.
112. W. D. Cornell, P. Cieplak, C. I. Bayly and I. R. Gouls, *J. Am. Chem. Soc.*, 1995, **117**, 5179.
113. K Lindorff-Larsenn, *Proteins: Struct., Funct., Bioinf.*, 2010, **78**, 1950.
114. M. Levitt and A. Warshel, *Nature*, 1975, **253**(5494), 694.
115. A. Liwo, M. R. Pincus, R. J. Wawak, S. Rackovsky and H. A. Scheraga, *Protein Sci.*, 1993, **2**(10), 1715.
116. A. Liwo, M. R. Pincus, R. J. Wawak, S. Rackovsky, S. Ołdziej and H. A. Scheraga, *J. Comput. Chem.*, 1997, **18**(7), 874.

117. G. G. Maisuradze, P. Senet, C. Czaplewski, A. Liwo and H. A. Scheraga, *J. Phys. Chem. A*, 2010, **114**(13), 4471.
118. M. Makowski, E. Sobolewski, C. Czaplewski, S. Ołdziej, A. Liwo and H. A. Scheraga, *J. Phys. Chem. B*, 2008, **112**(36), 11385.
119. M. Makowski, A. Liwo, E. Sobolewski and H. A. Scheraga, *J. Phys. Chem. B*, 2011, **115**(19), 6119.
120. M. Makowski, A. Liwo and H. A. Scheraga, *J. Phys. Chem. B*, 2011, **115**(19), 6130.
121. N. Basdevant, D. Borgis and T. Ha-Duong, *J. Phys. Chem. B*, 2007, **111**(31), 9390.
122. S. Izvekov and G. A. Voth, *J. Phys. Chem. B*, 2005, **109**(7), 2469.
123. A. Liwo, P. Arłukowicz, C. Czaplewski, S. Ołdziej, J. Pillardy and H. A. Scheraga, *Proc. Natl. Acad. Sci. U. S. A.*, 2002, **99**(4), 1937.
124. A. Liwo, M. Khalili, C. Czaplewski, S. Kalinowski, S. Ołdziej, K. Wachucik and H. A. Scheraga, *J. Phys. Chem. B*, 2007, **111**(1), 260.
125. L. Monticelli, S. K. Kandasamy, X. Periole, R. G. Larson, D. P. Tieleman and S. J. Marrink, *J. Chem. Theory Comput.*, 2008, **4**(5), 819.
126. M. R. Betancourt, *J. Phys. Chem. B*, 2009, **113**(44), 14824.
127. H. Wang, C. Junghans and K. Kremer, *Eur. Phys. J. E. Soft Matter*, 2009, **28**(2), 221.
128. M. Praprotnik, S. Matysiak, L. D. Site, K. Kremer and C. Clementi, *J. Phys.: Condens. Matter*, 2007, **19**, 292201.
129. T. Ha-Duong, N. Basdevant and D. Borgis, *Chem. Phys. Lett.*, 2009, **468**(1-3), 79.
130. S. O. Yesylevskyy, L. V. Schäfer, D. Sengupta and S. J. Marrink, *PLoS Comput. Biol.*, 2010, **6**(6), e1000810.
131. T. Lazaridis and M. Karplus, *Proteins Structure Function and Genetics*, 1999, **35**(2), 133.
132. C. Levinthal, *J. Chem. Phys.*, 1968, **65**(1), 44.
133. C. B. Anfinsen, *Science*, 1973, **181**(96), 223.
134. M. Karplus and A. Sali, *Curr. Opin. Struct. Biol.*, 1995, **5**(1), 58.
135. J. N. Onuchic, Z. Luthey-Schulten and P. G. Wolynes, *Annu. Rev. Phys. Chem.*, 1997, **48**, 545.
136. N. D. Socci, J. N. Onuchic and P. G. Wolynes, *Proteins*, 1998, **32**(2), 136.
137. S. Bromberg and K. A. Dill, *Protein Sci.*, 1994, **3**(7), 997.
138. H. H. Gan, A. Tropsha and T. Schlick, *Proteins*, 2001, **43**(2), 161.
139. K. A. Dill, S. Bromberg, K. Yue, K. M. Fiebig, D. P. Yee, P. D. Thomas and H. S. Chan, *Protein Sci.*, 1995, **4**(4), 561.
140. J. D. Bryngelson and P. G. Wolynes, *J. Phys. Chem.*, 1989, **93**(19), 6902.
141. C. Clementi, *Curr. Opin. Struct. Biol.*, 2008, **18**(1), 10.
142. Y. Zhou and M. Karplus, *Nature*, 1999, **401**(6751), 400.
143. B. Sharma, S. V. Bykov and S. A. Asher, *J. Phys. Chem. B*, 2008, **112**(37), 11762.
144. Z. Luo, J. Ding and Y. Zhou, *J. Chem. Phys.*, 2008, **128**(22), 225103.
145. P. C. Whitford, J. K. Noel, S. Gosavi, A. Schug, K. Y. Sanbonmatsu and J. N. Onuchic, *Proteins*, 2009, **75**(2), 430.

146. P. Derreumaux, *J. Chem. Phys.*, 1999, **111**, 2301.
147. J. Maupetit, P. Tuffery and P. Derreumaux, *Proteins*, 2007, **69**(2), 394.
148. J. F. St-Pierre, N. Mousseau and P. Derreumaux, *J. Chem. Phys.*, 2008, **128**(4), 045101.
149. Y. Chebaro, X. Dong, R. Laghaei, P. Derreumaux and N. Mousseau, *J. Phys. Chem. B*, 2009, **113**(1), 267.
150. N. Mousseau and P. Derreumaux, *Acc. Chem. Res.*, 2005, **38**(11), 885.
151. Y. Lu, P. Derreumaux, Z. Guo, N. Mousseau and G. Wei, *Proteins*, 2009, **75**(4), 954.
152. B. R. Jefferys, L. A. Kelley and M. J. Sternberg, *J. Mol. Biol.*, 2010, **397**(5), 1329.
153. J. Moult, K. Fidelis, A. Kryshtafovych, B. Rost and A. Tramontano, *Proteins*, 2009, **77**(S9), 1.
154. J. U. Bowie, R. Luthy and D. Eisenberg, *Science*, 1991, **253**(5016), 164.
155. D. T. Jones, W. R. Taylor and J. M. Thornton, *Nature*, 1992, **358**(6381), 86.
156. H. Zhou and Y. Zhou, *Proteins*, 2004, **55**(4), 1005.
157. S. Liu, C. Zhang, S. Liang and Y. Zhou, *Proteins*, 2007, **68**(3), 636.
158. Y. Hu, X. Dong, A. Wu, Y. Cao, L. Tian and T. Jiang, *PLoS ONE*, 2011, **6**(2), e17215.
159. R. Samudrala and J. Moult, *J. Mol. Biol.*, 1998, **275**(5), 895.
160. D. Rykunov and A. Fiser, *BMC Bioinf.*, 2010, **11**, 128.
161. L. Wroblewska and J. Skolnick, *J. Comput. Chem.*, 2007, **28**(12), 2059.
162. C. M. Summa and M. Levitt, *Proc. Natl. Acad. Sci. U. S. A.*, 2007, **104**(9), 3177.
163. J. U. Bowie and D. Eisenberg, *Proc. Natl. Acad. Sci. U. S. A.*, 1994, **91**(10), 4436.
164. K. M. Misura, D. Chivian, C. A. Rohl, D. E. Kim and D. Baker, *Proc. Natl. Acad. Sci. U. S. A.*, 2006, **103**(14), 5361.
165. D. T. Jones, *Proteins*, 2001, **Suppl 5**, 127.
166. D. T. Jones and L. J. McGuffin, *Proteins*, 2003, **53 Suppl 6**, 480.
167. Y. Zhang and J. Skolnick, *Proc. Natl. Acad. Sci. U. S. A.*, 2004, **101**(20), 7594.
168. Y. Zhang, A. K. Arakaki and J. Skolnick, *Proteins*, 2005, **61 Suppl 7**, 91.
169. S. Ołdziej, C. Czaplewski, A. Liwo, M. Chinchio, M. Nanias, J. A. Vila, M. Khalili, Y. A. Arnautova, A. Jagielska, M. Makowski, H. D. Schafroth, R. Kaźmierkiewicz, D. R. Ripoll, J. Pillardy, J. A. Saunders, Y. K. Kang, K. D. Gibson and H. A. Scheraga, *Proc. Natl. Acad. Sci. U. S. A.*, 2005, **102**(21), 7547.
170. P. Tian, *Chem. Soc. Rev.*, 2010, **39**(6), 2071.
171. B. Kuhlman and D. Baker, *Proc. Natl. Acad. Sci. U. S. A.*, 2000, **97**(19), 10383.
172. B. Kuhlman, G. Dantas, G. C. Ireton, G. Varani, B. L. Stoddard and D. Baker, *Science*, 2003, **302**(5649), 1364.
173. X. Fu, J. R. Apgar and A. E. Keating, *J. Mol. Biol.*, 2007, **371**(4), 1099.
174. I. W. Davis, W. B. Arendall, D. C. Richardson and J. S. Richardson, *Structure*, 2006, **14**(2), 265.

175. F. Lauck, C. A. Smith, G. F. Friedland, E. L. Humphris and T. Kortemme, *Nucleic Acids Res.*, 2010, **38**(Web Server issue), W569.
176. A. Bazzoli, A. G. Tettamanzi and Y. Zhang, *J. Mol. Biol.*, 2011, **407**(5), 764.
177. D. Baker, *Protein Sci.*, 2010, **19**(10), 1817.
178. H. Frauenfelder, P. W. Fenimore and R. D. Young, *IUBMB Life*, 2007, **59**(8–9), 506.
179. B. J. Grant, A. A. Gorfe and J. A. McCammon, *Curr. Opin. Struct. Biol.*, 2010, **20**(2), 142.
180. T. R. Strick, M. N. Dessinges, G. Charvin, N. H. Dekker, J. F. Allemand, D. Bensimon and V. Croquette, *Rep. Prog. Phys.*, 2003, **66**, 1.
181. C. Bustamante, Y. R. Chemla, N. R. Forde and D. Izhaky, *Annu. Rev. Biochem.*, 2004, **73**, 705.
182. M. Karplus and J. A. McCammon, *Nat. Struct. Mol. Biol.*, 2002, **9**(9), 646.
183. A. Laio and M. Parrinello, *Proc. Natl. Acad. Sci. U. S. A.*, 2002, **99**(20), 12562.
184. Y. Sugita and Y. Okamoto, *Chem. Phys. Lett.*, 1999, **314**(1-2), 141.
185. D. Tobi and I. Bahar, *Proc. Natl. Acad. Sci. U. S. A.*, 2005, **102**(52), 18908.
186. F. Tama and Y. H. Sanejouand, *Protein Eng.*, 2001, **14**(1), 1.
187. I. Bahar, B. Erman, R. L. Jernigan, A. R. Atilgan and D. G. Covell, *J. Mol. Biol.*, 1999, **285**(3), 1023.
188. I. Bahar and R. L. Jernigan, *Biochemistry*, 1999, **38**(12), 3478.
189. O. Keskin, *J. Biomol. Struct. Dyn.*, 2002, **20**(3), 333.
190. I. Navizet, R. Lavery and R. L. Jernigan, *Proteins*, 2004, **54**(3), 384.
191. M. K. Kim, R. L. Jernigan and G. S. Chirikjian, *J. Struct. Biol.*, 2003, **143**(2), 107.
192. C. Atilgan, Z. N. Gerek, S. B. Ozkan and A. R. Atilgan, *Biophys. J.*, 2010, **99**(3), 933.
193. W. Zheng and M. Tekpinar, *BMC Struct. Biol.*, 2009, **9**, 45.
194. I. Navizet, F. Cailliez and R. Lavery, *Biophys. J.*, 2004, **87**(3), 1426.
195. S. Sacquin-Mora, E. Laforet and R. Lavery, *Proteins*, 2007, **67**(2), 350.
196. S. Sacquin-Mora, P. Sebban, V. Derrien, B. Frick, R. Lavery and C. Alba-Simionesco, *Biochemistry*, 2007, **46**(51), 14960.
197. S. Sacquin-Mora and R. Lavery, *Biophys. J.*, 2006, **90**(8), 2706.
198. D. Ming, Y. Kong, Y. Wu and J. Ma, *Proc. Natl. Acad. Sci. U. S. A.*, 2003, **100**(1), 104.
199. P. Maragakis and M. Karplus, *J. Mol. Biol.*, 2005, **352**(4), 807.
200. Z. Yang, P. Májek and I. Bahar, *PLoS Comput. Biol.*, 2009, **5**(4), e1000360.
201. K. Okazaki, N. Koga, S. Takada, J. N. Onuchic and P. G. Wolynes, *Proc. Natl. Acad. Sci. U. S. A.*, 2006, **103**(32), 11844.
202. V. Tozzini, J. Trylska, C. E. Chang and J. A. McCammon, *J. Struct. Biol.*, 2007, **157**(3), 606.
203. V. Tozzini, W. Rocchia and J. A. McCammon, *J. Chem. Theory Comput.*, 2006, **2**(3), 667.

204. D. Alemani, F. Collu, M. Cascella and M. Dal Peraro, *J. Chem. Theory Comput.*, 2010, **6**(1), 315.
205. C. M. Deane, Ł. Salwiński, I. Xenarios and D. Eisenberg, *Mol. Cell. Proteomics*, 2002, **1**(5), 349.
206. P. Aloy and R. B. Russell, *Trends Biochem. Sci.*, 2002, **27**(12), 633.
207. G. Franzot and O. Carugo, *J. Struct. Funct. Genomics*, 2003, **4**(4), 245.
208. P. Aloy and R. B. Russell, *Nat. Biotechnol.*, 2004, **22**(10), 1317.
209. S. Pitre, M. Alamgir, J. R. Green, M. Dumontier, F. Dehne and A. Golshani, *Adv. Biochem. Eng./Biotechnol.*, 2008, **110**, 247.
210. P. J. Kundrotas, Z. Zhu and I. A. Vakser, *Nucleic Acids Res.*, 2010, **38**(Database issue), D513.
211. A. Tovchigrechko, C. A. Wells and I. A. Vakser, *Protein Sci.*, 2002, **11**(8), 1888.
212. M. Kurcinski and A. Koliński, *J. Mol. Model.*, 2007, **13**(6-7), 691.
213. P. Aloy, H. Ceulemans, A. Stark and R. B. Russell, *J. Mol. Biol.*, 2003, **332**(5), 989.
214. P. Aloy, M. Pichaud and R. B. Russell, *Curr. Opin. Struct. Biol.*, 2005, **15**(1), 15.
215. P. Aloy and R. B. Russell, *Proc. Natl. Acad. Sci. U. S. A.*, 2002, **99**(9), 5896.
216. H. Chen and J. Skolnick, *Biophys. J.*, 2008, **94**(3), 918.
217. T. Kortemme, L. A. Joachimiak, A. N. Bullock, A. D. Schuler, B. L. Stoddard and D. Baker, *Nat. Struct. Mol. Biol.*, 2004, **11**(4), 371.
218. E. L. Humphris and T. Kortemme, *PLoS Comput. Biol.*, 2007, **3**(8), e164.
219. L. A. Clark and H. W. van Vlijmen, *Proteins*, 2008, **70**(4), 1540.
220. O. Sharabi, C. Yanover, A. Dekel and J. M. Shifman, *J. Comput. Chem.*, 2011, **32**(1), 23.
221. E. Katchalski-Katzir, I. Shariv, M. Eisenstein, A. A. Friesem, C. Aflalo and I. A. Vakser, *Proc. Natl. Acad. Sci. U. S. A.*, 1992, **89**(6), 2195.
222. H. A. Gabb, R. M. Jackson and M. J. Sternberg, *J. Mol. Biol.*, 1997, **272**(1), 106.
223. D. W. Ritchie and G. J. Kemp, *Proteins*, 2000, **39**(2), 178.
224. D. W. Ritchie and V. Venkatraman, *Bioinformatics*, 2010, **26**(19), 2398.
225. J. G. Mandell, V. A. Roberts, M. E. Pique, V. Kotlovyi, J. C. Mitchell, E. Nelson, I. Tsigelny and L. F. Ten Eyck, *Protein Eng.*, 2001, **14**(2), 105.
226. A. Heifetz, E. Katchalski-Katzir and M. Eisenstein, *Protein Sci.*, 2002, **11**(3), 571.
227. M. Zacharias, *Proteins*, 2005, **60**(2), 252.
228. S. Fiorucci and M. Zacharias, *Proteins*, 2010, **78**(15), 3131.
229. E. J. Gardiner, P. Willett and P. J. Artymiuk, *Proteins: Struct., Funct., Genet.*, 2001, **44**(1), 44.
230. J. J. Gray, S. E. Moughon, T. Kortemme, O. Schueler-Furman, K. Misura, A. V. Morozov and D. Baker, *Proteins: Struct., Funct., Genet.*, 2003, **52**(1), 118.

231. R. L. Dunbrack, *Curr. Opin. Struct. Biol.*, 2002, **12**(4), 431.
232. N. Andrusier, E. Mashiach, R. Nussinov and H. J. Wolfson, *Proteins*, 2008, **73**(2), 271.
233. M. Zacharias, *Curr. Opin. Struct. Biol.*, 2010, **20**(2), 180.
234. K. Bastard, C. Prévost and M. Zacharias, *Proteins*, 2006, **62**(4), 956.
235. S. Y. Huang and X. Zou, *Proteins*, 2007, **66**(2), 399.
236. L. Jiang, B. Kuhlman, T. Kortemme and D. Baker, *Proteins*, 2005, **58**(4), 893.
237. H. Behringer, A. Degenhard and F. Schmid, *Phys. Rev. Lett.*, 2006, **97**(12), 128101.
238. J. Taktikos and H. Behringer, *Phys. Rev. E: Stat., Nonlinear, Soft Matter Phys.*, 2009, **79**(4 Pt 1), 041908.
239. K. E. Gottschalk, H. Neuvirth and G. Schreiber, *Protein Eng Des Sel*, 2004, **17**(2), 183.
240. H. Hwang, B. Pierce, J. Mintseris, J. Janin and Z. Weng, *Proteins*, 2008, **73**(3), 705.
241. S. Liu, Y. Gao and I. A. Vakser, *Bioinformatics*, 2008, **24**(22), 2634.
242. J. Janin, *Mol. Biosyst.*, 2010, **6**(12), 2351.
243. S. Sacquin-Mora, A. Carbone and R. Lavery, *J. Mol. Biol.*, 2008, **382**(5), 1276.
244. P. L. Kastritis and A. M. Bonvin, *J. Proteome Res.*, 2010, **9**(5), 2216.
245. T. Yoshikawa, K. Tsukamoto, Y. Hourai and K. Fukui, *J. Chem. Inf. Model.*, 2009, **49**, 693.
246. S. Engelen, L. A. Trojan, S. Sacquin-Mora, R. Lavery and A. Carbone, *PLoS Comput. Biol.*, 2009, **5**(1), e1000267.
247. P. L. Kastritis, I. H. Moal, H. Hwang, Z. Weng, P. A. Bates, A. M. Bonvin and J. Janin, *Protein Sci.*, 2011, **20**(3), 482.
248. P. Smialowski, P. Pagel, P. Wong, B. Brauner, I. Dunger, G. Fobo, G. Frishman, C. Montrone, T. Rattei, D. Frishman and A. Ruepp, *Nucleic Acids Res.*, 2010, **38**(Database issue), D540.
249. X. W. Chen, J. C. Jeong and P. Dermyer, *Nucleic Acids Res.*, 2011, **39**(Database issue), D750.
250. A. Ben-Hur and W. S. Noble, *BMC Bioinf.*, 2006, **7 Suppl 1**, S2.
251. M. Beck, M. Topf, Z. Frazier, H. Tjong, M. Xu, S. Zhang and F. Alber, *J. Struct. Biol.*, 2011, **173**(3), 483.
252. F. Alber, M. F. Kim and A. Sali, *Structure*, 2005, **13**(3), 435.
253. F. Alber, F. Förster, D. Korkin, M. Topf and A. Sali, *Annu. Rev. Biochem.*, 2008, **77**, 443.
254. M. Topf and A. Sali, *Curr. Opin. Struct. Biol.*, 2005, **15**(5), 578.
255. F. Alber, S. Dokudovskaya, L. M. Veenhoff, W. Zhang, J. Kipper, D. Devos, A. Suprapto, O. Karni-Schmidt, R. Williams, B. T. Chait, M. P. Rout and A. Sali, *Nature*, 2007, **450**(7170), 683.
256. F. Alber, S. Dokudovskaya, L. M. Veenhoff, W. Zhang, J. Kipper, D. Devos, A. Suprapto, O. Karni-Schmidt, R. Williams, B. T. Chait, A. Sali and M. P. Rout, *Nature*, 2007, **450**(7170), 695.
257. H. D. Nguyen, V. S. Reddy and C. L. Brooks, *Nano Lett.*, 2007, **7**(2), 338.

258. A. J. Olson, Y. H. Hu and E. Keinan, *Proc. Natl. Acad. Sci. U. S. A.*, 2007, **104**(52), 20731.
259. G. S. Ayton and G. A. Voth, *Biophys. J.*, 2010, **99**(9), 2757.
260. R. P. Sear, *J. Phys.: Condens. Matter*, 2005, **17**, S3587.
261. S. R. McGuffee and A. H. Elcock, *J. Am. Chem. Soc.*, 2006, **128**(37), 12098.
262. S. R. McGuffee and A. H. Elcock, *PLoS Comput. Biol.*, 2010, **6**(3), e1000694.
263. M. S. Cheung, D. Klimov and D. Thirumalai, *Proc. Natl. Acad. Sci. U. S. A.*, 2005, **102**(13), 4753.
264. X. Li, I. H. Moal and P. A. Bates, *Proteins*, 2010, **78**(15), 3189.

CHAPTER 10

Generalised Multi-level Coarse-grained Molecular Simulation and its Application to Myosin-V Movement

WILLIAM R. TAYLOR* AND ZOE KATSIMITSOULIA

Division of Mathematical Biology, MRC National Institute for Medical Research, The Ridgeway, Mill Hill, London NW7 1AA, U.K
*Email: wtaylor@nimr.mrc.ac.uk

10.1 Introduction

Physical methods for the determination of macromolecular structure have improved greatly over the past decade, resulting in a wealth of large molecular structures including both protein and nucleic acid. Most, if not all of these structures are dynamic, either in their assembly (and disassembly) such as cytoskeletal fibrils and viruses or often have large motions associated with their function, such as the F1/F0 ATP synthetase complex. Indeed for the class of molecular motors, motion is the essential aspect of their function.[1,2]

With the exception of real-time scanning atomic force microscopy,[3] which is still not a widely used method, most of the physical techniques, such as electron microscopy or X-ray crystallography, provide only snap-shots of discrete states of a dynamic system.[4,5] This can be augmented by data from a wide variety of physical methods, such as spectroscopy or single particle measurements,[6,7] that can provide dynamic information against which the different 'snap-shots' can be interpreted. However, given the coordinates for different states, it is still

RSC Biomolecular Sciences No. 23
Innovations in Biomolecular Modeling and Simulations: Volume 1
Edited by Tamar Schlick

Published by the Royal Society of Chemistry, www.rsc.org

natural to want to visualise how the system progresses from one to the other, or given a set of components, to wonder how they might assemble into a complex.

An impression of how complex systems behave can be gleaned by interpolating between states using a 'geometry engine' of the type more often used in animating cartoon characters in popular games and movies. Depending on the constraints of the data, these can to some extent be predictive but more typically, simply recapitulate the relationship between the existing structures in a more accessible manner. At the other extreme to this top-down approach, molecular dynamics methods can be used to animate large structures from the bottom up: that is, by summing the interactions of individual atoms into increasingly larger motions.[8–10]

Both the top-down and bottom-up approaches have their problems. The top-down approach may not have a geometry engine of sufficient quality to ensure that the system does not wander into un-natural configurations between known states, which is a problem if there is a large difference between the states (or just one known state). Conversely, with the bottom-up approach, the repeated summation of small atomic forces into large motions is likely to be a divergent process, giving little confidence in the uniqueness of the final configuration.

The ideal solution to both these problems is to combine the two approaches. However, this raises a new set of questions as to the best way that this can be done and there is a variety of approaches depending on whether the aim is mainly to predict or recapitulate. This type of modelling is broadly referred to as coarse-grained (CG) as the constraints from the data are typically imposed at a level above atomic interaction.[11–13] The definition of levels is essentially arbitrary but for proteins and RNA there is a good succession in which each level incorporates roughly an order of 10 elements from the lower level giving a progression form atoms to residues/bases to secondary structures to domains to multi-domain/chain macromolecules or assemblies. This follows the conventional structural hierarchy of primary, secondary, tertiary and quaternary, with the allowance that in the latter, it is often arbitrary whether the units are linked as domains or exist as distinct chains.[14]

As structural constraints are imposed at levels above atomic, they becomes less general until the highest level may be specific to a particular known state of a large system. For example, atomic interactions are the same in all molecules, whether nucleic acid or protein. Interactions at the monomer level require knowledge of the chemical structure of four bases for RNA/DNA and twenty amino acids for proteins. At the secondary structure level, both RNA and protein become very different and the interaction (packing) of the secondary structures can only be described because of the large numbers of structures in which they have been observed. The interaction has shifted from being fundamental to being empirical. At the domain level constraints on the fold of the chain must be derived largely from a specific fold and at the top level of a multi-domain chain or multi-chain assembly, constraints derive from a unique structure.

If only one of these levels of constraint is utilised, then the others become poorly determined. As discussed above, using only atomic interaction leads to uncertainty at the higher levels and using just high-level information (say, domain/domain distances) leaves the atomic detail unspecified. Taken by itself, any intermediate level will leave those above and below less well defined. The ideal solution clearly is not to neglect any level but to impose constraints at every level, ranging from the generic at the lowest level to the specific at the highest. Shifts from one set of high-level constraints to another can then be used to model the transition from one state of the system to another while preserving integrity through the lower levels.

In this chapter, we will give an overview of a new generalised coarse-grained simulation method that we have developed along the lines described above with a view to simulating the motion of large macromolecular systems and, in particular, molecular motors.

10.2 Implementation

10.2.1 Overview

As outlined in the introduction, the nature of the macromolecular structure data is clearly hierarchic so it is natural to turn to a hierarchic data structure to capture it. With a view to generality, it was our aim that the structure of each unit in the hierarchy at any level should be identical with the distinction between levels being made only through externally specified (user) parameters. The most fundamental aspects to be specified include the shape of the unit, its linkage to other units (such as whether it is in a chain or multiply bonded) and how units interact with each other. For ease of description, a unit contained in a higher level unit will be called its 'child' with the higher level referred to as the 'parent' of the lower level unit. Units with the same parent are therefore 'siblings' and if they form a chain, are designated 'sister' (preceeding) and 'brother' (following). Anthropomorphic terms based on family relationships will be used through.

Interactions can be distinguished as inter- and intra-level. Within a level (intra) these will either be repulsive (active for a pair of units in collision) or attractive (active between a bonded pair). However, there are no bonds between levels (unless individually specified by the user) and no bumps between levels. Inter-level interaction consists only of coordination of motion and containment. Coordinate movement means that when a parent moves, all its children move too, which by implication, continues down through all successive generations. In the opposite direction, the centroid of the children determines the position of the parent. This relationship also implies indirect interaction between the levels so that when children in different families collide then their parents will also experience a lesser repulsion. Containment was implemented as a simple kick-back to any children that had 'strayed' beyond the shape boundary of their parent.

The motion of each unit is purely random with a fixed step-size defined in the user-specified parameter file. Every move is accepted whether it violates bond geometry of leads to collisions.[15] Clearly this would lead to a degradation in the molecular structure so independently of this imposed motion, the bond lengths and local chain geometry (if there is a chain) are continually refined towards their initial configuration. Similarly, units in collision are also corrected. All these processes run concurrently (implemented as separate threads) and in addition a user specified process also runs in parallel in which the directed

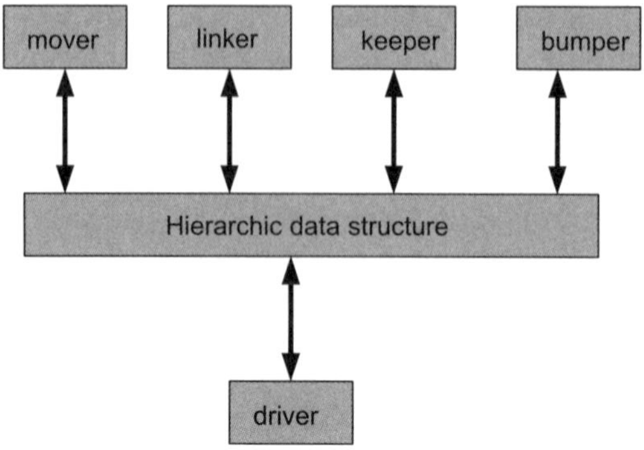

Figure 10.1 The structure of the program is viewed as independent processes acting on a single common `hierarchic data structure`. The `mover` routine gives every object a random displacement. The `linker` routine maintains bonded distances and cross-links. The `keeper` routine restrains children to stay inside their parents. The `bumper` routine detects collisions and corrects steric violations. The `driver` is a user supplied routine that implements the application specific aspects of the dynamics. Each routine is independent and runs as a parallel process (threads) interacting only with the common data.

elements of the model are specified. All of these processes operate on a single representation of the coordinates, so care has been taken to ensure that undefined states do not arise in one process that would disrupt another. In general, this can be avoided with each process working on a temporary copy of the coordinates it needs then writing these in one step back into the structure.

The overall structure of the implementation is shown in Figure 10.1 along with the names of the processes that will be referred to below. The two processes that maintain the integrity of the molecular structure are the collision detection and correction process and the process that maintains the specified links in the chain: respectively, called the `bumper` and the `linker` which will be considered first.

10.2.2 Steric Exclusion

It is intended that the implementation should be applied to very large systems so any collision detection based on a full pairwise algorithm would be impractical. This is commonly avoided in molecular (and other) dynamics programs through the use of a neighbour list in which each atom maintains a list of its current neighbours and checks only these for collision. This has the unavoidable problem that the list must be revised periodically. We adopted a similar approach except that we used the hierarchic structure of the data to provide built-in neighbour lists where any unit only checks its siblings for collisions. As each family is usually in the order of 10–100 units, this would not be a large task to compute in a pairwise manner, however, we use a faster approach based on ranked lists of units that are maintained for each dimension (X,Y,Z). As sorting avoids quadratic operations in the number of objects, this is much faster for large families.[13]

When two units are in collision, they are repelled only if they have no children. Again a single (user specified) kick-back step is applied equally to each unit along the line connecting their centres. If the units have children, then collisions are found between their joint families within a restricted range along the line connecting the two parent centres. If there are no grandchildren, repulsion is again implemented along the line connecting their two centres, otherwise the process is repeated at each lower level until the lowest atomic† level is reached.

As mentioned above, parents of different families will automatically adjust their position indirectly to the repulsion between their families as the centroid of the children will move slightly apart. This is often a small effect, and before it becomes significant, the children can become bunched at the collision interface which has the reverse effect of bringing the parents (and hence their children) even closer together. To avoid this, on the return path from the traversal of the family hierarchy (*i.e.* revisiting the parents of the colliding children), the parents themselves are given a small direct displacement proportional to the number of their children that were in collision. If only the positions of the atomic level units were observed, this would have the appearance of a repulsive 'field' as there would seem to be 'action at a distance' across a family. Alternatively, it can be imagined that the children are embedded in a soft parental jelly-like matrix. Computationally, the approach means that in any collision at a high level, there will be fewer low-level collisions generated which saves on computational expense.

This approach to the treatment of collisions has an additional effect in that it is relatively insensitive to the shape of the colliding objects, which in our implementation can be spheres, ellipsoids or tubes (discussed in more detail below). If we assume that the atomic level consists of spheres, then different shapes at higher levels are primarily defined by the distribution of their children within them, which in turn is determined by the shape within which the family is confined. The only discrepancy comes through the point at which units at the parental level detect collision as, for computational simplicity, this is kept as an isotropic test at the radius of their (user defined) bumping sphere. Objects that are long/thin tubes or extremely prolate ellipsoids which are fully contained in a bumping sphere can therefore be considered in collision before any of their family members come in contact. At worst, however, this is just a slight waste of computer time as the count of colliding children will be zero and the parents will not respond.

10.2.3 Polymers and Cross-linking

10.2.3.1 Specifying Chain Connectivity

Unless liquids are being modelled, the links between units (which are not distinguished from bonds) are the components that impart the greatest structure. For biological polymers, links between just adjacent units along a chain,

† The term "atomic" only means the lowest level in the hierarchy, which in the protein applications discussed, is the residue level (based on the alpha-carbon position).

combined with steric exclusion, are sufficient to define a basic model. However, even this, apparently simple, imposition of structure leads to complications in a hierarchic model. If the atomic level is a linear chain, then so too are all higher levels but this is not so if each atomic family forms a separate or a circular chain. Then higher levels can be unlinked (*e.g.* a liquid of cyclic peptides) or otherwise linked in their own way. The linkage polymer state of each level can be specified by the user independently but chain interdependencies are checked internally and imposed by the program.

The structural hierarchy (family structure) is not determined by chain connectivity but should be based on groups of units that will tend to move together as they are all acted upon by their parent's transform operations. At the lower level, such groupings will typically consist of consecutive units along a chain (such as an α-helix), however, at the domain level and secondary structure level for RNA, families of units will also be composed of sequentially discontinuous segments. Computationally, this requires some book-keeping to keep note of which members are linked across families. To facilitate this, each unit holds a record of its preceding unit, referred to as its 'sister', and its following unit, referred to as its 'brother'; both of which may specify any unit in any family at the same level. Together, the sisters and brothers constitute two linked-lists, with brothers running from the start of the chain to the end and sisters in the opposite direction.

As chain connectivity is not restricted by family groupings, its path at the next higher level is not necessarily linear and can be branched. This means that each unit may have more than one brother or sister which is equivalent to branches in the chain in both directions. In general, this specifies a network and to deal with the associated 'book-keeping', each unit holds a stack of its brothers and sisters. Following a chain is therefore not simple and when listing a chain in 'sequential order', the lists of brothers and sisters are followed recursively from any given starting unit.

10.2.3.2 Inter-chain Cross-links

Polymer chain links are such a common feature of biological macromolecules that the capacity to encode them was included as a general feature in the data structure of each unit. Inter-chain cross-links, which are less ubiquitous, were allocated only as requested by the user in the data file that specifies the model. For any level, a fixed number of links could be specified, not all of which need necessarily be used. If no linking capacity was specified then computer memory was not allocated.

Although links can be individually specified, some automated features were incorporated to ease the burden of assigning the local cross-links associated with secondary structure, both in proteins and RNA. For proteins, two types of secondary structure can be defined: the α-helix and β-sheet. The former is purely local and two links were automatically set to the relative chain positions +4 and −3 of the ideal length found in proteins. Similarly, two local links were made along a β-strand to the +2 and −2 positions. However, each strand in a sheet makes non-local links which can be specified by data provided in the coordinate input specification which is automatically generated by a separate program that calculates the definition of secondary structures. Similarly, for RNA, the base-pairings are pre-calculated by a separate program.

At a higher level, links between secondary structures can be allocated within a given range of interaction. Unlike the secondary structures which are pre-calculated, these links are computed as part of the data input process.

10.2.4 Geometric Regularisation

Steric exclusion combined with the range of linkage described above can generate a relatively stable structure. However, given a background of 'thermal' noise, any less constrained parts of the structure will be free to diverge from their starting configuration under the given distance constraints. Typically, this involves twisting and shearing, which can generate large motions with little violation of the specified distances, which in principle, cannot constrain chirality. A general mechanism, based only on local angles and distances was provided to reduce these distortions and was applied equally to all levels that form a chain.

In a chain segment of five units (designated: b2,b1,c0,a1,a2), six distances were recorded from the starting configuration in the upper half of the matrix of pairwise distances excluding adjacent units. Three angles were also recored as b1-c0-a1 and the torsion angles around b1-c0 and c0-a1. These local distances were continually refined, as were the angles. Distances can be regularised with little disruption, however, refining torsion angles can sometimes lead to an error propagation with dramatic effects. To limit the potential for this the torsion angles were dialled-up exactly to generate new positions for b2 (b2′) and a2 (a2′). These were used to form a basis-set of unit length vectors along: $x = a2 - b2'$, $y = c0 - (a2' + b2')/2$, with z mutually orthogonal. Starting from the centroid of the five points, the coefficients of an equivalent basis-set defined on the original positions were applied to the new basis-set to generate the new coordinate positions. The result is a compromise between angle and position that remains stable over repeated application.

Although only local information is used, its application over all levels leads to a global effect and indeed is sufficient by itself to recapitulate a large structure. As the procedure was designed to correct defects caused by the addition of random motion (caused by `mover` in Figure 10.1), it was not implemented as an independent parallel process but included in `mover` and applied after the coordinates had been displaced, so keeping a balance between disruption and correction. A final feature was included to allow for the necessary requirement that in a dynamic model, the starting configuration of the structure should not be exclusively maintained. This was accommodated by periodically shifting the target distances and angles towards those found in the current configuration. The overall effect of this procedure is to provide a buffering effect against random motion and is similar to giving rigidity to the structure but still allowing movement under a persistent 'force'. In the current implementation, this shift is by 1% once in roughly every 100 activations of `mover`. This can be adjusted depending on the application.

10.2.5 Shape Specification

Three basic shapes were implemented: cylinder, ellipsoid and sphere. Although the sphere is a special instance of an ellipsoid, there are implementation details, described below, that make them distinct. Each shape type by itself has elements of symmetry that can make their orientation arbitrary, however, this symmetry is broken when a unit contains children in an irregular configuration. Thus each unit needs to have an associated reference frame that determines its orientation and is acted on by rotational operations. For a unit in a chain, the current reference frame is based on the direction from its sister to brother (X) with the Y direction as the projection of the unit's position onto this line in an orthogonal direction and Z as their mutual perpendicular. A consequence of this is that flexing of the chain does not preserve the end-point distances between consecutive cylinders or ellipsoids along the chain.

The length of cylinders and ellipsoids is set by reading in two end-points from the coordinate input data which have been pre-calculated from the inertial axes of the point-set that comprises the current unit (say, a secondary structure element or a domain). As well as the length, the line linking these end-points specifies the axis that corresponds with the X direction in the internal reference frame. The two end-points are then set within the data-structure that defines each unit as two points equidistant from the central point along the X direction. While the length of a unit is determined by these end-points, this is different from the size of each unit which is set generically for every unit on a given level by a value specified in the parameter file that describes the model. For a sphere, this is the only value that is needed and specifies the radius. For a cylinder, it also specifies the radius which is the thickness of the tube. For an ellipsoid, the end-points specify the length along the X axis and the size parameter specifies the other two axes. Therefore all ellipsoids are radially symmetric around X, giving a progression from oblate (disc) through spherical to prolate (cigar). Ignoring scalene ellipsoids excludes only long flat discs which are not common shapes for secondary structures or domains.

10.2.6 Implementation

10.2.6.1 Time and Memory Allocation

The adoption of a common data structure for each node in the hierarchy can lead to the allocation of memory for variables that are seldom, if ever, used. For example, the data structure allows for a general shape type which includes the coordinates of the end-points for tubes and ellipsoids yet if the object is a sphere, which it commonly is on the most populated atomic level, then space is wasted. Fortunately, with a reasonable workstation or laptop, memory is seldom a limitation for the system and tests have been made using over a million allocated nodes.

With a parallel implementation (using threads), the time allocated to the different processes can present scheduling considerations. A simple solution was adopted in which the call-back loop of each process was interleaved with a sleep call which suspended the process for a fixed period of time (currently 0.1 sec.). Within each process, higher priority was allocated to branches in the hierarchy that were in an active state, such as undergoing collision or close to a component that had been selected as being of special interest (such as the myosin molecule in the example considered below).

10.2.6.2 Visualisation

Objects were visualised in a simple viewer with all levels except the atomic being rendered as transparent according to the shape they had been given. Objects in a chain were linked by a thin tube which for spheres ran along the centre-centre direction and so was always normal to the spherical surface. For cylinders and ellipsoids, the linker tube ran from a sphere placed on the end-points, which for cylinders had the same radius as the cylinder (producing sausage-like objects) and for ellipsoids, was only slightly larger than the linker tube. This provides a visual distinction between spheres and spherical ellipsoids.

10.3 Example Application

The program described in the previous section corresponds to a general geometry engine but does not contain any implementation of forces that are more typically found in molecular dynamics (MD) or Monte Carlo (MC) programs. As it stands, the objects simulated will simply shake and diffuse in a random way under the constraints of the imposed steric exclusion, linkage constraints and shape envelope confinement. For anything of biological interest to occur, such as a transition from one state to another, then high level directions must be imposed. These can range from being completely determined, where the starting and final states are fully specified and only the pathway between is of interest or the change can be applied to just a part of the system and the interest is to see how the full system adjusts.

To illustrate how such a system can be implemented in more concrete terms, we will take the example of a molecular motor, specifically myosin-V on an actin filament,[6,16,17] in which large directed changes are applied to a component of the system (myosin) to drive it from one defined (bound) state to another then back to the original state but bound to a different actin molecule along the actin filament.[3] The system will be introduced in two parts: firstly, by describing how the model was set-up as a hierarchical data structure, then secondly, how the dynamics were introduced. The first part is done through data files that have been alluded to in the previous section, however, the second part requires direct run-time interaction with the simulation and this is implemented as a specific user-defined routine called the `driver` (Figure 10.1) which interacts only with the common data structure and executes as an independent parallel process.

10.3.1 Model Construction

We will introduce the data-structure from the top down, starting with file (`actmyo.run`) that specifies the two main components: a myosin-V dimer and the actin filament (Figure 10.2).

actmyo.run

```
PARAM myosin.model
PARAM actin.model
END
GROUP 2
MODEL 0
INPUT myosin.dimer.dat
MODEL 1
INPUT actin.linear.dat
```

The above run file directs the INPUT from two files for the myosin dimer (`myosin.dimer.dat`) and an actin filament (`actin.linear.dat`) that constitute two units (GROUPs) at the highest level. Each group is preceded by a specification of the parameter set (MODEL) that they will use (0 and 1) which corresponds to the PARAMeter files: `myosin.model` and `actin.model`, respectively. These files consist of columns of numbers with each column specifying the values for the different parameters at each level in the hierarchy. The file `myosin.model` consists of seven columns:

myosin.model

```
   0,     0,     0,     3,     1,     2,     1
9999,  1000,   500,   140,    70,    24,     5
   0,   100,   100,    50,    20,    20,    11
   0,     0,     0,     1,     2,     6,     4
   0,     0,     0,     1,     3,     1,     1
   0,     0,     1,     1,     1,     1,     1
   0,     1,     1,     1,     1,     1,     0
   0,     0,     1,     5,     5,     5,     5
   0,     0,     0,     1,     0,     0,     3
```

where the first is the state of the 'world' and the following six define six levels in the hierarchy. For example; the top line specifies the shape associated with each

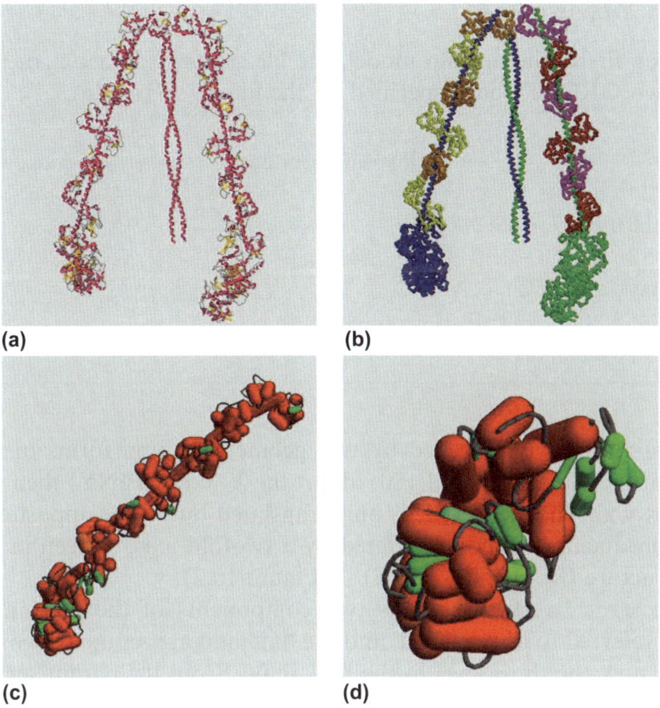

Figure 10.2 Myosin-V structure. The structure of dimeric myosin-V (`2DFS`), determined by single-particle cryo-electron microscopy is shown (a) with secondary structures represented in cartoon style (using RASMOL) with α-helix coloured pink and β-strands yellow. (b) The same structure is shown as a virtual α-carbon backbone with the two heavy chains coloured cyan and green and their associated light-chains in alternating yellow/orange and red/magenta, respectively. Secondary structure line-segments ('sticks') are shown as green tubes for β-strands and thicker red tubes for α-helices. (c) For the full structure of heavy and light chains (excluding the coiled-coil C-terminus) and (d) for the globular foot domain. (The amino-terminus lies in the all-β SH3 domain to the right)

level with: 1 = sphere, 2 = cylinder, 3 = ellipsoid and 0 being a virtual sphere that is not rendered. The following two lines specify the size of each object and its bumping radius (both in arbitrary units). The equivalent lines for the `actin.model` file show a slightly different structure using different values:

actin.model (part)

```
   0,     0,     1,     3,    1,    2,    1
9999,  3000,   120,   120,   70,   24,    5
   0,   400,   100,     0,   70,   20,   11
          :
```

10.3.1.1 Myosin

Considering firstly the myosin model, the file `myosin.dimer.dat` contains another two GROUPs both specified by the file `myosin.dat`:

myosin.dimer.dat

```
GROUP  2
SPINY  180.0
TRANS  55.0     -250.0     -450.0
INPUT  myosin.dat
TRANS  -55.0    -250.0     -450.0
INPUT  myosin.dat
```

The first INPUT is preceded by two geometrical transforms in which the contents of the file are rotated 180° about the Y axis (SPINY) then translated (TRANS x y z). The second file is only translated but in the opposite direction along X producing two copies related by a twofold axis, as seen in the X-ray crystal structure (PDB code: 2DFS).[18] (Figure 10.2).

The file `myosin.dat` contains two components of the myosin molecule which are referred to as the 'head' and the 'tail' and are set up as two GROUPs in what is now the third level in the hierarchy. From the parameter file `myosin.model` it can be seen (column 4) that these objects are the first to have a defined shape and will behave and be rendered as ellipsoids.

myosin.dat

```
GROUP  2
TRANS  -31.0   136.5   59.0
INPUT  myosin.head.dat
INPUT  myosin.tail.dat
```

The file `myosin.head.dat` specifies the structure of the myosin head-group which is the globular kinase domain of the protein while the file `myosin.tail.dat` specifies the extended alpha-helical tail with its associated calmodulin-like light chains (sometimes referred to as the lever-arm or the 'leg'). Although these files contain coordinate data from the same protein structure (`1DGS`), they have been processed separately which puts their centroid to the origin and a translation (TRANS) was applied to the head group to restore their proper relative positions. By specifying the head and the tail as separate groups (units), they can be operated on independently by geometric transforms allowing the `driver` routine to recreate a large relative motion between them called the 'power-stroke' in which the tail swings through a large angle.

The myosin head group was split into seven domains as described previously[25] all of which are contained in distinct files: `head.dom[1-7].dat`. The level-3 unit that contains the domains was defined as an ellipsoid and its

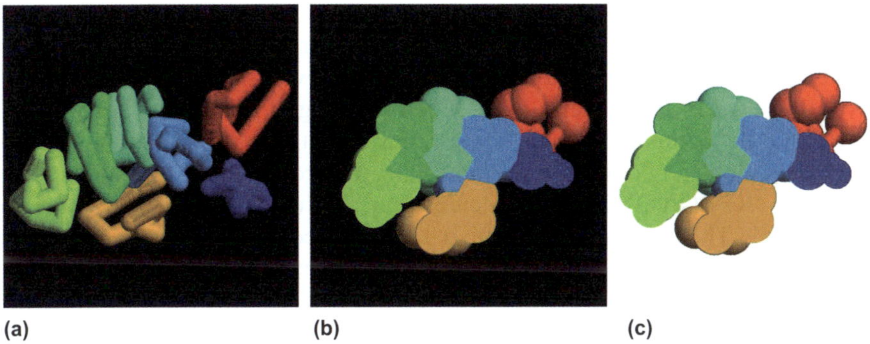

Figure 10.3 (a) SSEs allocated to produce seven domains: 3 core domains (light-blue, cyan, green), two binding domains (yellow, orange) and two ankle domains (blue, red). (b) The SSEs represented in Figure 10.2 are redrawn with space-filling spheres placed on each SSE end-point and sliced through by a plane that would also contain the 'leg' extension. (c) As in part (b) but with domain boundaries sketched and domain codes added: A1–2 (ankle), B1–2 (binding) and C1–3 (core). The direction of the X and Y axes of the internal reference frame are indicated by arrows.

end-points are defined on the GROUP definition line (along with the number of children it contains) (Figure 10.3).

myosin.head.dat

```
GROUP   7    -1.2   -42.6   -37.2   4.32   34.44   39.0
INPUT   head.dom1.dat
INPUT   head.dom2.dat
INPUT   head.dom3.dat
INPUT   head.dom4.dat
INPUT   head.dom5.dat
INPUT   head.dom6.dat
INPUT   head.dom7.dat
REBOND  134   169
REBOND  275   320
REBOND  411   429
REBOND  495   412
REBOND  428   276
REBOND  292   541
REBOND  667   496
REBOND  540   293
REBOND  319   135
REBOND  168   668
```

The path of the chain through the seven head-group domains does not correspond simply to the domain order. Since each domain group is defined automatically by compactness, (giving sets of units that should move together), it is necessary to specify the chain path through the domains. This is done using the identity of units at the atomic level, sequentially numbered over the scope of the current group. Each rewiring of the units is specified by a REBOND command which states that the unit with the first identity number should be bonded to the unit with the second identity number. For example; 'REBOND 134 169' specifies that residue 134 should now link to residue 169. The resulting loose end at 135 is picked-up by the later command 'REBOND 319 135'. These connections can be specified 'by-hand' but are written automatically by the preprocessing program that defines the domains.

Each domain file now takes us down to the lowest (atomic) level at which the actual X,Y,Z, coordinates are encountered. To avoid the proliferation of many small files, the two lowest levels (secondary structures and residues) are defined together with the first GROUP command stating that there are nine secondary structures in the group and the second (lower level) GROUP command specifying six ATOMS (atomic-level units) in each subgroup. The coordinate data (from the PDB file) is given at the atom level along with the secondary structure state encoded in the final column as: 1 = loop, 2 = beta and 3 = alpha. As the secondary structures are defined in the parameter file as cylinders, they are given end-point coordinates on the GROUP command line. If these are omitted, as with the first loop segment, default end-points are generated inside the program.

head.dom1.dat

```
GROUP 9
GROUP 6
ATOM     1    CA    GLY A     1    -3.537 -37.198 -20.771 1.00 1.00
ATOM     2    CA    GLY A     2    -1.470 -34.014 -20.628 1.00 1.00
ATOM     3    CA    GLY A     3     1.589 -35.231 -18.657 1.00 1.00
ATOM     4    CA    GLY A     4     3.165 -37.036 -21.663 1.00 1.00
ATOM     5    CA    GLY A     5     6.875 -37.568 -22.449 1.00 1.00
ATOM     6    CA    GLY A     6     8.601 -34.333 -23.626 1.00 1.00
GROUP 7       6.15  -31.57  -21.01   12.09 -24.98   -5.29
ATOM     7    CA    GLY A     7     6.079 -32.117 -21.857 1.00 2.00
ATOM     8    CA    GLY A     8     7.638 -29.779 -19.295 1.00 2.00
ATOM     9    CA    GLY A     9     6.751 -28.831 -15.716 1.00 2.00
ATOM    10    CA    GLY A    10     7.995 -26.626 -12.874 1.00 2.00
:
SHEET
BETA    44    36
BETA    43    37
BETA    37    43
BETA    36    44
BETA    36    23
BETA    23    36
```

The links between β-strands in a SHEET are specified as pairings on the BETA records at the end of the file.

The structure of the tail component follows along similar lines and will not be described in detail.

10.3.1.2 Actin

The highest level actin specific file, `actin.linear.dat` describes a segment of an actin filament which consists of repeated actin molecules related by helical symmetry with a rotation of $-167°$ and 55 Å translation. Because of the way they interact with the myosin, actins were taken in pairs (called a dimer) giving a shift between dimers of 23° and 110Å‡. These relationships could be encoded by separate rotate and translate commands but as helical symmetry is common, a combined HELIX command was created specifying the two components together:

actin.linear.dat

```
GROUP   16
INPUT   actin.dimer.dat
HELIX   0.0     0.0    -5.8    26.0
INPUT   actin.dimer.dat
HELIX   0.0     0.0    -11.6   52.0
INPUT   actin.dimer.dat
HELIX   0.0     0.0    -17.4   78.0
INPUT   actin.dimer.dat
HELIX   0.0     0.0    -23.2   104.0
INPUT   actin.dimer.dat
:
```

Each `actin.dimer.dat` file simply introduces another level in the hierarchy and maintains the same symmetry around the fibre axis but because the dimer centre lies on the axis, only a shift along Z is needed at the higher level.

actin.dimer.dat

```
GROUP   2
HELIX   -3.55   -0.45     0.0     -0.0
INPUT   actin.one.dat
HELIX    3.55    0.45   -2.9   -167.0
INPUT   actin.one.dat
```

The actin molecule consists of four domains forming a flat disc which was naturally encoded as an oblate ellipsoid. This was (hand) specified by an axis of length 10 (-5 to 5) which relative to the size of the domain specified in

‡NB: the internal coordinates in the data structure are not Ångstroms.

`actin.model` gives an excessive axial ratio which is set to the maximum allowed ratio of 5.

actin.one.dat

```
GROUP    4      -5.0  0.0  5.0  0.0  0.0
INPUT    actin.dom1.dat
INPUT    actin.dom2.dat
INPUT    actin.dom3.dat
INPUT    actin.dom4.dat
RELINK   32    127
RELINK   172    33
RELINK   90    173
RELINK   216   281
RELINK   372   217
RELINK   280    91
ENDS     372   126
```

As with the myosin head domains, the chain path through the actin domains must be relinked. This also entails the creation of a new terminal position which is specified by the ENDS command. (The equivalent myosin command falls in the tail segment).

Each domain file (`actin.dom[1-4].dat`) is similar to the equivalent myosin head group domain files and introduce no novel features.

10.3.2 Driver Construction

The `driver` routine encodes the dynamic aspect of the myosin motor. This includes specifying the actin/myosin bound state along with motion at the hinge points between the myosin head and tail and the myosin dimer.

10.3.2.1 Myosin-V Processive 'Walking'

The structure of the actin/myosin bound complex is known from electron microscopy studies which have been refined by energy minimisation.[19–22] From any given starting configuration, the actin lying closest to the myosin head was identified and the myosin molecule moved towards it. When the myosin came within a predefined distance, its orientation was also refined. Together these operations 'docked' the myosin molecule into a configuration relative to the actin that corresponded to the known structure. When in this position (referred to as 'tightly bound'), the orientation of the myosin tail relative to the head was rotated about an axis that corresponded to what can be inferred from the structures of myosin with tails in different positions. This motion, known as the 'power-stroke' can only occur if the other half of the myosin dimer is unbound.[23,24] In this situation, the free myosin will be carried along towards a new region of the actin filament where it will then search for a new binding site.

Cycling through these states leads to a processive motion of myosin along the actin filament.[§]

The mechanics of the myosin walking motion can be decomposed into three distinct components: the myosin can be bound or unbound to actin, the head can be either in a pre- or post- power-stroke position and the two halves of the myosin dimer can swivel around their dimer interface. There is no explicit communication between the binding states of the two myosin molecules, except what can be communicated through their dimer interface. This has the form of an alpha-helical coiled-coil but little is known of how it responds under stress (tension) or how it affects the diffusion search of an unbound head. By contrast, the power-stroke (PS) transition is well characterised by many structural studies and the forward stroke (from pre- to post-PS positions) occurs only when the myosin is bound to actin whereas the reverse stroke (cocking the trigger) occurs in the unbound state. These transitions are all driven by an ATP hydrolysis cycle (summarised in Figure 10.4), the details of which will not be discussed here.

The deterministic progression of the myosin walking cycle lends itself well to being described as a finite state machine in which a transition to one state opens the options to progress to others. From a theoretical viewpoint, it is of interest to investigate how simple this machine can be to result in processive motion.

10.3.2.2 The Myosin Dimer Hinge

The most independent component of the myosin machine is the dimeric interface which, by analogy to walking motion, will be referred to as the 'hip' joint. As little is known about its structure or dynamics, it was modelled simply as a constraint to hold the ends of the legs at a fixed distance.

Walking motion requires movement of the legs about the hip-joint (Figure 10.5) and rather than rely on the generic diffusion built into the geometry engine, an additional motion was included in the `driver` routine to give any unbound myosin a rotational displacement about the hip. This provides an example of how the `driver` routine can utilise structural information across levels in the hierarchy as the rotation is applied to the whole myosin molecule (level-2) whereas the axis is determined by the domain positions at the end of the tail in both molecules (level-4 in separate sub-trees).

While a faster rate of driven motion reduces the search time for the free myosin head to find a new binding site without disrupting the position of the bound head, it also gives less time for the generic collision detection algorithm to avert clashes between the myosin molecules and the actin filament. To rectify this, a specific high-level check was made on the moving myosin based on the distance of the head from the actin filament (the actin dimer centre) and the other head group. The tail was also checked but as this is an elongated

[§]Note that this processive, or proper walking, motion of myosin-V differs from that of muscle myosin (II) in which the actin myosin contact is only transient. The latter is more akin to a bank of rowers in which each myosin oar is dipped into the actin river.

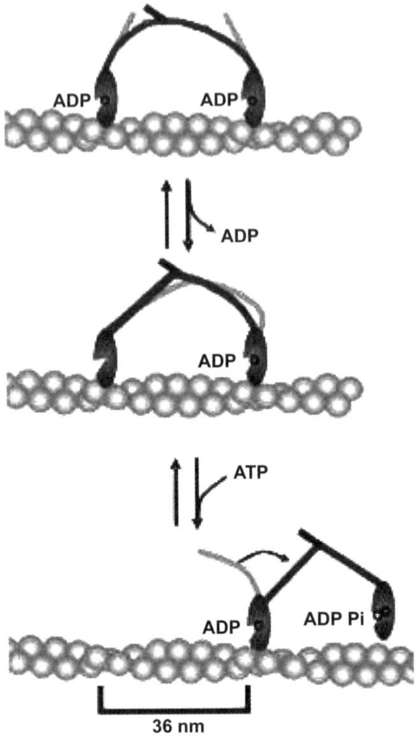

Figure 10.4 (a) Both myosins bind ADP and are attached to actin (creating some bending strain from the relaxed 'leg' positions shown in light grey). (b) With loss of ADP in the left leg, the power stroke can progress in the right leg. (c) On ATP binding, the left leg is released from actin and swings to the right and with ATP hydrolysis, it to returns to the pre-power-stroke conformation. (Reproduced from Reedy (2000) with permission).[23]

substructure, the closest approach of the ellipsoid major axes was monitored and if these fell below a fixed cutoff (set at the same distance as the hip joint) then the two myosins were separated.

10.3.2.3 The Actin/Myosin Binding

The distance of the myosin head from each actin dimer centre was monitored and when this fell within a given range, the myosin was moved closer to the selected actin. This choice is made afresh every time the `driver` routine is activated, so the target actin can vary as the configuration of molecules changes. This approach mode, referred to as 'loose binding', which includes no orientation component, continues until a shorter threshold is reached at which point the closer actin molecule is selected and the myosin is orientated and translated towards its known binding position. The orientation component is calculated based on the internal reference frames of the actin and the myosin head by applying the rotation matrix and translation vector that reproduces the docked complex. A fully bound state is declared if this transformation can be

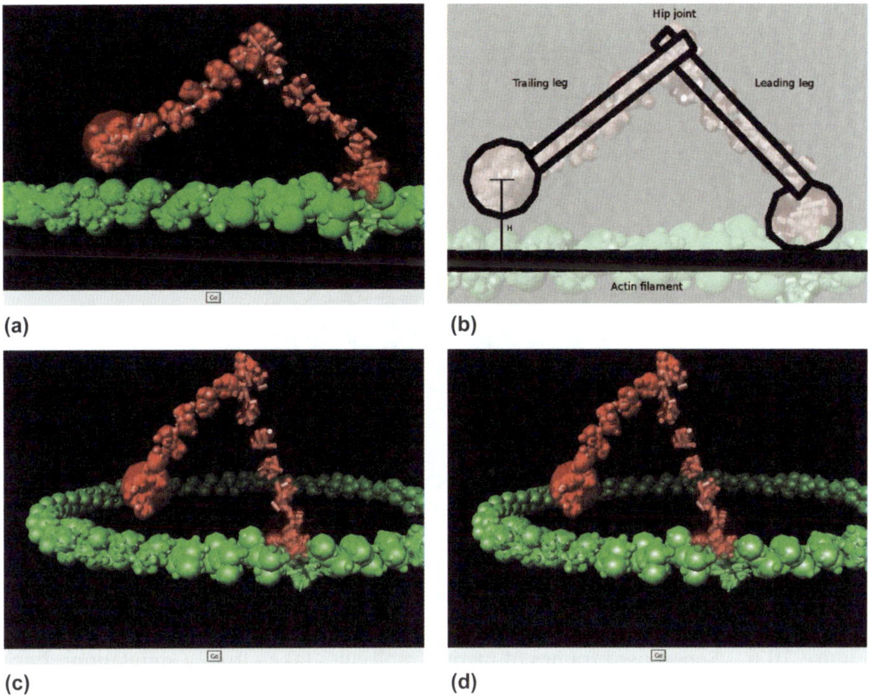

Figure 10.5 (a) The myosin-V dimer model (red) is shown bound to the actin filament model (green). Secondary structure elements (SSEs) are depicted as cylinders (with large/small diameters for α/β) connected by fine lines. The translucent spheres show the higher level groupings of SSEs into domains. (b) Annotates part (a) with the myosin **leading leg** bound tightly to the actin in pre-power-stroke conformation. The myosin **trailing leg** has just detached from the **actin filament** (solid line) and is now free to pivot about the **hip-joint** between the two legs. The height (**H**) of the free myosin above the filament determines when it is recaptured, initially into a loose-binding mode. Parts c and d capture the myosin in a similar conformation showing the circuit of 104 actin molecules (52 dimers with 8 half-repeats) over which the myosin can move.

applied and the end of the tail is placed within the range of the hip-joint to the other myosin.

10.3.2.4 The Power-stroke

The power-stroke (PS) consists of a large swinging motion of the tail relative to the head which is easily encoded in the `driver` routine as a rotation about an axis that lies within the head (Figure 10.6). Structural studies have associated the swivel point with a particular α-helix and the coordinates of this helix centre were taken as the hub. The rotation axis is less well defined but can also be inferred from the known structures and this was calculated and set as a fixed axis in the `driver` routine. Similarly, the extent of the left and right swing were preset with reference to the internal reference frame of the myosin head.

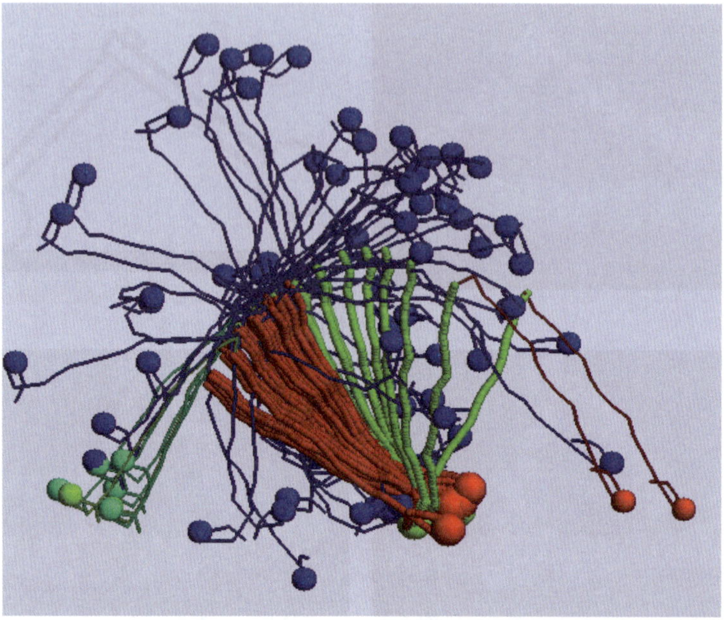

Figure 10.6 Superposed frames of myosin molecules taken from a simulationi. Each myosin is represented in simplified form by a line connecting the mid-points of sequential domains in the tail segment which is drawn thicker for a tightly bound myosin. The head groups are represented by a sphere. The colours represent: active power-stroke (green), post-power-stroke conformation (red), unbound (blue) and loosely bound (cyan).

Given these constraints, a simple mechanism was encoded that maintained a position at the end of the swing range depending on the bound state of the myosin. If the myosin was unbound the tail angle was incremented until it attained the pre-PS position and only when the myosin became fully bound, was this motion reversed towards the post-PS position. These two states were used to impose an additional condition on binding: that the myosin can only select an actin for binding when it is in the pre-PS position. This means that the post-PS state must detach from the actin and revert to the pre-PS position before it can rebind. This introduces a refractory period that depends on how fast the tail can swing which has a physical correspondence to the release of ADP and the subsequent binding of ATP.

10.3.3 Simulation

The specification of the `driver` process introduced a number of parameters that will not be considered in detail here. These were tested and roughly optimised to generate processive motion along the actin filament at as high a speed as could be achieved (Figure 10.7). To avoid end effects, the actin filament was bent into a circular 'race' track and the performance of the system was monitored by the lap-time for the myosin, combined with a penalty for myosin runners that fell off the track (*i.e.* both heads simultaneously detached).

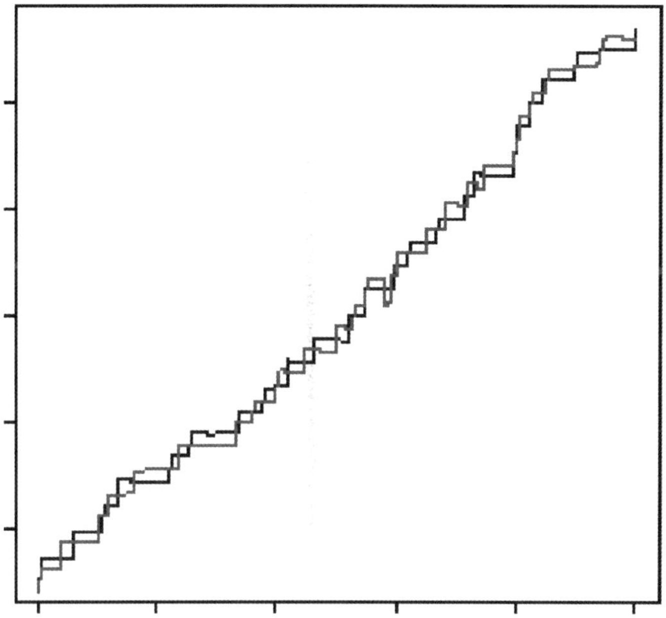

Figure 10.7 Step traces for myosin simulations showing a staircase of overlapping 'steps'. The X-axis is time and the Y-axis actin position in the filament. Trajectories for the two myosin heads are in black and grey.

Although the time scale is arbitrary, this simulation generated behaviour that was remarkably similar to that observed for real myosin-V motion by single particle studies. The major divergence from reality was the source of the rate-limiting step in the stepping cycle, which in the real system is the time for release of ADP whereas in our simulation, the search time for the next binding site was rate limiting.

Various tests were performed on how rigid the myosin tail (legs) had to be and how strictly the hip-joint separation had to be maintained. If the legs were too weak or the hip too wide, then a state could be attained in which both myosins were tightly bound in the post-PS position. If this occurred then there was no frustration in the system and both myosins could remain stably bound forever, resulting in no motion. At the other extreme, if the legs were too stiff and the hip joint maintained strictly at a short separation, then the myosin pair did not have sufficient flexibility to attain simultaneous binding of the two heads, leading to one head always bound with the other always free or in a loosely bound state to various nearby actins – again resulting in no motion.

10.4 Conclusions

In this chapter, we have described a general multi-scale modelling and simulation system that can be easily customised to any large system both in terms of setting-up the model geometry and in directing the behaviour of the simulation.

The use of a multi-level approach to the treatment of collisions means that, at higher levels, the system does not depend strongly on the choice of size or shape assigned to each level. Thus, crude shape approximations do not determine the direct behaviour of the system but just alert the lower level to check for collisions. The propagation of this cascade reaches down to the atomic level where eventually the 'hard-sphere' repulsion is implemented giving a more realistic representation of the true excluded volume.

Unlike a conventional MD or MC simulation, the approach we have described cannot be expected to generate predictive behaviour from a summation of atomic interactions. It relies instead on a hierarchy of user or data defined constraints, such as the definition of secondary structures and their packing interactions. However, these constraints can be incomplete and the missing constraints will allow freedom for change in the system that will be stochastically explored through the underlying random motion applied at every level.

The example we provided of the actin/myosin motor is typical of the scale of application that the system can accommodate, both in terms of number of atoms and size of displacement. While there are many aspects of the behaviour of the system that can be adjusted to reproduce realistic behaviour, we see its use mainly as a way of setting up the physical constraints for a complex system that can then be used to test if a hypothesis about the dynamics of the system is consistent with its physical structure.

Acknowledgements

ZK was supported by the Oxford IRC in bionanotechnology and WRT by the Medical Research Council (UK). Claudia Veigel and Justin Molloy are thanked for valuable discussion and comments.

References

1. J. A. Spudich, How molecular motors work, *Nature*, 1994, **372**(6506), 515–8.
2. R. D. Astumian, Making molecules into motors, *Sci. Am.*, 2001, **285**, 56–64.
3. N. Kodera, D. Yamamoto, R. Ishikawa and T. Ando, Video imaging of walking myosin v by high-speed atomic force microscopy, *Nature*, 2010, **468**, 72–76.
4. P. D. Coureux, H. L. Sweeney and A. Houdusse, Three myosin v structures delineate essential features of chemo-mechanical transduction, *EMBO J.*, 2004, **23**(23), 4527–37.
5. S. Y. Bershitsky, M. A. Ferenczi, N. A. Koubassova and A. K. Tsaturyan, Insight into the actin-myosin motor from X-ray diffraction on muscle, *Front. Biosci.*, 2009, **14**, 3188–3213.
6. J. N. Forkey, M. E. Quinlan and M. A. Shaw, J. E. Corrie, and Y. E. Goldman, Three-dimensional structural dynamics of myosin v by single-molecule fluorescence polarization, *Nature*, 2003, **422**(6930), 399–404.
7. C. Veigel, S. Schmitz, F. Wang and J. R. Sellers, Load-dependent kinetics of myosin-V can explain its high processivity, *Nat. Cell Biol.*, 2005, **7**, 861–869.

8. T. Schlick, *Molecular modeling and simulation: an inter-disciplinary guide*, Springer, New York and London, 2002.
9. T. P. Burghardt and J. Y. Hu, and K. Ajtai, Myosin dynamics on the millisecond time scale, *Biophys. Chem.*, 2007, **131**, 15–28.
10. T. Kawakubo, O. Okada and T. Minami, Dynamic conformational changes due to the atp hydrolysis in the motor domain of myosin: 10-ns molecular dynamics simulations, *Biophys. Chem.*, 2009, **141**, 75–86.
11. S. Izvekov and G. A. Voth, A multiscale coarse-graining method for biomolecular systems, *J. Phys. Chem. B*, 2005, **109**, 2469–2473.
12. Z. Zhang, J. Pfaendtner, A. Grafmuller and G. A. Voth, Defining coarse-grained representations of large biomolecules and biomolecular complexes from elastic network models, *Biophys. J.*, 2009, **97**, 2327–2337.
13. W. R. Taylor and Z. Katsimitsoulia, A soft collision *detection algorithm* for simple Brownian dynamics. *Compu. Biol. Chem.,* 2010, **34**, 1–10. doi.10.1016/jcompbiolchem.2009.11.003.
14. W. R. Taylor, and A. Aszoódi, *Protein Geometry, Classification, Topology and Symmetry*, Institute of Physics, Currently published by CRC Press, 2005.
15. Z. Katsimitsoulia and W. R. Taylor, A hierarchic algorithm for simple Brownian dynamics, *Compu. Biol. Chem.*, 2010, **34**, 71–79. doi.10. 1016/j.compbiolchem.2010.01.001.
16. A. D. Mehta, R. S. Rock, M. Rief, J. A. Spudich, M. S. Mooseker and R. E. Cheney, Myosin-v is a processive actin-based motor, *Nature*, 1999, **400**(6744), 590–3.
17. E. M. De La Cruz, A. L. Wells, S. S. Rosenfeld, E. M. Ostap and H. L. Sweeney, The kinetic mechanism of myosin v, *Proc. Natl. Acad. Sci. U. S. A.*, 1999, **96**(24), 13726–31.
18. J. Liu, D. W. Taylor, E. B. Krementsova, K. M. Trybus and K. A. Taylor, Three-dimensional structure of the myosin V inhibited state by cryoelectron tomography, *Nature*, 2006, **442**, 208–211.
19. K. C. Holmes, D. Popp, W. Gebhard and W. Kabsch, Atomic model of the actin filament, *Nature*, 1990, **347**, 44–49.
20. L. F. Chen, H. Winkler, M. K. Reedy and M. C. Reedy, and A., T. K., Molecular modeling of averaged rigor crossbridges from tomograms of insect flight muscle, *J. Struct. Biol.*, 2002, **138**, 92–104.
21. T. Kawakubo, O. Okada and T. Minami, Molecular dynamics simulations of evolved collective motions of atoms in the myosin motor domain upon perturbation of the ATPase pocket, *Biophys. Chem.*, 2005, **115**, 77–85.
22. M. Lorenz, and K. C. Holmes, *The actin-myosin interface*, *Proc. Natl. Acad. Sci. U. S. A.*, 2010, **107**, 12529–12534.
23. M. C. Reedy, Visualizing myosin's power stroke in muscle contraction, *J. Cell Sci.*, 2000, **113**, 3551–3562.
24. R. Dominguez, Y. Freyzon, K. M. Trybus and C. Cohen, Crystal structure of a vertebrate smooth muscle myosin motor domain and its complex with the essential light chain: visualization of the pre-power stroke state, *Cell*, 1998, **94**, 559–571.
25. W. R. Taylor and Z. Katsimitsoulia, A coarse-grained molecular model for actin-myosin simulation, *J. Mol. Graphics Modell.*, 2010, **29**(2), 266–279.

CHAPTER 11

Top-down Mesoscale Models and Free Energy Calculations of Multivalent Protein-Protein and Protein-Membrane Interactions in Nanocarrier Adhesion and Receptor Trafficking

JIN LIU,[a,†] NEERAJ J. AGRAWAL,[b] DAVID M. ECKMANN,[a,c] PORTONOVO S. AYYASWAMY[d] AND RAVI RADHAKRISHNAN*[a]

[a] Department of Bioengineering, University of Pennsylvania, Philadelphia, PA 19104, USA; [b] Department of Chemical and Biomolecular Engineering, University of Pennsylvania, Philadelphia, PA 19104, USA; [c] Department of Anesthesiology and Critical Care, University of Pennsylvania, Philadelphia, PA 19104, USA; [d] Department of Mechanical Engineering and Applied Mechanics, University of Pennsylvania, Philadelphia, PA 19104, USA
*Email: rradhak@seas.upenn.edu

[†] Current address: School of Mechanical and Materials Engineering, Washington State University, Pullman, WA 99164, USA

RSC Biomolecular Sciences No. 23
Innovations in Biomolecular Modeling and Simulations: Volume 1
Edited by Tamar Schlick

Published by the Royal Society of Chemistry, www.rsc.org

11.1 Introduction

Experimental technologies such as x-ray crystallography and nuclear magnetic resonance are well established at the atomic resolution (1–10 Å) and optical microscopy methods are well established at the micron resolution. However the mesoscale resolution (1–200 nm), which is the most relevant for functional signaling modules, has been elusive. Multiscale modeling can complement experimental technologies in order to access the mesoscale (10–100 nm). Traditional multiscale modeling involves bottom-up approaches of systematically coarse-graining the atomistic description, in order to access the mesoscale. Here, we describe a top-down approach, in which models are constructed at the mesoscale based on phenomenological interaction potentials, and the parameters are determined directly by independent biophysical experimentation. In prior work published in the literature, such an approach has been extensively employed, and specific choices of the governing equations have been validated based on experimental studies.[1–9] The top-down strategy is already proven to be a viable avenue for pursuing models that provide physical insight as well as for enabling direct contact with experiments. In this "applications" focused chapter, we illustrate this approach on two biologically relevant problems: (1) adhesion of nanocarriers to cells mediated by multivalent receptor-ligand interactions in targeted drug delivery; (2) internalization of cell surface receptors in cells via the biological process of endocytosis. Below, we provide brief overviews on the two applications to highlight the overall motivation, experimental challenges, and the need for quantitative modeling. We describe details of our model formulations and results in Sections 11.2 and 11.3 and conclude this chapter in Section 11.4 by providing an outlook for future studies.

Targeted drug delivery using functionalized nanocarriers (*i.e.* carriers coated with specific targeting ligands) represents a promising approach in therapeutic applications. However, targeting of nanocarriers (NCs) to endothelial cells (ECs) remains an important design challenge in biomedical science. The use of functionalized NCs offers a range of tunable design parameters (*i.e.* size, shape, *etc.*) and a high-dimensional tunable parameter space needs to be spanned to determine optimal design.[10] Challenges inherent to design include: (i) molecular and geometric parameters surrounding receptor-ligand interactions and NCs, (ii) accurate characterization of hydrodynamics, (iii) physico-chemical barriers for NC uptake/arrest/internalization, and (iv) uncertainty in targeting environment, to name a few. Binding affinity or association constant (K_a) is a well-defined physico-chemical measure of the efficiency of NC targeting. This quantity has been experimentally measured and often employed in guiding the rational design of functionalized NCs.[11–13] Despite such previous studies on NC binding, a comprehensive understanding of the determinants of NC binding to EC is still limited. The challenge, from a modeling perspective and one we address in Section 11.2, is to predict how the binding affinity depends on experimentally tunable parameters; such a computational frame-work can directly aid in the optimization of NC design.

Curvature-driven processes in cell membranes are of considerable interest to intracellular trafficking, organelle homeostasis, and biogenesis. Several recently discovered protein membrane-binding domains have been postulated to assemble in a process that is driven by membrane curvature and membrane tension and in the process induce local deformations of the bilayer.[14] During the process of endocytosis, clustering of proteins with Epsin N-terminal homology (ENTH) domain in regions of background mean curvature[15] have been reported. The mechanism by which clathrin sustains curvature in the bud region is another open question in clathrin-mediated endocytosis (CME). Since individual heavy chain interactions are comparable to thermal fluctuations, it is difficult to rationalize how the clathrin coat can impose its intrinsic curvature on the membrane.[16–18] The emerging view is that membrane tubulating proteins (*e.g.* epsins) are incorporated as part of the growing coat through clathrin or adaptor protein 2 (AP-2) mediated interactions with the CLAP (clathrin/AP-2) domain on epsin, in order to stabilize the membrane curvature. Efforts to determine the interaction map of CME-related proteins using established methods in structural biology and biochemistry have yielded successful results.[19–37] Yet, crucial to our understanding of endocytosis, and still elusive, is the nature of spatial and temporal localization of these proteins. Through a mesoscale model described in Section 11.3, we provide insights at the 1–200 nm resolution, which aid in our understanding of the structural and energetic aspects of curvature driven processes.

11.2 Mesoscale Model for Functionalized Nanocarrier Targeting to Cells in Drug Delivery

From a modeling perspective, numerical estimation of the binding affinities (or the absolute free energy of binding) is a significant challenge. The calculation of binding affinities necessarily involves the determination of absolute binding free energies, which requires extensive sampling over conformational degrees of freedom as well as the determination of various (translational and rotational) entropy changes upon binding. Recently, inspired by the framework of Woo and Roux[38] on the calculation of binding affinity between a flexible ligand and a receptor, we have developed a mesoscale model to compute the absolute binding free energy of binding of NC to EC.[39] The mesoscale model pertains to spherical NC functionalized with antibodies and binding to antigens on EC surface amidst fluid flow and glycocalyx, see Figure 11.1. We discuss in the following sections, how the physical and geometrical parameters for the mesoscale model and interactions are derived from the independent experiments; this includes the properties of the flow, NC size and shape, antibody surface coverage on NC, glycocalyx resistance, and antigen-antibody interaction. Then, using the parameters derived from experiments and molecular dynamics simulations (MD) in our mesoscale model, we compute the binding affinities of NC to

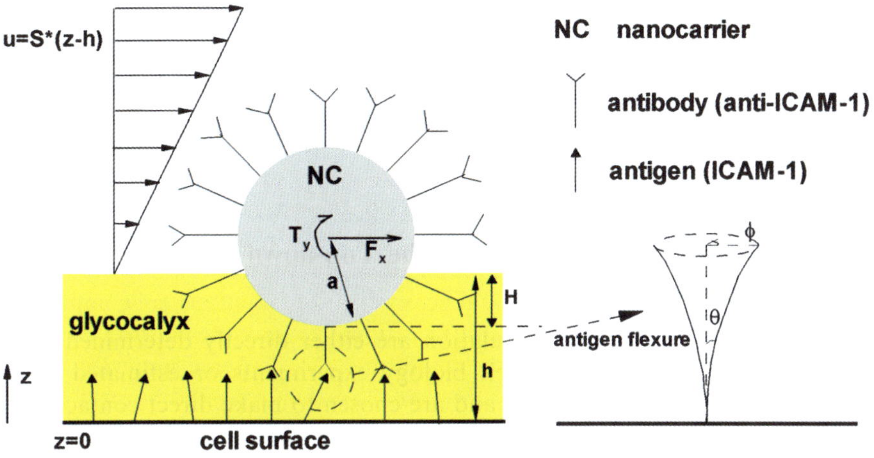

Figure 11.1 A two-dimensional schematic of the mesoscale model for NC binding to EC. The adhesion is mediated through multivalent interactions between antibodies (ligands) on NC and ICAM-1s (receptors) on the EC surface. The ICAM-1 flexure is taken into account by allowing it to bend and rotate in θ and ϕ. Simple 1-D shear flow and glycocalyx are also included as shown. Figure reproduced from reference 39.

EC (described below) and compare the model predictions with experimental data. Using this model, we have investigated the effects from various tunable and design parameters under physiological conditions and compared model predictions quantitatively with corresponding experimental measurements.

As shown in Figure 11.1, we adopt a top-down modeling approach and characterize our system using phenomenological interaction potentials.[39] The NC is treated as a rigid sphere with uniformly distributed antibodies on its surface. The interactions between NC and cell surface are mimicked by the interactions between antibodies and antigens (which are, in the current model confined to a flat surface). The interactions are considered through the Bell model,[40] *i.e.* through the reaction free energy: $\Delta G_r(d) = \Delta G_0 + 0.5kd^2$, where d represents the distance between the reaction sites of the interacting antibody and antigen, ΔG_0 is the free energy change at equilibrium state ($d = 0$) and k is the interaction bond force constant. We also account for the antigens' flexure by allowing them to bend and rotate in θ and ϕ space (see Figure 11.1). Under the assumption of small flexural deformations ($y_L \approx L\theta$), we model the flexure of an antigen as bending a beam from equilibrium (upright) position; thus the bending energy due to flexure can be calculated as (see reference 41 for details):

$$\Delta G_f(\theta) = (2EI/L)\theta^2, \tag{11.1}$$

where EI is the flexural rigidity of receptors, L represents the antigen length and y_L is the distance of the antigen tip from its equilibrium upright position.

Since the antigen flexural movement is highly orientation dependent, a configurational-biased sampling technique[42] is implemented in our model to improve the efficiency of sampling of the configurations of flexural movement. Other additions to the model include the effect of shear flow and the presence of glycocalyx, which have been considered using models and experiments reported earlier.[43–47]

11.2.1 Parameter Estimation in the Top-down Modeling Strategy

The free parameters in the simulation are either directly determined from independent biophysical and cell biology experiments or estimated using molecular dynamics simulations and are chosen to make direct contact with the experimental system in reference 11. The receptor (antigen) parameters are chosen to mimic ICAM-1, and the ligand (antibody) parameters are chosen to mimic the murine anti-ICAM-1 antibody which binds specifically to ICAM-1. For the interactions between antibody and ICAM-1, Muro *et al.*[11] reported the equilibrium free energy change between antibody and ICAM-1 to be -7.98×10^{-20} J at 4 °C, which we set as ΔG_0 in the Bell model of our simulations. We obtain the bond spring constant $k=1000\,\text{dyn/cm}$ by fitting rupture force distribution data reported from single-molecule force spectroscopy.[41,48] Both ΔG_0 and k are assumed to be temperature independent based on which we derive the value of the reactive compliance γ (distance along the reaction coordinate to reach the transition state or point of rupture) to be $\sim 0.4\,\text{nm}$, which agrees very well with experimental evaluations.[48,49] The saturation antibody coverage has been experimentally determined[11] to be 220 per NC with a diameter of 100 nm. The ICAM-1 surface density on endothelial cell is set at 2000 ICAM-1molecules/μm^2 to be consistent with experimental reports of the available binding sites per endothelial cell.[11] We have employed a flexural rigidity of 7000 $\text{pN}\cdot\text{nm}^2$ in our mesoscale model; this value falls between 12 000–24 000 $\text{pN}\cdot\text{nm}^2$ estimated for the ICAM-1 dimer and 800 $\text{pN}\cdot\text{nm}^2$ for the ICAM-1 monomer in prior coarse-grained molecular dynamics calculations,[50] and fits within the spectrum of known biomolecule and receptor rigidities (for *e.g.* the *EI* for glyco-proteins is reported as 700 $\text{pN}\cdot\text{nm}^2$ and that for actin filaments as $15-70\times10^3$ $\text{pN}\cdot\text{nm}^2$).

11.2.2 Calculation of the Absolute Binding Free Energy

For binding of ligands (**L**) (or NC) to receptors (**R**), the binding process described as $\mathbf{L}+\mathbf{R}\leftrightarrow\mathbf{LR}$ and the binding affinity (or association constant) K_a is defined as:

$$K_a=\frac{[\mathbf{LR}]}{[\mathbf{L}][\mathbf{R}]}=\frac{p_1[\mathbf{R}]_{tot}}{[\mathbf{L}]p_0[\mathbf{R}]_{tot}}=\frac{1}{[\mathbf{L}]}\times\frac{p_1}{p_0}. \tag{11.2}$$

Here [**L**], [**R**] and [**LR**] are concentrations of each species. Here, p_0 and p_1 are the fraction of receptors with no ligand and one ligand bound respectively, so that $[\mathbf{R}]=p_0[\mathbf{R}]_{tot}$ and $[\mathbf{LR}]=p_1[\mathbf{R}]_{tot}$, where $[\mathbf{R}]_{tot}$ is the total receptor concentration. The association constant is expressed in terms of the ratio of the integral of configurational degrees of freedoms in the bound state to the unbound state:[38]

$$K_a=\frac{1}{[\mathbf{L}]}\times\frac{\int_{bound}d\mathbf{1}d\mathbf{X}e^{-\beta U_{bound}}}{\int_{unbound}d\mathbf{1}d\mathbf{X}e^{-\beta U_{unbound}}},\tag{11.3}$$

where U_{bound} and $U_{unbound}$ are the total potential energies of the system at bound and unbound states, $\beta=1/k_BT$ in which k_B is the Boltzmann constant and T is the absolute temperature. Here, **1** represents all the degrees of freedom associated with the ligand (NC), **X** represents the degrees of freedom for the remaining molecules (receptors), and $[\mathbf{L}]=1/V_{unbound}$. The integral associated with the unbound state ($U_{unbound}=0$) is over translational degrees of freedom yielding the volume, $V_{unbound}$, and over rotational degrees of freedom yielding a factor of $8\pi^2$ in 3-dimensions. The final form of the binding affinity for interactions between NC and cell surface can be expressed as (see references 39 and 51):

$$K_a=\frac{1}{[\mathbf{L}]}\times T_1\times T_2\times T_3,\tag{11.4}$$

where T_1 represents the entropy loss from the bounded receptors and it can be calculated as:

$$T_1=\frac{A_{R,b}^{(1)}\times A_{R,b}^{(2)}\times\ldots\times A_{R,b}^{(N_b)}}{A_{R,ub}^{(1)}\times A_{R,ub}^{(2)}\times\ldots\times A_{R,ub}^{(N_b)}}.\tag{11.5}$$

Here $A_{R,b}^{(n)}$ is the accessible surface area to the nth receptor in the bound state and $A_{R,ub}^{(n)}$ is the area in the unbound state. Term T_2 accounts for the NC rotational entropy loss due to binding:

$$T_2=\frac{(N_{ab}/N_b)\Delta\omega}{8\pi^2},\tag{11.6}$$

where N_{ab} is the number of antibodies per NC, N_b is the total number of bonds in equilibrium state, $\Delta\omega$ is the rotational volume of the NC in the bound state which can be determined from the distributions of Euler angles as described in reference 52. The term T_3 accounts for NC translational entropy loss,

$$T_3=\frac{A_{NC,b}\int e^{-\beta W(z)}dz}{A_{NC,ub}l_z},\tag{11.7}$$

where $A_{NC,b}$ is the area for the NC translational movement in the bound state, $A_{NC,ub}$ and $A_{NC,ub}l_z$ are the area and volume accessible to the NC in the unbound state, and $W(z)$ is the potential of mean force (PMF).

11.2.3 Calculation of Potential of Mean Force (PMF) of NC Binding to EC

Metropolis Monte Carlo (MC) moves are employed for bond formation/breaking, NC translation and rotation, and antigen translation. The simulations are run in parallel on multiple processors with different realizations of the same physical system. An adaptive step size for NC translation/rotation and antigen diffusion is implemented to ensure a Metropolis acceptance rate of 50%. To calculate the PMF $W(z)$, we choose a reaction coordinate z along which we perform umbrella sampling with harmonic biasing potentials. The umbrella sampling is performed with window size of $\Delta z = 0.05\,\text{nm}$ and the harmonic biasing potential in each window is chosen to be $0.5k_u(z - z_{0,i})^2$, where $0.5k_u(\Delta z)^2 = 1.0 \times 10^{-20}$ J, k_u is the harmonic force constant and $z_{0,i}$ is the location of the center of window i. The NC is slowly moved to the cell surface by updating $z_{0,i}$. A total of 200 million Monte Carlo steps are performed in each window and the histogram is stored only when there exists at least one bond. All the relevant parameters including the window size Δz, strength of the biasing potential k_u and the sampling size in each window have been tested to ensure convergence. The weighted histogram analysis method (WHAM) algorithm[53] is used to unbias and combine the histograms in different windows to form a complete PMF ($W(z)$) profile using a tolerance factor of 10^{-6}. PMF profiles for each system are averaged over four independent realizations.

In reference 39 we reported the PMF profile for NCs with diameter of 100 nm and the number of antibody $N_{ab} = 162$ (which corresponds to 74% of saturation coverage). Figure 11.2(a) shows the typical PMF profile at temperature of 27 °C, as clearly indicated, three firm bonds form ($N_b = 3$) with a characteristic free energy well around $-32\,k_BT$ in equilibrium state when projected along the

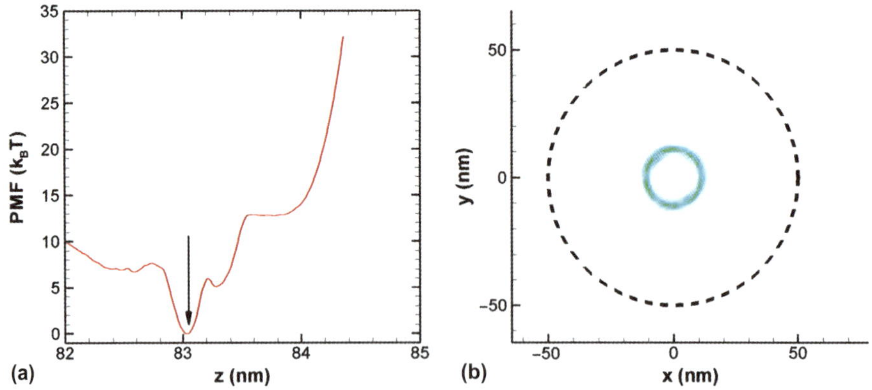

Figure 11.2 Typical PMF ($W(z)$) profile (a) and the averaged spatial distribution of bound antibody-ICAM-1 pairs at the equilibrium state (b) indicated by the arrow in (a) for antibody surface coverage of 74%. Figure reproduced from reference 39.

reaction coordinate z. The PMF change is smaller for the second and third bonds compared to the first as a result of ICAM-1 flexure and differences in loss in translational and rotational entropies of NC as well as ICAM-1. The multivalency or the number of bonds formed at equilibrium is affected by the bond reaction free energy, bond spring constant, ICAM-1 bending rigidity, antibody surface coverage, and NC size. Figure 11.2(b) shows the spatially averaged distribution of the bound ICAM-1 relative to the center of NC, and this is an annulus distribution, based on which the all the area terms $A_{R,b}{}^{(i)}$ described above can be estimated: the outer envelope of the bond distribution in (b) defines the minimal patch on the EC surface for complete NC binding. In particular, the first bound ICAM-1 can freely access the area within the outer circle with outer radius $r_0 = 12.5$ nm, yielding $A_{R,b}{}^{(i)} = \pi r_0{}^2$, the second bound ICAM-1 can only access the area within the annulus between the outer and inner circles (inner radius $r_i = 9.7$ nm) with corresponding $A_{R,b}{}^{(2)} = \pi(r_0{}^2 - r_i^2)$, and the third bound ICAM-1 is restricted to a patch with accessible area $A_{R,b}{}^{(3)} = (r_0 - r_i)^2$. In contrast, each of these ICAM-1 molecules in the unbound state can access a surface area determined by the ICAM-1 density. Similarly, the translational area accessed by a bound NC is $A_{NC,b} = (r_0 - r_i)^2$, and as stated above the rotational mobility of bound NC, ($\Delta\omega$) is estimated from the Euler angle distributions. Finally, the computed binding affinity is estimated as $K_a = 5.9 \times 10^{10}\,\text{nm}^3$ and the dissociation constant is $K_d = 1/K_a = 28$ pM. In a similar fashion,[51] the binding affinity between single antibody and ICAM-1 can also be calculated and as listed in Table 11.1, both the interactions of NC-cell surface (for 100 nm NCs) and antibody-ICAM-1 are consistent with corresponding experimental measurements under similar conditions, see below.

11.2.4 Comparison with Experiments

In Table 11.1, we compared the dissociation constants (K_d) computed from our model with *in vitro* measurements[11] under similar conditions for both interactions of NC-cell and single antibody-ICAM-1. Both interactions agree within reasonable uncertainties. We also studied the effect of the antibody surface coverage (σ_s) on NC,[39] and interestingly we observed a threshold at about 45% coverage, below which the NC binding precipitously dropped below that of single antibody-ICAM-1 interactions. Intriguingly, this threshold was almost exactly reproduced in *in vivo* measurements of the endothelial targeting of NCs in the lungs of mice at different antibody coverages; detailed description of the comparison can be found in reference 39. The individual PMF profiles at

Table 11.1 The dissociation constants for different interactions; comparison between model predictions and experimental measurements.

Dissociation Constant ($K_d = 1/K_a$)	*Model*	In vitro *Experiment*
Antibody-ICAM-1 interaction	3.2 nM	8.5 nM
NC-cell interaction	28 pM	77 pM

different σ_s revealed an integer reduction in the multivalency (from three to two) at the threshold, and as a consequence, the binding free energy reduced by about $6\,k_BT$. The significance of this result is that for 100 nm NCs, there exists an optimal antibody coverage (at ~45%), above which the NCs can bind very effectively to endothelial cells. We further demonstrated the accuracy of our model by comparing our model predictions with AFM force measurements. The AFM rupture force measurements of NC unbinding yielded a mean rupture force of 316 pN over 89 trials; we predicted a mean rupture force of ~215–230 pN derived from our computed PMF profiles, which again, is in close agreement with experiments (see reference 39 for details). Furthermore, our most recent reports of the effects of particle size, glycocalyx and shear flow,[51] have all yielded model predictions in close agreement with experimental observations. In reference 51, we have extended our analysis to include the effects of glycocalyx and shear flow (hydrodynamics) on binding between NC and cell surface. The results indicated that a glycocalyx layer of thickness of 55~70 nm only reduces the binding affinity by about 2-fold although a glycocalyx layer of thickness of 100 nm can reduce the binding by a factor of 500. This exponential dependence was confirmed in our simulations with different particle sizes (100 nm *vs.* 200 nm particles), and helps to rationalize the large differences between *in vitro*[51] and *in vivo*[54] experimental results of glycocalyx-mediated NC binding. The effect of flow was also investigated. From the PMF profiles at different shear rates, it was found that the free energy of binding first decreases below a threshold S then increases. As a result, the corresponding dissociation constant (K_d) shows a biphasic trend with shear rate, *i.e.* below a threshold shear rate the flow actually enhances NC binding by reducing K_d, while above the threshold the flow reduces the binding by promoting NC detachment. Consistent with experimental measurements of particle rolling velocities,[55] we have shown that this biphasic behavior was insensitive to the particle size. Our computational results strongly suggest that multivalent interactions are the key to this shear enhanced NC binding phenomenon, which has been extensively studied experimentally due to its important implications in many disease related processes.[56]

11.3 Mesoscale Model for Protein-Induced Membrane Curvature

Since curvature-driven processes like endocytosis manifest as a result of cooperative protein-protein as well as protein-membrane interactions that span multiple length and time scales, an integration of models for membrane dynamics with models for cellular signaling is necessary.[57,58] Several bottom-up approaches to study membrane phase and dynamical behavior have been described involving both atomistic as well as coarse-grained models.[59–64] Such efforts have been successful in delineating the interactions between the membrane and membrane-bound biomolecules,[59,65,66] in describing micelle formation and vesicle fusion,[67–73] and in characterizing the elastic properties of membranes.[74] In contrast, top-down approaches have included continuum field-based mesoscale models[72,75–78] to describe applications at the mesoscale,[79] which have inspired us to formulate the mesoscale model for induced membrane curvature.

11.3.1 Membrane Curvature Model and Conformational Sampling

11.3.1.1 Helfrich Elastic Bending Free Energy for Fluid Membranes

Following the Helfrich formulation,[80] we model the membrane energy E as:[17,81]

$$E = \int_S \left\{ \frac{\kappa}{2} (2M - H_0)^2 + 2\kappa' K + \sigma \right\} dA \tag{11.8}$$

Here, κ and κ' are membrane elastic bending rigidities, $M = (c_1 + c_2)/2$ is the mean curvature of the membrane, c_1 and c_2 are the two principal radii of curvatures, H_0 is the spontaneous imposed (or intrinsic) curvature of the membrane due to curvature-inducing proteins, $K = c_1 c_2$ is the Gaussian curvature, σ is the frame tension, and A is the total membrane area. For topologically invariant surfaces (membrane patches) characterized by constant Gaussian rigidity, the Gauss curvature term contributes just a constant to the free energy (owing to the Gauss-Bonnet theorem)[80] and is dropped for this reason from our analysis. For small deformations of the membrane, the shape of the membrane can be represented in the Monge gauge ($z = z(x,y)$), converting the Hamiltonian (not including the Gaussian rigidity term) into the more computationally efficient form:

$$E = \iint \left[\frac{\kappa}{2} \left(\nabla^2 z - H_0\right)^2 + \left(\frac{\kappa}{4} H_0^2 + \frac{\sigma}{2} \right) (\nabla z)^2 \right] dxdy. \tag{11.9}$$

In our membrane simulations, the membrane surface is discretized by a triangulation system consists of $(N+1)^2$ vertices connected by $(3N^2 + 2N)$ links, here N is the number of links in each side of the square membrane. The total number of triangles is $(2N^2)$. Therefore, the total Helfrich energy for bending elasticity associated with mean curvature (*i.e.* dropping the Gaussian curvature and the frame tension contributions) can be calculated by summation over all the vertices:

$$H = \frac{\kappa}{2} \sum_{v=1}^{(N+1)^2} A(v) \left[\left(c_1(v) + c_2(v) - H_0(v) \right)^2 \right]. \tag{11.10}$$

Here, $A(v)$ is the average surface area around vertex v. The two principal curvatures at vertex v are computed very efficiently through the transformation between the global Cartesian frame and local Darboux frame (see reference 82

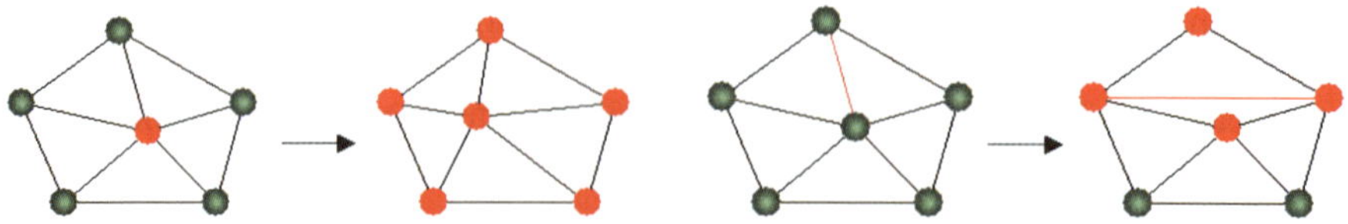

Figure 11.3 Illustration of vertex shift and link flip moves left: the vertex shift move, and right: link flip move in the Helfrich MC method.

for a detailed description). The inclusions influence membrane topology by inducing spontaneous curvature H_0^e. The total Helfrich energy of the membrane is minimized and the equilibrium properties of the membrane-inclusion system are evaluated using a novel Monte Carlo (MC) technique. The MC steps are divided into two classes:

Vertex shift: as indicated in Figure 11.3, the vertex (red circle in the left) is randomly selected and an attempt is made to move to new position. In this step, the mean curvatures and therefore the Helfrich energies on all the connecting vertices (red circles in the right) are updated, based on which the attempt is accepted or rejected. The step-size of the vertex movement is adjusted to ensure the acceptance rate $\sim$30–50%.

Link flip: as indicated in Figure 11.3, the link (red line in the left) connecting two vertices is randomly selected and an attempt is made to flip to the pair of opposite vertices. During this step, the mean curvatures and therefore the Helfrich energies on four related vertices (red circles in the right) are recalculated, based on which the attempt is accepted or rejected. For each of the movements, the new system energy is computed and the acceptance probability is based on the Metropolis criteria by comparing with old system energy. The duration of the simulation is measured in Monte Carlo steps (MCs) and each step contains $(N+1)^2$ vertex shifts and $(3N^2+2N)$ link flips.

11.3.1.2 Diffusion of Membrane Bound Curvature Inducing Proteins (Epsins)

We model the membrane curvature induced by (curvature-inducing) proteins as a Gaussian function with a range b_i and magnitude $C_{0,i}$:

$$H_{0,i}^e(x,y) = C_{0,i}e^{\left[(x-x_i)^2+(y-y_i)^2\right]/b_i^2}, \tag{11.11}$$

where x_i and y_i are the x and y coordinates of the ith inclusion on the membrane. When multiple proteins are present on the membrane, the resulting protein-induced membrane curvature is calculated as:

$$H_0^e(x,y) = \sum_{i=1}^{N_e} H_{0,i}{}^e, \tag{11.12}$$

where N_e is the number of proteins present on the membrane. We chose $b_i = 8.3$ nm while the value of $C_{0,i}$ was set to be 0.075 nm^{-1}, consistent with the properties of epsin (ENTH domain) interacting with membranes, as described in our previous work.[83] The protein diffusion is modeled through the simulation of probabilistic hopping along the membrane surface. In each diffusion step, while we fix the membrane profile, we allow the protein to translate continuously on the $x-y$ plane to a new position, then the energy change associated with such movement is calculated and this movement is rejected or accepted according to Metropolis criteria. We self-consistently vary the magnitude of the protein translation step in order to ensure a 30% rate of acceptance. In addition to the mobile curvature inducing proteins described above, a background curvature induced in the membrane (such as due to presence of a clathrin coat) is modeled as:[17]

$$H_0^c(x,y) = C_0\Gamma(r_0), \tag{11.13}$$

where $\Gamma(r_0)$ is a function that is unity within a circular domain (centered at zero) of radius r_0 and zero otherwise; r_0 is the linear extent (radius) of the curvature-field induced by clathrin and projected on the $x-y$ plane; and C_0 is the magnitude of clathrin-induced curvature. Finally the net intrinsic curvature at any position is then expressed as:

$$H_0(x,y) = max\left[H_0^e(x,y), H_0^c(x,y)\right], \tag{11.14}$$

where, the *max* operator denotes the maximum of the two values.

11.3.2 Calculation of Relative Free Energies of Curved Membranes

We have recently developed a computational methodology for calculating relative free energies for fluid membranes subject to spatially dependent intrinsic curvature fields using the method of thermodynamic integration.[17] To demonstrate the applicability of the method, we consider a model system of membrane deformations caused by a static (*i.e.* non-diffusing) heterogeneous curvature field. The model is characterized by a radially symmetric mean curvature field on the membrane over a localized region characterized by a linear extent r_0 (see Figure 11.4 and Eq. 11.13). We calculate the free energy change of the membrane as a function of the extent of the curvature field (r_0) as well as the magnitude of the curvature field (C_0) using the method of thermodynamic integration (TI) along C_0. Considering C_0 to be a coupling parameter that can be varied independently, the derivative of the Helmholtz free energy $F = -k_BT \ln Q$ (where Q is the partition function), can be written as:

$$\frac{\partial F}{\partial C_0} = \left\langle \frac{\partial E}{\partial C_0} \right\rangle_{C_0}. \tag{11.15}$$

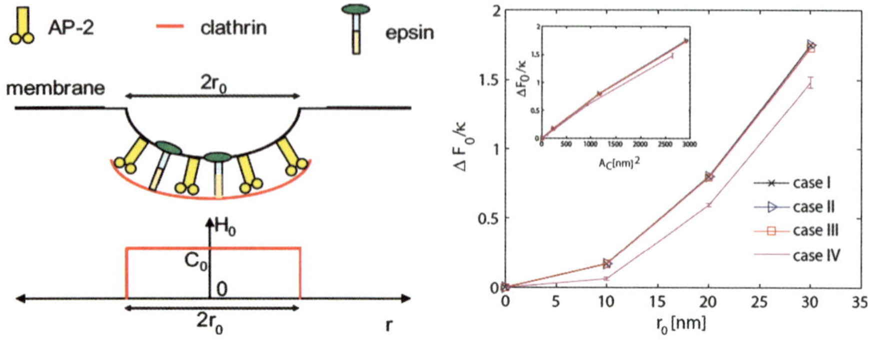

Figure 11.4 **Left:** Depiction of the intrinsic curvature field in our model for the presence of the clathrin-coat assembly. The lower panel depicts the form of spontaneous curvature function H_0, corresponding to a region of induced curvature of linear extent r_0. **Right:** Computed membrane free-energy change as a function of r_0. Error bars, where not visible, are smaller than the symbols. The four cases differ in the parameters (L/nm, κ/k_BT, $\sigma/(\mu N/m)$) considered in the simulations, where L is the total length of the square membrane patch: case I = (250,20,0), case II = (250,50,0), case III = (250,50,3), case IV = (500,50,3). Figure reproduced from reference 17.

In the Monge gauge, the above identity can be integrated to yield,

$$F(C_0, r_0) - F(0, r_0) = \int_0^{C_0} \left\langle \Gamma(r_0)\kappa \iint \left[-(\nabla^2 z - C_0\Gamma(r_0)) + \left(\frac{C_0}{2}\right)(\nabla z)^2 \right] dxdy \right\rangle_{C_0} dC_0. \tag{11.16}$$

In Figure 11.4, we show the results of the numerically integrated free energy (based on Eq. 11.16) as a function of r_0, for a fixed value of $C_0 = 0.04\ \text{nm}^{-1}$ (magnitude of the intrinsic curvature in clathrin coated vesicular bud formation). In doing so, we have used the identity that $F(0,r_0)$ is independent of r_0 and evaluate $F(C_0,r_0) - F(0,r_0)$ for different values of r_0. Hence, by explicitly computing the free-energy changes we are able to quantify change in entropy associated with increasing membrane deformation. Based on our results, two important outcomes have emerged: first, by explicitly computing the free-energy changes and by quantifying the loss of entropy associated with increasing membrane deformation, we show that the membrane stiffness increases with increasing intrinsic field, thereby, renormalizing the membrane bending rigidity. Second, we show that the entropy of the membrane decreases with increasing magnitude of r_0. These results help to quantify the free-energy change when a planar membrane deforms under the influence of curvature-inducing proteins at a finite temperature.[84]

11.3.3 Application to Clathrin-Mediated Endocytosis

Using our simulation methodology, we first analyzed the equilibrium behavior of bilayer membranes under the influence of curvature-inducing proteins in the Monge approximation.[17,81] This study revealed how the protein diffusion and cooperativity associated with a multitude of protein–membrane interactions elicits rearrangements and new topographies in the membrane phase. Based on our model simulations, we predicted that, depending on the density of membrane-bound proteins and the degree to which a single protein can induce intrinsic mean curvature in the membrane, a range of membrane phase behaviors can be observed including two different modes of vesicle-bud nucleation and repressed membrane undulations. A state diagram as a function of experimentally tunable parameters to classify the underlying states is provided in Figure 11.5.

Based on the state behavior depicted in Figure 11.5, we explored the role of cooperative protein-protein and protein-membrane interactions in the ubiquitous endocytic pathway in mammalian cells, namely the CME, see Figure 11.6.[83,85] In our model for CME, the epsins bind to the lattice of a growing clathrin coat through the interaction of the CLAP domain of epsin with the clathrin triskelion. This way, multiple epsins localized spatially and orientationally templated on the clathrin coat collectively play the role of a curvature inducing capsid. In addition, epsin serves as an adapter in binding the clathrin coat to the membrane through the interaction of its ENTH domain with the phosphotidylinositol (4,5) bisphosphate (PI(4,5)P2 or PIP2) molecules on the membrane. By employing the Helfrich methodology outlined above, we

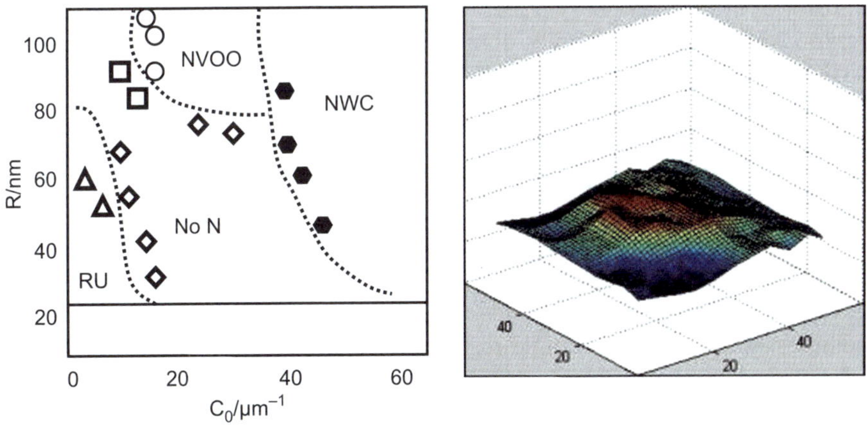

Figure 11.5 **Left:** for different sets of parameters, the calculated global state diagram describing membrane behavior in parameter space. Abbreviations: RU: Repressed undulations; No N: No nucleation; NVOO: Nucleation via bond orientational ordering with cooperativity; NWC: Nucleation without cooperativity. Here, $R = 2b_i$, and b_i, C_0 are defined earlier. Figure reproduced from reference 81.

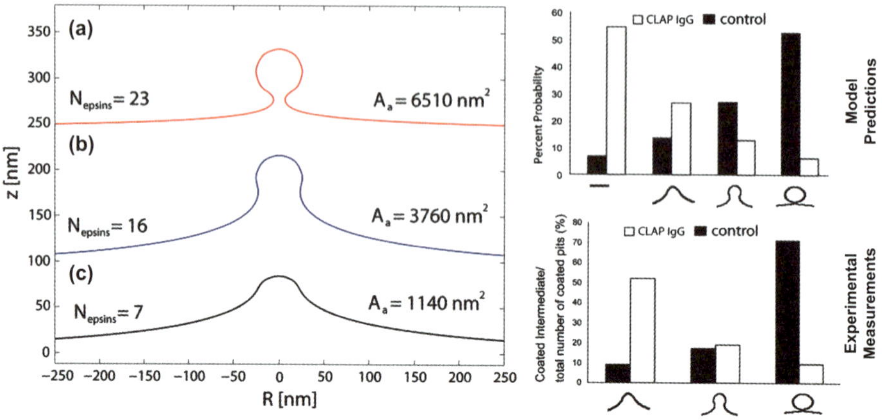

Figure 11.6 **Left:** membrane deformation profiles under the influence of imposed curvature of the epsin shell model for three different coat areas; here $\kappa = 20\ k_BT$. For the largest coat area, the membrane shape is reminiscent of a clathrin-coated vesicle. **Right:** calculated (top) and experimentally measured (bottom) probability of observing a clathrin-coated vesicular bud of given size in WT cells (filled) and CLAP IgG injected cells (unfilled). Figure reproduced from reference 83.

have addressed how the shapes and energetics of vesicular-bud formation in a planar membrane is stabilized by presence of the epsin/clathrin assembly, see Figure 11.6. We suggest that there exists a critical size of the coat above which a vesicular bud with a constricted neck resembling a mature vesicle is stabilized. Our model also provides estimates for the number of epsins involved in stabilizing a coated vesicle (Figure 11.6, left) and without any direct fitting, reproduces the experimentally observed shapes of vesicular intermediates as well as their probability distributions quantitatively, in wildtype as well as CLAP IgG injected neuronal cell experiments (Figure 11.6, right).

11.4 Future Outlook

From a methodological vantage point, the top-down mesoscale strategy we have employed can be further enriched by considering a variety of additional interactions. For example, models treating membrane as a pair of slightly compressible monolayers bound together with non-instantaneous lipid density relaxation have been proposed.[86] Minimal models for cytoskeleton fortified membranes have also been developed[87] and applied to erythrocyte deformation.[88–92] This model includes shear elasticity of the cytoskeleton in combination with the Helfrich Hamiltonian. Based on the Helfrich description, the dynamics of the elastic membrane sheet in the overdamped limit including hydrodynamic coupling to surrounding solvent and arbitrary external forces have been investigated in previous studies.[84,92,93] The infinitely thin elastic sheet assumption has also been relaxed and the inter-layer friction and slippage between the lipid monolayers have been incorporated.[94,95] Mechanistic models for cell membrane deformation and vesicle budding in the cellular context based on the elastic free energy formulations have also been proposed.[96] These studies have further motivated the development of models for protein diffusion in ruffled surfaces[97–101] and the simultaneous diffusion of protein and membrane dynamics.[15,72,102–110] From the application stand-point, an exciting avenue of future work is to synergize the mesoscale model with recent advances in live-cell fluorescent imaging (which have allowed for accurate recording of the dynamics of specific tagged molecules *in vitro*[108,111–113]) and directly compare spatial localization of these proteins in three dimensions[14,114] between model and experiment. The inclusion of membrane in the NC model is an obvious next step, which will enable the exciting study of NC internalization in ECs. Predictive simulations or guides for future NC experimentation may be envisioned including the effect of ICAM-1 density on targeting in normal *vs.* inflamed endothelium in cardiovascular disease models, nanocarrier internalization subsequent to binding mediated by membrane mechanics, and the effect of particle size and shape in binding and internalization, and targeting NC to complex disease states such as cancer. With these suggested extensions to our model, we believe that our current mesoscale modeling platform can be further enhanced to guide pharmacological design.

Acknowledgements

We acknowledge financial support from NSF grants CBET-0853389, CBET-0853539, and NIH grants NIBIB-1R01EB006818, NHLBI-1R01HL087036. Computational resources were provided in part by the National Partnership for Advanced Computational Infrastructure (NPACI) under the allocation grant MRAC MCB060006.

References

1. A. K. Chakraborty, M. L. Dustin and A. S. Shaw, *Nature Immunology*, 2003, **4**, 933–936.
2. S. E. Lee, Y. Hori, J. T. Groves, M. L. Dustin and A. K. Chakraborty, *Trends in Immunology*, 2002, **23**, 500.
3. S. Raychaudhuri, A. K. Chakraborty and M. Kardar, *Physical Review Letters*, 2003, **91**, 208101.
4. K. H. Lee, A. R. Dinner, C. Tu, G. Campi, S. Raychaudhuri, R. Varma, T. N. Sims, W. R. Burack, H. Wu, J. Wang, O. Kanagawa, M. Markiewicz, P. M. Allen, M. L. Dustin, A. K. Chakraborty and A. S. Shaw, *Science*, 2003, **302**, 1218–1222.
5. S. J. Lee, Y. Hori and A. K. Chakraborty, *Proc. Natl. Acad. Sci. USA*, 2003, **100**, 4383–4388.
6. A. K. Chakraborty, *Nature Immunology*, 2002, **3**, 895–896.

7. S. J. Lee, Y. Hori, J. T. Groves, M. L. Dustin and A. K. Chakraborty, *Trends in Immunology*, 2002, **23**, 492–499.
8. S. Y. Qi, J. T. Groves and A. K. Chakraborty, *Proc. Natl. Acad. Sci. USA*, 2001, **98**, 6548–6553.
9. A. K. Chakraborty, *Science's Stke*, 2002, **122**, PE10.
10. A. Khademhosseini and R. Langer, *Chem. Eng. Prog.*, 2006, **102**, 38–42.
11. S. Muro, T. Dziubla, W. Qiu, J. Leferovich, X. Cui, E. Berk and V. R. Muzykantov, *J. Pharmacol. Exp. Ther.*, 2006, **317**, 1161–1169.
12. J. B. Haun and D. A. Hammer, *Langmuir*, 2008, **24**, 8821–8832.
13. K. Ho, Y. Lapitsky, M. Shi and M. S. Shoichet, *Soft Matter*, 2009, **5**, 1074–1080.
14. S. Saffarian, E. Cocucci and T. Kirchhausen, *PLoS Biol*, 2009, **7**, e1000191.
15. A. Roux, G. Koster, M. Lenz, B. Sorre, J. B. Manneville, P. Nassoy and P. Bassereau, *Proc Natl Acad Sci U S A*, **107**, 4141–4146.
16. N. J. Agrawal, J. Weinstein and R. Radhakrishnan, *Molecular Physics*, 2008, **106**, 1913–1923.
17. N. J. Agrawal and R. Radhakrishnan, *Phys Rev E Stat Nonlin Soft Matter Phys*, 2009, **80**, 011925.
18. M. M. Kozlov, *Nature*, 2007, **447**, 387–389.
19. J. Hirst, D. A. Sahlender, S. Li, N. B. Lubben, G. H. H. Borner and M. S. Robinson, *Traffic*, 2008, **9**, 1354–1371.
20. F. T. Huang, A. Khvorova, W. Marshall and A. Sorkin, *Journal of Biological Chemistry*, 2004, **279**, 16657–16661.
21. D. Loerke, M. Mettlen, D. Yarar, K. Jaqaman, H. Jaqaman, G. Danuser and S. L. Schmid, *Plos Biology*, 2009, **7**, 628–639.
22. J. Rejman, V. Oberle, I. S. Zuhorn and D. Hoekstra, *Biochemical Journal*, 2004, **377**, 159–169.
23. N. Rusk, P. U. Le, S. Mariggio, G. Guay, C. Lurisci, I. R. Nabi, D. Corda and M. Symons, *Current Biology*, 2003, **13**, 659–663.
24. S. Semerdjieva, B. Shortt, E. Maxwell, S. Singh, P. Fonarev, J. Hansen, G. Schiavo, B. D. Grant and E. Smythe, *Journal of Cell Biology*, 2008, **183**, 499–511.
25. D. Vanden Broeck and M. J. S. De Wolf, *Biotechniques*, 2006, **41**, 475–484.
26. S. Vergarajauregui, A. San Miguel and R. Puertollano, *Traffic*, 2006, **7**, 686–698.
27. F. Soulet, D. Yarar, M. Leonard and S. L. Schmid, *Molecular Biology of the Cell*, 2005, **16**, 2058–2067.
28. C. X. Zhang, A. E. Y. Engqvist-Goldstein, S. Carreno, D. J. Owen, E. Smythe and D. G. Drubin, *Traffic*, 2005, **6**, 1103–1113.
29. A. Benmerah, C. Lamaze, B. Begue, S. L. Schmid, A. Dautry-Varsat and N. Cerf-Bensussan, *Journal of Cell Biology*, 1998, **140**, 1055–1062.
30. S. Confalonieri, A. E. Salcini, C. Puri, C. Tacchetti and P. P. Di Fiore, *Journal of Cell Biology*, 2000, **150**, 905–911.
31. S. D. Conner and S. L. Schmid, *J. Cell Biol.*, 2003, **162**, 773–780.

32. A. A. de Melker, G. van der Horst, J. Calafat, H. Jansen and J. Borst, *Journal of Cell Science*, 2001, **114**, 2167–2178.
33. F. Irie, M. Okuno, E. B. Pasquale and Y. Yamaguchi, *Nature Cell Biology*, 2005, **7**, 501–U569.
34. T. W. Koh, V. I. Korolchuk, Y. P. Wairkar, W. Jiao, E. Evergren, H. L. Pan, Y. Zhou, K. J. T. Venken, O. Shupliakov, I. M. Robinson, C. J. O'Kane and H. J. Bellen, *Journal of Cell Biology*, 2007, **178**, 309–322.
35. S. Pennock and Z. X. Wang, *Molecular and Cellular Biology*, 2008, **28**, 3020–3037.
36. S. Sigismund, T. Woelk, C. Puri, E. Maspero, C. Tacchetti, P. Transidico, P. P. Di Fiore and S. Polo, *PNAS*, 2005, **102**, 2760–2765.
37. S. vanDelft, C. Schumacher, W. Hage, A. J. Verkleij and P. Henegouwen, *Journal of Cell Biology*, 1997, **136**, 811–821.
38. H. Woo and B. Roux, *Proc. Natl. Acad. Sci. USA*, 2005, **102**, 6825–6830.
39. J. Liu, G. Weller, B. Zern, P. S. Ayyaswamy, D. M. Eckmann, V. R. Muzykantov and R. Radhakrishnan, *Proc. Natl. Acad. Sci. USA*, 2010, **107**, 16530–16535.
40. G. I. Bell, M. Dembo and P. Bongrand, *Biophys. J.*, 1984, **45**, 1051–1064.
41. N. J. Agrawal and R. Radhakrishnan, *J. Phys. Chem. C*, 2007, **111**, 15848–15856.
42. D. Frenkel and B. Smit, *Molecular simulation: from algorithms to applications*, Academic, Orlando, FL, 2001.
43. S. Weinbaum, X. Zhang, Y. Han, H. Vink and C. Cowin, *Proc. Natl. Acad. Sci. USA*, 2003, **100**, 7988–7995.
44. M. L. Smith, D. S. Long, E. R. Damiano and K. Ley, *Annu. Biophys. J.*, 2003, **85**, 637–645.
45. E. R. Damiano, D. S. Long and M. L. Smith, *J. Fluid Mech.*, 2004, **512**, 1–19.
46. A. J. Goldman, R. G. Cox and H. Brenner, *Chem. Engn. Sci.*, 1967, **22**, 653–660.
47. A. W. Mulivor and H. H. Lipowsky, *Am. J. Physiol. Heart Circ. Physiol.*, 2002, **283**, H1282–H1291.
48. X. Zhang, E. Wojcikiewicz and V. T. Moy, *Biophys. J.*, 2002, **83**, 2270–2279.
49. W. Hanley, O. McCarty, S. Jadhav, Y. Tseng, D. Wirtz and K. Konstantopoulos, *J Biol Chem*, 2003, **278**, 10556–10561.
50. J. Liu, R. P. Bradley, D. M. Eckmann, P. S. Ayyaswamy and R. Radhakrishnan, *Current Nanoscience*, 2011, **7**, 727–735.
51. J. Liu, N. J. Agrawal, A. Calderon, P. S. Ayyaswamy, D. M. Eckmann and R. Radhakrishnan, *Biophys. J.*, 2011, **101**, 319–326.
52. J. Carlsson and J. Aqvist, *J. Phys. Chem.* ~ *B*, 2005, **109**, 6448–6456.
53. B. Roux, *Comput. Phys. Commun.*, 1995, **91**, 275–282.
54. A. W. Mulivor and H. H. Lipowsky, *Am J Physiol Heart Circ Physiol*, 2002, **283**, H1282–1291.
55. T. Yago, J. Wu, C. D. Wey, A. G. Klopocki, C. Zhu and R. P. McEver, *J. Cell Biol.*, 2004, **166**, 913–923.

56. E. B. Finger, K. D. Puri, R. Alon, M. B. Lawrence, U. H. von Andrian and T. A. Springer, *Nature*, 1996, **379**, 266–269.
57. U. Seifert, *Advances in Physics*, 1997, **46**, 13–137.
58. G. H. Fredrickson, V. Ganesan and F. Drolet, *Macromolecules*, 2002, **35**, 16–39.
59. G. A. Khelashvili, S. A. Pandit and H. L. Scott, *J Chem Phys*, 2005, **123**, 34910.
60. S. W. Chiu, E. Jakobsson, R. J. Mashl and H. L. Scott, *Biophys J*, 2002, **83**, 1842–1853.
61. S. O. Nielsen, C. F. Lopez, G. Srinivas and M. L. Klein, *J Phys-Condens Mat*, 2004, **16**, R481–R512.
62. L. Saiz, S. Bandyopadhyay and M. L. Klein, *Bioscience Rep*, 2002, **22**, 151–173.
63. A. P. Alivisatos, P. F. Barbara, A. W. Castleman, J. Chang, D. A. Dixon, M. L. Klein, G. L. McLendon, J. S. Miller, M. A. Ratner, P. J. Rossky, S. I. Stupp and M. E. Thompson, *Adv Mater*, 1998, **10**, 1297–1336.
64. S. Bandyopadhyay, M. Tarek and M. L. Klein, *Current Opinion in Colloid & Interface Science*, 1998, **3**, 242–246.
65. S. A. Pandit, D. Bostick and M. L. Berkowitz, *Biophys J*, 2003, **85**, 3120–3131.
66. T. Murtola, E. Falck, M. Patra, M. Karttunen and I. Vattulainen, *J Chem Phys*, 2004, **121**, 9156–9165.
67. K. Katsov, M. Muller and M. Schick, *Biophysical Journal*, 2006, **90**, 915–926.
68. K. Katsov, M. Muller and M. Schick, *Pramana-J Phys*, 2005, **64**, 1127–1134.
69. M. Schick, K. Katsov and M. Mueller, *Biophysical Journal*, 2005, **88**, 66A–66A.
70. M. Muller, K. Katsov and M. Schick, *J Polym Sci Pol Phys*, 2003, **41**, 1441–1450.
71. M. Schick, M. Mueller and K. Katsov, *Biophysical Journal*, 2002, **82**, 543A–543A.
72. B. J. Reynwar, G. Illya, V. A. Harmandaris, M. M. Muller, K. Kremer and M. Deserno, *Nature*, 2007, **447**, 461–464.
73. A. Grafmuller, J. Shillcock and R. Lipowsky, *Phys Rev Lett*, 2007, **98**, 218101.
74. S. J. Marrink and A. E. Mark, *Journal of Physical Chemistry B*, 2001, **105**, 6122–6127.
75. J. Liu, M. Kaksonen, D. G. Drubin and G. Oster, *Proceedings of the National Academy of Sciences of the United States of America*, 2006, **103**, 10277–10282.
76. J. Liu, Y. Sun, D. G. Drubin and G. F. Oster, *Plos Biology*, 2009, **7**, e1000204.
77. P. D. Blood and G. A. Voth, *Proceedings of the National Academy of Sciences of the United States of America*, 2006, **103**, 15068–15072.
78. A. Arkhipov, Y. Yin and K. Schulten, *Biophys J*, 2009, **97**, 2727–2735.

79. S. A. Safran, *Adv Phys*, 1999, **48**, 395–448.
80. W. Helfrich, *Z. Naturforsch. C*, 1973, **28**, 693–703.
81. N. Agrawal, J. Weinstein and R. Radhakrishnan, *Mol Phys*, 2008, **106**, 1913–1923.
82. N. Ramakrishnan, P. B. Sunil Kumar and J. H. Ipsen, *Phys. Rev. E*, 2010, **81**, 041922.
83. N. J. Agrawal, J. Nukpezah and R. Radhakrishnan, *PLoS Comput Biol*, 2010, **6**, e1000926.
84. U. Seifert and S. A. Langer, *Europhys. Lett.*, 1993, **23**, 71–76.
85. V. Ramanan, N. J. Agrawal, J. Liu, S. Engles, R. Toy and R. Radhakrishnan, *Integrative Biology*, 2011, **3**, 803–815.
86. B. T. Stokke, A. Mikkelsen and A. Elgsaeter, *European Biophysics Journal with Biophysics Letters*, 1986, **13**, 203–218.
87. B. T. Stokke, A. Mikkelsen and A. Elgsaeter, *European Biophysics Journal with Biophysics Letters*, 1986, **13**, 219–233.
88. C. L. L. Lawrence and L. H. B. Frank, *Physical Review Letters*, 2004, **93**, 256001.
89. H. A. Pinnow and W. Helfrich, *Eur. Phys. J. E*, 2000, **3**, 149–157.
90. W. Cai and T. C. Lubensky, *Phys Rev E*, 1995, **52**, 4251.
91. N. Gov, *Physical Review Letters*, 2004, **93**, 268104–268104.
92. S. A. Shkulipa, W. K. den Otter and W. J. Briels, *The Journal of Chemical Physics*, 2006, **125**, 234905–234911.
93. W. K. den Otter and S. A. Shkulipa, *Biophys. J.*, 2007, **93**, 423–433.
94. H. Gao, W. Shi and L. B. Freund, *Proc. Natl. Acad. Sci. U.S.A.*, 2005, **102**, 9469–9474.
95. J. Liu, M. Kaksonen, D. G. Drubin and G. Oster, *PNAS*, 2006, **103**, 10277–10282.
96. N. S. Gov, *Physical Review E (Statistical, Nonlinear, and Soft Matter Physics)*, 2006, **73**, 041918.
97. F. Divet, G. Danker and C. Misbah, *Phys Rev E*, 2005, **72**, 041901.
98. F. Divet, T. Biben, I. Cantat, A. Stephanou, B. Fourcade and C. Misbah, *Europhys Lett*, 2002, **60**, 795–801.
99. A. Naji and F. L. H. Brown, *The Journal of Chemical Physics*, 2007, **126**, 235103–235116.
100. E. Reister-Gottfried, S. M. Leitenberger and U. Seifert, *Physical Review E (Statistical, Nonlinear, and Soft Matter Physics)*, 2007, **75**, 011908–011911.
101. E. Atilgan and S. X. Sun, *The Journal of Chemical Physics*, 2007, **126**, 095102.
102. K. Bohinc, D. Lombardo, V. Kraljiglic, M. Fosnaric, S. May, F. Pernus, H. Hagerstrand and A. Iglic, *Cell Mol Biol Lett*, 2006, **11**, 90–101.
103. A. Iglic, B. Babnik, K. Bohinc, M. Fosnaric, H. Hagerstrand and V. Kralj-Iglic, *J Biomech*, 2007, **40**, 579–585.
104. H. T. McMahon, and J. L. Gallop, *Nature*, 2005, **438**, 590–596.
105. T. Baumgart, S. Das, W. W. Webb and J. T. Jenkins, *Biophys J*, 2005, **89**, 1067–1080.

106. T. Baumgart, S. T. Hess and W. W. Webb, *Nature*, 2003, **425**, 821–824.
107. B. R. Capraro, Y. Yoon, W. Cho and T. Baumgart, *J Am Chem Soc*, **132**, 1200–1201.
108. D. A. Christian, A. Tian, W. G. Ellenbroek, I. Levental, K. Rajagopal, P. A. Janmey, A. J. Liu, T. Baumgart and D. E. Discher, *Nat Mater*, 2009, **8**, 843–849.
109. J. E. Hinshaw and S. L. Schmid, *Nature*, 1995, **374**, 190–192.
110. S. Muro, R. Wiewrodt, A. Thomas, L. Koniaris, S. M. Albelda, V. R. Muzykantov and M. Koval, *J Cell Sci*, 2003, **116**, 1599–1609.
111. F. Gentile, C. Chiappini, D. Fine, R. C. Bhavane, M. S. Peluccio, M. M. Cheng, X. Liu, M. Ferrari and P. Decuzzi, *J Biomech*, 2008, **41**, 2312–2318.
112. M. Lakadamyali, M. J. Rust and X. W. Zhuang, *Cell*, 2006, **124**, 997–1009.
113. J. Z. Rappoport, *Biochemical Journal*, 2008, **412**, 415–423.
114. S. Saffarian and T. Kirchhausen, *Biophysical Journal*, 2008, **94**, 2333–2342.

CHAPTER 12

Studying Proteins and Peptides at Material Surfaces

JUN FENG,[a] GILLIAN C. LYNCH[b] AND B. MONTGOMERY PETTITT*[a,b]

[a] Department of Biology and Biochemistry, University of Houston, Houston, TX 77204-5001, USA; [b] Department of Chemistry and Institute for Molecular Design, University of Houston, Houston, TX 77204-5003, USA
*Email: pettitt@uh.edu

12.1 Introduction

The microarray is a device composed of a collection of biomolecules, such as oligonucleotides, peptides/proteins, carbohydrates etc. tethered to a solid substrate. It allows thousands of experimental tests to be performed at the same time in a high-throughput fashion. A well-known example is DNA microarrays which primarily target gene expression profiling, genetic variation, and single nucleotide polymorphism (SNP) genotyping. That well-established technology and has played a major role in genomic research.[1,2] With the completion of numerous genome sequences, analyses of proteins, and their interaction networks are required to harvest the information contained in the genomes. The protein microarray has provided an invaluable tool for large-scale analysis of protein-protein, protein-nucleotide, and other molecular interactions in the post-genomic era. Theory and simulation provide understanding to better interpret the wide variety of data possible and improve experimental design.

Because of the versatility of proteins, protein microarrays have been widely used in proteomic research, including protein expression profiling, detection

RSC Biomolecular Sciences No. 23
Innovations in Biomolecular Modeling and Simulations: Volume 1
Edited by Tamar Schlick

Published by the Royal Society of Chemistry, www.rsc.org

of molecular interactions, enzymatic activity assays, and identification of post-translational modifications. Consider a small sampling of the breadth. In a classic work, Zhu *et al.*[3] constructed a yeast proteome microarray to probe protein-protein and protein-lipid interactions and identified many new calmodulin and phospholipid interacting proteins. Similarly those authors[4] also developed a protein microarray to assay the phosphorylation specificity of 119 yeast protein kinases using 17 different substrates. Novel activities of protein kinases were identified and common amino acid residues near catalytic sites of many of tyrosine phosphorylating enzymes were found. Others[5] used an enzyme-linked immunosorbent assay (ELISA)-based antibody microarray to assess multiple cytokine expression levels from tissue culture media and from patients' sera. Detection of cytokine levels can reach as low the pg/ml concentration. Ivanov *et al.*[6] designed antibody microarrays to capture target proteins of post-translational modifications from mammalian cell extract. Tyrosine phosphorylation, ubiquitination, as well as acetylation, in response to different treatments were examined and compared with samples from control cells.

Protein microarrays are obviously useful for a variety of applications, including drug target discovery, disease biomarker detection, and clinical diagnostics. A critical example is the G-protein coupled receptors that are commonly regarded as the single most important class of drug targets for current therapeutics. High throughput screening of such disease and therapeutic related proteins with chemical compound libraries using microarrays greatly facilitates lead compound identification and validation in pharmaceutics.[7,8] Microarray-based immunoassay also has application in the acceleration of the diagnoses of diseases. Joos *et al.*[9] spotted eighteen autoantigens known to be serologic biomarkers of several autoimmune diseases on a microarray format and detected autoantibodies in autoimmune sera in a highly sensitive and specific manner. A number of applications of protein microarrays have been reviewed in the literature in the last decade and here we wish to consider a more restricted set of problems.[10–17]

In a typical protein microarray experiment, capture protein/peptide probes are immobilized on a solid support, such as a glass or gold-coated slide, or bead. Samples such as cell extract or sera are mixed and incubated with the immobilized probes to search for possible molecular interactions. Finally, the binding of probes with target proteins or other molecules can be detected for instance, by an antibody tagged with a fluorescent label, enzyme-catalyzed chemiluminescence, or other strategies. Despite the high throughput and parallel screening capability, several technical challenges remain in the design and interpretation of protein microarrays.[18–21] The availability of antibodies or other capture proteins with high affinity and specificity is one limitation for the broader applications of this microarray technology. Second, because of the complex nature of the proteome, a dynamic concentration range of proteins may exist in a given sample. Also, detecting target proteins at a picomolar or even femtomolar scale in backgrounds many decades higher is challenging. Due to the intrinsic nature of proteins, the attachment of capture peptides/proteins on microarray supports without compromising their stability and functionality is another technological bottleneck. DNA biophysics and its interactions with surfaces is dominated by electrostatics. Proteins display differential solubilities along their sequence and as a result surfaces can induce unexpected and unwanted global conformational changes.

Theoretical and computational studies for protein arrays focus on consideration of the influence of the liquid-solid interface on the fundamental behavior of the proteins. Since recognition and binding are known to be affected at surfaces, computational techniques must include relevant forces, length scales and time scales to answer the questions of response. This paper focuses on the effects produced by microarray surface chemistry, protein immobilization strategies. We first consider experimental set ups and issues to give a background for the needed theoretical studies. We then review computations and simulations of relevance to the interactions of material surfaces with proteins/peptides in aqueous solution.

12.2 Experimental Surface Chemistry and Immobilization Strategies

Widely utilized in DNA microarray fabrication, glass and silicon substrates are also commonly used substrates for protein microarray supports. These materials are relatively chemically inert to biological molecules and ideal for optical detection, but need to be covered with molecules which both passivate the surface and maintain the proteins in the proper state for attachment to avoid probe denaturation. Much of the early literature for proteins on surfaces is under the rubric of fouling. The mechanism of protein immobilization[20,22,23] can be classified into three categories: physical absorption, covalent binding, and noncovalent bioaffinity binding, with a variety of surface chemistries available in each category (Table 12.1). Some of these represent interesting challenges for simulation.

Table 12.1 List of commonly used surface chemistry for protein immobilization.

Immobilization strategies	*Surface chemistry*	*Protein functional groups*
physical absorption	agarose/polyacrylamide gel	none
	nitrocellulose	none
	poly(vinylidene difluoride)	none
	poly-lysine	none
covalent binding	aldehyde	amines
	epoxide	amines/hydroxyls
	succinimidyl ester	amines
	maleimide	thiols
	pyridyil disulfide	thiols
	vinyl sulfone	thiols
	boronic acid	diols of carbohydrate moiety
noncovalent bioaffinity binding	avidin	biotin tag
	Ni-nitriloacetic acid	His tag
	protein A/protein G	IgG Fc reagion
	oligonucleotide	ssDNA tag

Proteins can be spontaneously absorbed on most surfaces by interplay of enthalpic and entropic contributions including the displaced solvent's contributions.[24] In fact, physical adsorption of proteins has long been utilized in Enzyme-linked immunosorbent assay (ELISA) and Western blot methods. The nanoporous surfaces and membranes such as agarose, polyacrylamide and nitrocellulose embed proteins in three-dimensional matrices and potentially offer a higher capacity (density) for protein attachment than other more strictly two-dimensional strategies. Glass slides coated with polylysine,[25] polyvinylidene difluoride (PVDF),[26] agarose,[27] polyacrylamide,[28] or nitrocellulose.[29] are routinely utilized in protein microarrays. Despite the simplicity for protein immobilization, some drawbacks are present for physical adsorption. For instance, high background signals emanating from non-specific binding to nitrocellulose can undermine the detection assay. Moreover, physically entrapped proteins are often weakly attached and can be removed by harsh washing conditions. Alternatively, these surfaces can be derivatized with reactive groups[27,28,30] allowing covalent cross linkages that endure stringent washing conditions to remove non-specific interactions in the steps following the binding assay.

There exist a number of covalent strategies for protein immobilization on a 2-D microarray chip. Glass is often derivatized with organosilanes such as 3-aminopropyl triethoxysilane, 3-mercaptopropyl trialkoxysilane or 3-glycidoxypropyl trimethoxysilane to introduce chemically reactive groups for protein attachment. The resulting aldehyde,[31,32] epoxy,[4] active ester,[33,34] and other functional groups[35] coated surfaces can interact with hydrophilic functional groups on the side chains of amino acids. Such functional groups include the epsilon amine in lysine, hydroxyl groups in serine, threonine or tyrosine, and the thiol of cystein. The drawback is that many of these residues are present more than once in a protein molecule and any side chain of these residues can react with the surface functional group, often resulting in a variety of orientations of these capture proteins.

In addition, many proteins, including antibodies, hormones, and enzymes, are glycosylated during *in vivo* post-translational modification and their carbohydrate moieties are often suitable for immobilization, generally without affecting their biological activity. Immobilization of glycoproteins via their carbohydrate components is commonly used.[36,37] Cyclic saccharides containing *cis*-diols readily associate with boronic acid activated surfaces. This approach is used by Zayats *et al.*[38] to immobilize flavoenzymes like lactate dehydrogenase, malate dehydrogenase, and glucose oxidase for biosensor fabrication. Moreover, a functional aldehyde group can be introduced into a glycoprotein by oxidizing the carbohydrate moiety by a periodate treatment. Knezevic *et al.*[39] used this method to immobilize lipase on an amine-bearing surface with a high retention of hydrolytic activity.

Photochemical coupling has been widely used in DNA microarray fabrication.[40,41] Surface patterning and immobilization of proteins, peptide fragments, and other biomolecules can also take advantage of this methodology.[42–48] Covalent coupling by azido compounds, which generate highly reactive nitrene radicals upon excitation and readily insert into various chemical bonds (C=C, C–H, O–H, N–H, N–O), is one of the most commonly used photochemistry methods for photografting, cross-linking, photoaffinity labeling, and molecular conjugation. Studies of cell adhesion,[49] enzymatic activity,[50] and carbohydrate-protein recognition[51] have been reported on azido functionalized polymer surfaces. In the photoreaction, molecules bearing reactive groups are linked to photoactive reagents upon irradiation by light of the appropriate wavelength. Because only molecules deposited in the area exposed to light should be covalently attached, photochemical coupling can be spatially selective. Therefore, instead of using prepared molecules, biopolymers like DNA and peptide fragments can be directly synthesized on the chip through site-directed photo coupling. Gao and coworkers[42,52] reported parallel synthesis of addressable, combinatorial peptides on microarray platforms using digital photolithography. At each reaction cycle of peptide synthesis, an amino acid monomer was linked to the growing peptide through deprotection of the t-Boc group using photogenerated acids. Therefore, peptides of diverse sequences can be synthesized by light irradiation at selected sites.

Noncovalent protein immobilization based on bioaffinity principles offers an alternative approach. Capture proteins are linked to affinity tags by either chemical coupling or recombinant genetic engineering. There are a variety of bioaffinity tags, most of which fall into two categories: affinity ligand pairs and metal chelation. Avidin-biotin, known as a strong noncovalent binding pair with a dissociation constant (K_D) around 10^{-15} M has been exploited for the protein purification process, as well as biomolecular immobilization on microarrays.[53] Antibody immobilization can be achieved using protein A/protein G that specifically binds in the Fc constant region of IgG.[54,55] Nucleotide complementarity and hybridization, the strategy used for DNA microarrays can also be utilized in protein microarray fabrication. Proteins coupled with single-stranded DNA (ssDNA) molecules are immobilized by hybridizing with complementary oligonucleotides printed on the microarray.[56] An elastin-calmodulin scaffold is used to capture recombinant proteins containing an M13 tag, a 26-amino acid residue part of the myosin light chain kinase.[57] Genetically engineered recombinant proteins with polyhistine tags can be immobilized via a metal ion-chelated complex such as Cu^{2+}-iminodiacetic acid (IDA).[58] Reversible immobilization of calmodulin-fusion proteins has been achieved by calcium-mediated binding of calmodulin to its ligand phenothiazine.[59] Bioaffinity-based immobilization strategies usually orientate capture proteins in a uniform and specific manner, and are routinely implemented in the microarray surface design, along with other physical and chemical immobilization methods.

12.3 Surface Design

In protein microarrays, proteins/peptides are immobilized and further capture events all occur on the supporting surface. Since protein microarray screening and detection is based on the molecular recognition utilizing binding between antibody and antigen, or enzyme and ligand, or protein and protein, or protein and nucleotide, *etc*., the thermodynamic stability of the probe molecule is of critical importance and directs, and affects, the outcome of the microarray assay. The appropriate choice of microarray surface should allow retention of protein conformational stability and functionality to effectively present probes to agents in the solution phase and minimize non-specific binding.

In an endeavor to develop "tailor-made" surfaces for protein attachment, self-assembled monolayers (SAMs) have attracted significant attention and proved particularly useful.[60,61] The molecular architecture of a SAM, for example, the most commonly used alkanethiol SAM on gold substrate, is composed of densely packed alkyl chains. Their terminals can be derivatized to offer varied functional groups like –OH, $-NH_2$, –COOH, and –COOR for direct chemical coupling of the probe peptide/protein or as an interface to construct organic coatings incorporating a robust number of functional groups like thiol, aldehyde, or even streptavidin for protein attachment. The ordered structure of a SAM provides a regular and homogeneous environment for the bioanalytic assay. However, the closely packed molecular structure of a SAM can result in steric hindrance and affect molecular recognition. For instance, it is known that single-component biotin-terminated SAM has a poor streptavidin binding capacity due to steric constraints imposed by the tight biotin layer. This problem can be alleviated by fine-tuning the component of monolayers such as introducing spacer molecules to reduce repulsion. Perez-Luna *et al.*[62] reported that on SAMs composed of biotin and hydroxyl terminated thiols, there is a maximum binding of streptavidin when biotins are optimally spaced, corresponding to a surface mole fraction of 0.15 biotin. Usually the mixed SAMs approach is used to control the density of capture ligand and maximize binding capacity.

As mentioned earlier proteins have the potential to partially unfold or even denature at the solid-liquid interface.[63–65] Protein immobilization strategies must address the challenges posed by potential protein conformational instability. The ability to provide a hydrophilic environment that mimics the aqueous solution condition to preserve the native conformation of immobilized proteins is a desirable criterion for the choice of surface chemistry. Arenkov and coworkers[28] developed polymethoacrylamide gel-based protein microarrays and used photo-induced copolymerization to covalently attach proteins in the hydrogel; this provides a highly hydrophilic 3-D environment with high water content, but is stronger and more porous than the traditional polyacrylamide gel. Aside from those aforementioned 3-D hydrogels and porous membranes, a polyethylene glycol (PEG) surface is another useful option. The ethylene glycol groups have a high affinity for water molecules and hence form an aqueous microenvironment around the polymer. Studies have proved the ability of the PEG surface to stabilize proteins in a hydrophilic environment as well as to reduce nonspecific binding,[66–69] a problem that arises from some porous gels. Ruiz-Taylor *et al.*[70] designed monolayers of biotin-derivatized polylysine-grafted polyethylene glycol, and demonstrated its feasibility as a biocompatible surface in protein patterning. A large number of PEG-based surface designs have been successfully developed for microarray applications.[71–74]

Of course some proteins are unstable when isolated in aqueous solutions. One example is membrane proteins that serve as receptors, channels, and energy transducers, and are a particular class of proteins with important biological functionality. Unlike many other proteins, membrane proteins are vulnerable in the aqueous environment. Arraying membrane proteins require printing mixtures of the protein and its associated lipids. Fang *et al.*[75] were the first to fabricate membrane protein microarrays. The support membrane was formed by printing vesicular solutions of mixed phospholipids on glass surfaces modified with γ-aminopropylsilane and showed characteristics of long-range lateral fluidity and uniformity. They created arrays of G protein-coupled receptors (GPCRs) including adrenergic receptor (β1), the neurotensin receptor (NTR1), and the dopamine (D1) receptor for screening of ligands. Instead of using flat surfaces, Hong *et al.*[76] further developed G-protein-coupled receptor (GPCR) microarrays on porous glass to increase the accessibility of the supported membranes with membrane-bound protein complexes. The lipid bilayer is known to be unstable upon exposure to air due to the dehydration of the hydrophilic head groups, which can result in destruction and desorption from the surface.[77] To minimize this problem, Deng *et al.*[78] designed a fluidic and air stable supported lipid bilayer by covalent tethering of cholesterol, a naturally occurring component of cell membranes, to PEG-coated glass surface. The ordered cholesterol groups may be thought of as providing a free energy barrier that prevents the restructuring of the lipid bilayer when the top surface is dehydrated, and therefore leads to air stability. Membrane bound proteins, such as G protein-coupled receptors, represent the single most important class of drug targets. Despite the potential applications of membrane proteins in biomedical research and pharmaceutical industry, microarrays of membrane proteins are still a nascent technology that needs to be developed.

Immobilized proteins can display heterogeneous binding characteristics on the solid support.[79] Decreased binding affinity and lowered biological activity can arise upon immobilization and complicates assay interpretation. For instance, Mrksich *et al.*[80] measured the binding of bovine carbonic anhydrase on SAM terminated with a benzenesulfonamide ligand and reported an increased equilibrium dissociation constant, and hence a lowered binding affinity compared to one measured in solution. Consequently, maximizing proteins' biological activity and eliminating heterogeneity are other important aspects in surface construction to achieve sensitivity and reproducibility of bioanalytic assays. Presenting capture proteins in a desired orientation, reducing steric interaction, and increasing the accessibility of the active site are important considerations for chip design. For an antibody microarray, effective presentation of the antigen binding domain (the Fab fragments) toward the solution phase is critical for analyte-binding capacity. Peluso *et al.*[53] compared randomly versus specifically orientated capture agents based on both full-sized antibodies and Fab fragments. The performance of the specifically oriented IgG (immobilized through carbohydrates on the Fc region) and the Fab fragment (immobilized through the hinge region) in the immunoassay was consistently better, with up to 10-fold improvement, over their randomly oriented counterparts (immobilized through amine groups from lysine with multiple attachment points in the protein). Cha *et al.*[81] measured the catalytic activities of sulfotransferases immobilized on silicon surfaces with and without orientation. They found that compared to histine tagged enzymes with controlled orientations, the kinetics of the sulfuryl transfer reaction catalyzed by the randomly oriented ones on an epoxy or succinimidyl functionalized surface was 5–6 times slower, while the oriented ones performed almost the same as that of free enzyme molecules in solution.[82] assessed the binding ability of immobilized ribonuclease A (RNase A) with ribonuclease inhibitor protein (RI). Ellipsometric measurements indicated a 4-fold higher binding ability of RNase A immobilized with a preferred orientation (through an engineered cystein residue without interfering with the binding site of RNase A for RI) over RNase A in a random orientation (through amine coupling). These results demonstrate that the orientation of proteins immobilized on surfaces can influence their binding activities. Therefore, control of orientation is another important aspect in surface design.

Each protein molecule is unique, in terms of biological functionality, conformational stability, and physiochemical properties. The conformation and stability of a protein can be altered upon immobilization via interactions with the surfaces. The current state of the field indicates that there exists no universal way in preserving the structure and functionality of immobilized proteins.

The molecular structure of a protein ultimately determines its activity and functionality. Therefore, understanding the dynamics of protein structures at interfaces and the molecular interactions between proteins and solid surfaces has significant implications for protein microarray as well as other biotechnologies that require assembly of proteins at the material surface. Many experimental studies have been focusing on the interfacial phenomena in an effort to optimize surface construction and preserve protein functionality. They rely heavily on various spectroscopic techniques, which are valuable tools for evaluation of protein conformation, surface properties and protein-surface interactions. These analytical techniques and methods are reviewed in several articles.[83–85] Here, we focus on the interactions of proteins with material surfaces of varied chemistry and their impact on protein stability.

Studies of the kinetics and dynamics of protein adsorption on material surfaces have contributed to a major part of our knowledge of proteins at the solid-liquid interface. Early investigators using circular dichroism (CD) spectroscopy,[63,86] Fourier transform infrared spectroscopy,[87,88] total internal reflection florescence spectroscopy[89,90] and electron spin resonance spectroscopy[91] often observed conformational changes upon protein adsorption to solid surfaces. The degree of conformational change depends on the molecular flexibility of adsorbed proteins. "Soft" protein molecules such as myoglobin, hemoglobin, and bovine serum albumin exhibit remarkable conformational changes often leading to surface fouling, while "rigid" proteins like RNase A and horseradish peroxidase show relatively insignificant changes of conformation upon adsorption.[63]

Surface hydrophobicity plays a fundamental role in the protein-surface interaction and directly affects protein conformation. In many cases, the hydrophobic interaction is a major driving force for adsorption.[92] Using atomic force microscopy, Sethuraman *et al.*[93] measured the adhesion forces between seven globular proteins and a series of eight uncharged model surfaces. They found that proteins strongly interact with apolar surfaces while weak interactions are associated with polar surfaces. The attractive forces between all seven proteins and the solid substrates increase uniformly as the surface wettability decreases. A recent study by Sivaraman *et al.*[94] investigated the effect of surface chemistry on protein adsorption and conformational change. The structure of two model proteins, fibrinogen and albumin, adsorbed to SAM surfaces with hydrophobic, hydrophilic, and charged terminal groups was monitored by CD spectropolarimetry. The authors found that an increase in surface hydrophobicity causes an increase of both the amount of protein adsorbed and the degree of structural change.

Proteins are readily adsorbed onto hydrophobic surfaces. Interactions with hydrophobic surfaces often lead to the exposure of hydrophobic residues originally buried inside the native protein structure. Consequently, partial loss of secondary structure, denaturation, or even protein aggregation with fouling is reported on hydrophobic surfaces.[63,95] However, hydrophobic surfaces do not always cause protein unfolding. Under certain circumstances, a hydrophobic environment can promote secondary structure formation and protein activation. For example, LK14 is an amphiphilic 14-amino acid peptide composed of repeating leucine-lysine residues. It is known to form α-helical tetramers in solution and a monolayer at the water-air interface. Using CD measurements to study its structure on hydrophobic polystyrene, Mermut *et al.*[96] reported an α-helical structure for the adsorbed peptide. The system may have formed a monolayer and assumed a very ordered orientation with leucine residues interacting with the hydrophobic surface. Another example is the lipases, enzymes that hydrolyze triacylglycerols, and become activated from a closed to an open conformation at the lipid-water interface. Adsorption studies of lipases and the enzyme variants on hydrophobic surfaces[97,98] showed that, unlike other model proteins such as bovine serum albumin, lysozyme, and α-chymotrypsin, the secondary structure of the enzyme is stabilized. The hydrophobic surface also induces changes of tertiary structure to open conformations.

Compared to hydrophobic surfaces, hydrophilic surfaces, in general, adsorb relatively smaller quantities of the proteins and have less effect on surface-induced denaturation.[99–101] An interesting case is the PEG-coated surfaces. Whitesides and coworkers reported the ability of PEG to resist nonspecific protein adsorption.[68,102] Later, they screened a number of organic molecule functionalized SAMs by surface plasmon resonance (SPR) spectroscopy, and concluded four common properties present in the functional groups that are characteristic of protein resistance: hydrophilic, electrically neutral, hydrogen bond acceptors, but not hydrogen bond donors.[103] Many protein resistance material surfaces have also been discovered and designed, which include zwitterions-based surfaces such as phosphorylcholine,[104] polycarboxybetaine,[105] and other polymers,[106,107] carbohydrate derived surfaces such as galactose,[108] mannitol,[109] sorbitol,[110] and peptide-based surfaces.[111,112] The molecular mechanism of protein adsorption resistance has been extensively investigated, regarding various aspects of PEG and related polymers, including surface coverage, polymers chain length, polymer structure order, surface charge, and ionic strength.[113–117] Discrepancies in the mechanism of adsorption resistance still remain, but the molecular architecture of PEG polymers and surrounding solvation water is believed to play an essential role.

The interactions of proteins with charged surfaces and their adsorption originate from, and are dominated by, electrostatic attractions. In many cases, proteins nonspecifically bind to charged surfaces, regardless of the proteins net charge.[118–120] Glomm *et al.*[121] monitored the adsorption of eight different acidic and basic proteins onto a model anionic citrate-coated surface at equivalent bulk (solvent, ionic strength, and pH) and surface conditions by quartz crystal microbalance. Albeit to different degrees, adsorption is found to be independent of the global protein's charge. Instead, binding occurs through oppositely charged patches on the protein surface. On the other hand, some charged surfaces display charge selective protein uptake. Kusumo *et al.*[122] used SPR to measure the binding of proteins to cationic poly(2-(dimethylamino)ethyl methacrylate) (PDMAEMA) brushes grafted onto gold surfaces as a function of the grafted chain contour length and grafting density. The expected adsorption is observed for net negatively charged bovine serum albumin with an extremely high binding capacity. However, due to electrostatic repulsion, net positively charged lysozyme was completely rejected by the PDMAEMA brushes. Complete repulsion of proteins by surfaces with the same charge is not always observed. The authors argued that their observations may reflect the relatively small size of lysozyme and its high degree of conformational stability. The interactions between proteins and charged surfaces are also largely affected by solution pH and ionic strength. Solution pH, which modifies the charge carried by both the proteins and the surfaces, is often used to adjust the binding affinity. Because of electrostatic screening, the binding affinity of proteins with charged surfaces shows a reverse trend with a rise of solution ionic strength. Under high salt concentration, charged surfaces can become completely protein resistant.[123,124]

The stability of proteins upon interacting with charged surfaces clearly depends on the protein-surface pair. Change of secondary structure has been reported upon protein adsorption to charged surfaces.[94] But certain polyelectrolyte polymer brushes such as poly(alylamine hydrochloride)-, poly(acrylic acid)-, polystyrenesulfonate-, and polyethyleneimine- coated surfaces seem to stabilize the conformation of adsorbed proteins.[125–127] For poly(acrylic acid) polymers, adsorption studies of several proteins including bovine serum albumin, bovine β-lactoglobulin, and bovine pancreatic RNase A by Wittemann *et al.*[128] and hen egg white lysozyme, bovine serum albumin, bovine α-lactalbumin, and bovine insulin by Reichhart *et al.*[129] demonstrated almost full retention of their secondary structure on brushes. An enzymatic activity assay by Haupt *et al.*[130] reported the identical catalytic activities of immobilized glucoamylase and β-glucosidase in poly(acrylic acid) grafted colloidal particles as in solution. The mechanism of protein stabilization by a polyelectrolyte was postulated by Gibson[131–132] as the formation of an "electrostatic cage", which may limit structural fluctuations and anchor the protein in a more rigid conformation. Further discussions of proteins-polyelectrolyte systems can be found in a review by Cooper *et al.*[133]

The interactions of proteins with material surfaces are highly dependent on the individual nature of the protein and the surface involved. Protein-surface interactions are also affected by the surrounding solvent composition, such as pH and ionic strength. Generally, a hydrophilic and soft surface environment is favored for preserving the native conformation and maintaining the bioactivity of immobilized or embedded proteins.

12.4 Theoretical and Simulation Studies of Proteins at Surfaces

Direct observation of protein conformation or surface morphology by current spectroscopic techniques is not yet possible. Theoretical and simulation based studies offer a model based approach. Numerous simulations of proteins at material surfaces have added new insight into the interfacial behavior of proteins. Detailed descriptions of standard computer modeling and simulation methodologies can be found elsewhere.[134,135] Here we will concentrate on those of relevance to the systems this review is concerned with.

There has been a large number of colloidal scale modeling studies of protein-surface interactions, particularly the adsorption of proteins on different kinds of surfaces.[136–140] In order to lessen the demand on computational resources and time, colloidal scale modeling is coarse grained and often ignores the structural details of proteins. They often treat the protein as a rigid body composed of connected hard spheres with different charges and sizes. The model colloidal particle interacts with the surface through electrostatic and van der Waals interactions and the surrounding solvent is considered as a continuum medium with a specific dielectric constant. Colloidal scale modeling in the best cases is able to rationalize the adsorption kinetics and isotherms. Electrostatic interactions are often found to dominate the adsorption orientation on a charged surface. Jiang and coworkers[138,141] adopted a mesoscopic model in Monte Carlo simulations to study the adsorption and orientation of antibodies on charged surfaces. In their colloidal models, the antibodies were represented as Y-shaped 12-bead chains with different dipole moments. The effects of the sign and density of the surface charge, the solution pH and ionic strength on the orientation of the model antibodies were investigated. As expected the authors found the dipole moment of an antibody plays an important role on its orientation when on a charged surface. Their charge driven protein orientation hypothesis was further confirmed by experimental studies using SPR and time-of-flight secondary ion mass spectrometry.[142,143]

As previously mentioned, when a protein is attached to a surface, the kinetics and thermodynamic properties can be altered. To understand how the surface impacts the structure and thermal and mechanical folding behavior of immobilized proteins, Knotts *et al.*[144] studied a three-helix bundle peptide fragment using a GŌ-like reduced model attached to both attractive and neutral surfaces through either the C- or N-terminus. Compared to the free case, the authors found that the stability of the tethered peptide is reduced on both types of surface, thus lowering its melting temperature. Calorimetric cooperativity, as measured by van't Hoff analysis of calorimetric enthalpy ratios, similarly decreases. The effect of surface destabilization also depends upon the attachment site of the peptide.

The work by Friedel *et al.*[145] studied the effect of surface tethering on protein folding mechanisms. The authors modeled a four-stranded β-barrel protein by representing each amino acid as a single bead and tethered the protein to a surface through different residues. They found that both folding rates and stability are impacted differently by the surface, with dependences on both the length and location of the tether. These results showed that surface tethering can profoundly alter the folding mechanism and thermodynamic stability of proteins and also illustrated the sensitivity of the choice of the tethering point for the protein, which has important implications for protein immobilization strategies.

As mentioned above, proteins near hard surfaces may undergo conformational change, show large spreading, or even denaturation and fouling, particularly on hydrophobic and metallic surfaces. The interplay between the protein side chains of surface residues and the functional groups of the material surface affects protein conformation and stability and plays an important role on the interfacial behavior of immobilized proteins. Recent theoretical research takes into consideration the three-dimensional molecular architecture of proteins and uses (partially) atomic representations to model proteins at biomaterial surfaces.[141,146–156] Wilson *et al.*[147] studied the effect of surface chemistry on the adsorption of a fibronectin fragment spanning the 7–10th type III repeats ($FnIII_{7-10}$) on functionalized surfaces in an implicit aqueous environment. Their molecular dynamics simulation results showed distinctly different adsorption behavior for $FnIII_{7-10}$ on each surface, which is in qualitative agreement with the experimental studies. Zhou *et al.*[141] investigated the orientation and conformation of cytochrome *c* (Cyt-c) adsorbed on a carboxyl-terminated SAM by a combined Monte Carlo and molecular dynamics simulation approach. They found that the desired orientation of Cyt-c with its heme group perpendicular to the negatively charged surface is determined by the dipole of the protein. Clearly, severe conformational change could occur on a surface with high charge density.

A study of Cyt-c orientation on alkanethiol SAMs of various geometries was conducted by Trzaskowski *et al.*[154] using a combined computational docking-molecular dynamics method. Their results agreed with the previous simulation predictions of Zhou[139] and also showed that the orientation of the protein on the surface may be altered by manipulating not only the chemical properties of the surface or the protein but also the geometry of the surface. A Monte Carlo simulation study by Hsu *et al.*[151] suggested that the local hydropathy of the protein is a less important factor for the adsorption of human serum albumin on a hydrophilic hydroxyl-terminated SAM surface. Recently, a 5-ns Monte Carlo simulation of bovine β-Lactoglobulin (β-Lg) adsorbed to a stainless steel surface at acidic pH by Hagiwara *et al.*[156] indicated that the main interactions contributing to the adsorption of the positively charged β-Lg to the positively charged surface are van der Waals interactions.

In addition to those studies focusing on protein adsorption, Nordgren *et al.*[146] presented a collection of molecular dynamics simulation studies of Cyt-c attached to polar and nonpolar SAM surfaces via its unique surface cysteine residue. The authors found that hydroxyl-terminated SAM appears to interact more strongly with the protein than does the methyl-terminated SAM and induces a larger perturbation of the Cyt-c structure through hydrogen bonding.

We[149,157] have used all-atom molecular dynamics simulations to study the conformation of the tumor suppressor p53 epitope peptide tethered on a methyl-terminated microarray surface (Figure 12.1). We found that specific surface chemistry features determines the molecular interactions with the peptide residues. While conformational space visited by the tethered peptide largely overlapped with the one free in solution, surface tethering as well as the ions in the solution have changed both the thermodynamics and kinetics of the tethered peptide. These simulation results demonstrated that the chemical properties of the surface, as well as solution pH and ionic strength impact the protein-surface interactions, and manipulation of these factors could provide control over the structure and bioactivity of proteins at biomaterial surfaces.

The structure and dynamic properties of a biomaterial surface and the neighboring solvent is also amenable to modeling, which in turn gives us insight into the molecular events related to protein adsorption, desorption, orientation, and conformational rearrangement on these surfaces. One example

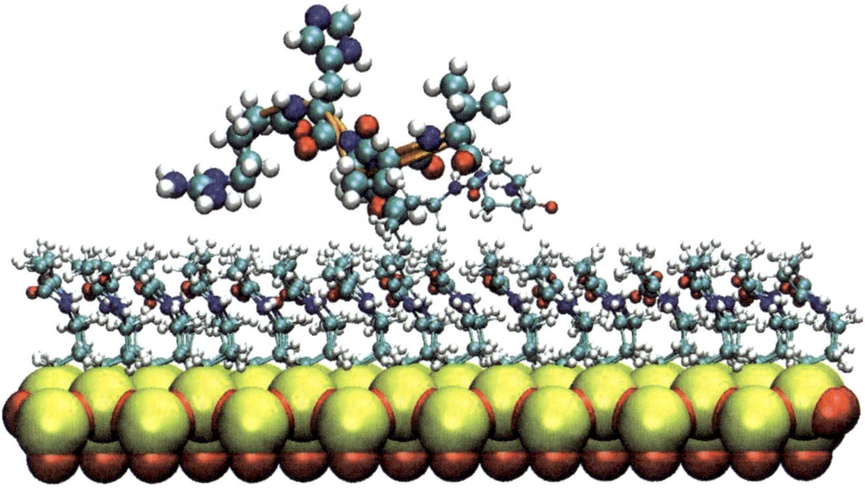

Figure 12.1 A snapshot of the p53 peptide RHSVV tethered on the microarray surface. The silica surface is covered with a monolayer of acetylated aminosilane linkers. The peptide is immobilized via C-terminus with a spacer Ahx (6-aminohexanoic acid)-βAla (3-aminopropanoic acid)-βAla bonded with the open aminosilane linker.

is PEG-coated surfaces, which are known to resist nonspecific protein adsorption and have been widely utilized in protein microarrays. In addition to numerous experimental studies, a number of theoretical modeling and simulations have also been performed to explore the interfacial properties of PEG and to uncover the underlying mechanism for protein resistance.[158–166] Using simplified oligo(ethylene glycol) (OEG) SAMs as a molecular model, an early *ab initio* calculation by Wang *et al.*[159] compared the structure and energetics of water interacting with helical and planer OEG strands and concluded that protein resistance of the OEG-terminated surface is a consequence of the stability of the interfacial water layer. Studies by Jiang and coworkers[163,165] supported the observation that the interfacial water layer above the OEG SAM surface, which is tightly bounded and forms a stable structure, is responsible for the protein-surface repulsive force and serves as a physical barrier to resist protein adsorption. However, a similar molecular modeling study by Ismail *et al.*[164] examined the interfacial structure and dynamics of water as a function of the OEG SAM coverage and did not find a significant difference in the interfacial water dynamics. Consequently, they suggested that the flexibility of the interface is a key factor in protein resistance. That is, an interface that is increasingly more like bulk water is more resistant to protein adsorption. In all cases, hydration water is believed to be essential for the non-fouling property of PEG/OEG surfaces. A detailed discussion of water structure and bioactivity at material surfaces is adressed by Vogler.[167,168]

Solvent transport properties, such as diffusion and viscosity of the interfacial water, are another active area of research. Micro and nanofluidic devices are notoriously hard to design with Navier-Stokes level fluid flow models. Many simulations have studied the diffusion coefficient of water near a biomolecule such as a DNA or protein. Water mobility is often found to be retarded near the molecular surface.[169–171] Pozhar and Gubbins's transport theory of dense, inhomogeneous fluids[172,173] predicts the sensitivity of the transport coefficient to structured factors of fluids at an interface, such as number density profiles and pair correlation functions. Their theory is further verified by a recent equilibrium molecular dynamics simulation study in our group.[155] We calculated solvent transport properties near a methyl-terminated microarray surface. Ordered water structure is formed near the hydrophobic surface. Both solvent diffusion and viscosity are closely related to the water density profile along the microarray surface (Figure 12.2). Solvent in the high density regions shows limited mobility and high viscosity while low solvent density regions correspond to low viscosity and high diffusion rate.

We showed solvent diffusion coefficients as well as components of the local viscosity in the vicinity of a functionalized microarray surface are non monotonic in the direction away from the surface and closely correlated with the solvent density profile.[155] Translational diffusion coefficients for solvent molecules in directions parallel and perpendicular to a peptide surface are both restricted, while water molecules moving in a parallel direction are slightly more mobile than in the perpendicular direction. The surface changes our expectations for the kinetics and transport. With a tethered peptide near the

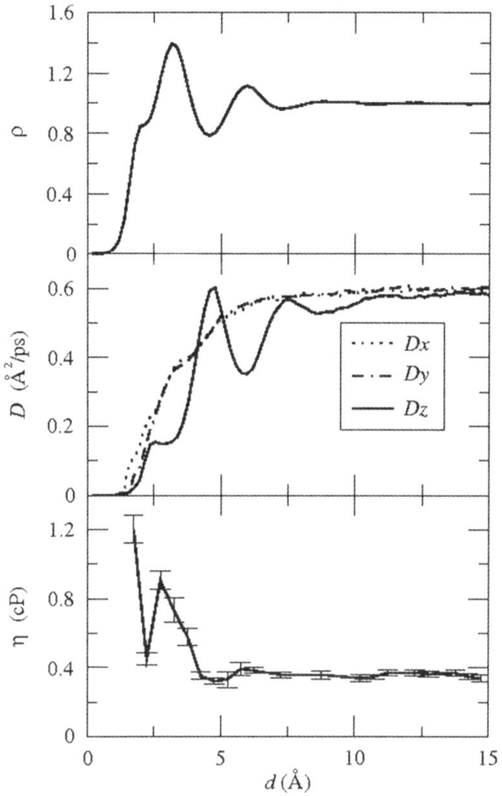

Figure 12.2 Solvent density (ρ, upper), parallel (*x*- and *y*-) and perpendicular (*z*-) diffusion (*D*, middle), and viscosity (η, lower) coefficients as a function of distance to the microarray surface. With permission of the authors, ref 155.

functionalized microarray surface, water diffusion decreases and viscosity increases when compared to an unfunctionalized microarray surface.

We note that nonequilibrium molecular dynamics simulations of fluids under sheer flow are typically utilized to calculate local pressure tensor magnitudes and position-dependent viscosity. We were able to use equilibrium molecular dynamics simulations to calculate local solvent viscosity components as a function of distance from the surface via an interpretation of the Green-Kubo formalism. The viscosity calculation results agree well with theoretical predictions of strong correlation with solvent density profile. At the microarray surface, solvent in the high density regions shows limited diffusion and high viscosity while low solvent density regions always correspond to low viscosity and high diffusion rate. This must be seen as in contrast to hard sphere results which may be due to a fundamental difference in mechanism for systems without sufficient attractions.

Molecular modeling and simulation is well suited to provide a testable atomic picture of proteins, surface morphology, solvent dynamics, and their interactions. However, empirical forced field-based simulation methods have been primarily designed to study biomolecules in a homogeneous environment such as in aqueous solution or in vacuum. Molecular modeling and simulation of proteins under an anisotropic condition like near a material surface is relatively new and parameters describing inter- and intra-molecular forces of both proteins and surfaces need to be verified. Moreover, the complicated process of reorientation and structural change of a protein upon interacting with a surface requires prolonged sampling of the conformational space. All-atom calculations of the large scale conformational rearrangement of a protein induced by surface interactions or binding with its ligand(s) are still computationally inconvenient. Nevertheless, we believe that with the continuing growth of computing power and simulation techniques, computer models will yield more accurate and quantitative information to understand the diverse behavior of proteins at material surfaces in solution.

12.5 Concluding Remarks

Envisioned as a type of "lab-on-a-chip", protein microarray technology has been rapidly developed and utilized in a wide variety of applications over the past decade. Spotting proteins or peptide molecules in a miniaturized array format for multianalytical screening has greatly reduced sample consumption and detection time compared to traditional bioassay techniques. However, immobilization from an isotropic solution onto an inhomogeneous supporting surface may result in conformational changes, unfolding and even loss of biological activity. The challenge is to retain the structural stability and functionality of the immobilized peptides/proteins and this demands understanding of the dynamic behavior of biomolecules interfacing with material surfaces. The relevance of protein-surface interactions and protein interfacial properties is being increasingly recognized, not only for the development of protein microarrays, but, also for a variety of bio- and nano-technology applications.[174] A multiplicity of molecular interactions exists between protein-material surface, protein-solvent, and material surface-solvent. Numerous experiments, as well as computer modeling and simulation studies, have provided us valuable information about the versatile interfacial phenomena. Yet, our understanding of proteins at the liquid-solid interface is still limited. Advances in analytical techniques and modeling methodologies will help to develop new designs and applications in interfacial biomolecular engineering.

Acknowledgements

Prof. Xiaolian Gao is thanked for many helpful discussions. We gratefully thank the National Institutes of Health (GM037657) and the Robert A Welch Foundation (E-1028) for partial financial support of this work. Some of our computational research was supported in part by the National Science Foundation through TeraGrid resources provided by the Pittsburgh

Supercomputing Center and the San Diego Supercomputer Center. Calculations were also performed in part using the Molecular Science Computing Facility (MSCF) in the William R. Wiley Environmental Molecular Sciences Laboratory, a national scientific user facility sponsored by the U.S. Department of Energy's Office of Biological and Environmental Research, located at the Pacific Northwest National Laboratory, operated for the Department of Energy by Battelle.

References

1. G. Ramsay, DNA chips: State-of-the-art, *Nature Biotechnology*, 1998, **16**(1), 40–44.
2. M. J. Heller, DNA microarray technology: Devices, systems, and applications, *Annual Review of Biomedical Engineering*, 2002, **4**, 129–153.
3. H. Zhu and M. Bilgin, *et al.*, Global analysis of protein activities using proteome chips, *Science*, 2001, **293**(5537), 2101–2105.
4. H. Zhu and J. F. Klemic, *et al.*, Analysis of yeast protein kinases using protein chips, *Nature Genetics*, 2000, **26**(3), 283–289.
5. R. P. Huang and R. Huang, *et al.*, Simultaneous detection of multiple cytokines from conditioned media and patient's sera by an antibody-based protein array system, *Analytical Biochemistry*, 2001, **294**(1), 55–62.
6. S. S. Ivanov and A. S. Chung, *et al.*, Antibodies immobilized as arrays to profile protein post-translational modifications in mammalian cells, *Mol. Cell Proteomics*, 2004, **3**(8), 788–795.
7. K. D. Kumble, Protein microarrays: new tools for pharmaceutical development, *Analytical and Bioanalytical Chemistry*, 2003, **377**(5), 812–819.
8. Y. L. Hong and B. L. Webb, *et al.*, G-protein-coupled receptor Microarrays for multiplexed compound screening, *Journal of Biomolecular Screening*, 2006, **11**(4), 435–438.
9. T. O. Joos and M. Schrenk, *et al.*, A microarray enzyme-linked immunosorbent assay for autoimmune diagnostics, *Electrophoresis*, 2000, **21**(13), 2641–2650.
10. E. T. Fung and V. Thulasiraman, *et al.*, Protein biochips for differential profiling, *Current Opinion in Biotechnology*, 2001, **12**(1), 65–69.
11. G. MacBeath, Protein microarrays and proteomics, *Nature Genetics*, 2002, **32**, 526–532.
12. H. Zhu and M. Snyder, Protein chip technology, *Current Opinion in Chemical Biology*, 2003, **7**(1), 55–63.
13. A. Lueking and D. J. Cahill, *et al.*, Protein biochips: a new and versatile platform technology for molecular medicine, *Drug Discovery Today*, 2005, **10**(11), 789–794.
14. D. Stoll and M. F. Templin, *et al.*, Protein microarrays: Applications and future challenges, *Curr. Opin. Drug Discovery Dev.*, 2005, **8**(2), 239–252.
15. S. F. Kingsmore, Multiplexed protein measurement: technologies and applications of protein and antibody arrays, *Nature Reviews Drug Discovery*, 2006, **5**(4), 310–320.

16. O. Stoevesandt and M. J. Taussig, *et al.*, Protein microarrays: High-throughput tools for proteomics, *Expert Review of Proteomics*, 2009, **6**(2), 145–157.
17. X. B. Yu and N. Schneiderhan-Marra, *et al.*, Protein microarrays for personalized medicine, *Clinical Chemistry*, 2010, **56**(3), 376–387.
18. P. Bertone and M. Snyder, Advances in functional protein microarray technology, *FEBS Journal*, 2005, **272**(21), 5400–5411.
19. K. Y. Tomizaki and K. Usui, *et al.*, Protein-detecting microarrays: current accomplishments and requirements, *ChemBioChem.*, 2005, **6**(5), 782–799.
20. F. Rusmini and Z. Zhong, *et al.*, Protein immobilization strategies for protein biochips, *Biomacromolecules*, 2007, **8**(6), 1775–1789.
21. A. A. Ellington and I. J. Kullo, *et al.*, Antibody-based protein multiplex platforms: Technical and operational challenges, *Clinical Chemistry*, 2010, **56**(2), 186–193.
22. D. S. Yeo and R. C. Panicker, *et al.*, Strategies for immobilization of biomolecules in a microarray, *Comb. Chem. High Throughput Screen*, 2004, **7**(3), 213–221.
23. T. Ekblad and B. Liedberg, Protein adsorption and surface patterning, *Curr. Opin. Colloid Interface Sci.*, 2010, **15**(6), 499–509.
24. J. D. Andrade and V. Hlady, Protein adsorption and materials biocompatibility: A tutorial review and suggested hypotheses, *Advances in Polymer Science*, 1986, **79**, 1–63.
25. B. B. Haab and M. J. Dunham, *et al.*, Protein microarrays for highly parallel detection and quantitation of specific proteins and antibodies in complex solutions, *Genome Biol.*, 2001, **2**(2), 1–13.
26. A. Lueking and M. Horn, *et al.*, Protein microarrays for gene expression and antibody screening, *Analytical Biochemistry*, 1999, **270**(1), 103–111.
27. V. Afanassiev and V. Hanemann, *et al.*, Preparation of DNA and protein micro arrays on glass slides coated with an agarose film, *Nucleic Acids Research*, 2000, **28**(12), E66.
28. P. Arenkov and A. Kukhtin, *et al.*, Protein microchips: use for immunoassay and enzymatic reactions, *Analytical Biochemistry*, 2000, **278**(2), 123–131.
29. N. A. Groathouse and A. Amin, *et al.*, Use of protein microarrays to define the humoral immune response in leprosy patients and identification of disease-state-specific antigenic profiles, *Infection and Immunity*, 2006, **74**(11), 6458–6466.
30. L. L. Lv and B. C. Liu, *et al.*, Construction of an antibody microarray based on agarose-coated slides, *Electrophoresis*, 2007, **28**(3), 406–413.
31. G. MacBeath and S. L. Schreiber, Printing proteins as microarrays for high-throughput function determination, *Science*, 2000, **289**(5485), 1760–1763.
32. A. Gordus and G. MacBeath, Circumventing the problems caused by protein diversity in microarrays: Implications for protein interaction

networks, *Journal of the American Chemical Society*, 2006, **128**(42), 13668–13669.

33. R. Benters and C. M. Niemeyer, *et al.*, Dendrimer-activated solid supports for nucleic acid and protein microarrays, *ChemBioChem.*, 2001, **2**(9), 686–694.
34. S. Anderson, Surfaces for Immobilization of N-Terminal Cysteine Derivatives via Native Chemical Ligation, *Langmuir*, 2008, **24**(24), 13962–13968.
35. P. K. Ajikumar and J. K. Ng, *et al.*, Carboxyl-Terminated Dendrimer-Coated Bioactive Interface for Protein Microarray: High-Sensitivity Detection of Antigen in Complex Biological Samples, *Langmuir*, 2007, **23**(10), 5670–5677.
36. H. Y. Hsiao and G. P. Royer, Immobilization of Glycoenzymes through Carbohydrate Side-Chains, *Archives of Biochemistry and Biophysics*, 1979, **198**(2), 379–385.
37. M. Saleemuddin and Q. Husain, Concanavalin A: A useful ligand for glycoenzyme immobilization – A review, *Enzyme and Microbial Technology*, 1991, **13**(4), 290–295.
38. M. Zayats and E. Katz, *et al.*, Electrical contacting of flavoenzymes and NAD(P)(+)-dependent enzymes by reconstitution and affinity interactions on phenylboronic acid monolayers associated with Au-electrodes, *Journal of the American Chemical Society*, 2002, **124**(49), 14724–14735.
39. Z. Knezevic and N. Milosavic, *et al.*, Immobilization of lipase from Candida rugosa on Eupergit (R) supports by covalent attachment, *Biochemical Engineering Journal*, 2006, **30**(3), 269–278.
40. X. L. Gao and E. LeProust, *et al.*, A flexible light-directed DNA chip synthesis gated by deprotection using solution photogenerated acids, *Nucleic Acids Research*, 2001, **29**(22), 4744–4750.
41. D. M. Dankbar and G. Gauglitz, A study on photolinkers used for biomolecule attachment to polymer surfaces, *Analytical and Bioanalytical Chemistry*, 2006, **386**(7–8), 1967–1974.
42. J. P. Pellois and X. Zhou, *et al.*, Individually addressable parallel peptide synthesis on microchips, *Nature Biotechnology*, 2002, **20**(9), 922–926.
43. K. L. Christman and H. D. Maynard, Protein micropatterns using a pH-responsive polymer and light, *Langmuir*, 2005, **21**(18), 8389–8393.
44. Y. Ito, Photoimmobilization for microarrays, *Biotechnology Progress*, 2006, **22**(4), 924–932.
45. K. H. Park and H. G. Park, *et al..*, Poly(dimethyl siloxane)-based protein chip for simultaneous detection of multiple samples: Use of glycidyl methacrylate photopolymer for site-specific protein immobilization, *Biosensors & Bioelectronics*, 2006, **22**(5), 613–620.
46. M. Duroux and E. Skovsen, *et al.*, Light-induced immobilisation of biornolecules as an attractive alternative to microdroplet dispensing-based arraying technologies, *Proteomics*, 2007, **7**(19), 3491–3499.

47. J. M. Alonso and A. Reichel, *et al.*, Photopatterned surfaces for site-specific and functional immobilization of proteins, *Langmuir*, 2008, **24**(2), 448–457.
48. D. M. He and W. Sun, *et al.*, Protein adsorbers from surface-grafted copolymers with selective binding sites, *Journal of Materials Chemistry*, 2009, **19**(2), 253–260.
49. Y. Ito and M. Nogawa, *et al.*, Photo-reactive polyvinylalcohol for photo-immobilized microarray, *Biomaterials*, 2005, **26**(2), 211–216.
50. J. S. Mecomber and R. S. Murthy, *et al.*, Photochemical functionalization of polymer surfaces for microfabricated devices, *Langmuir*, 2008, **24**(7), 3645–3653.
51. Z. C. Pei and H. Yu, *et al.*, Photogenerated carbohydrate microarrays, *ChemBioChem.*, 2007, **8**(2), 166–168.
52. J. P. Pellois and W. Wang, *et al.*, Peptide synthesis based on t-Boc chemistry and solution photogenerated acids, *Journal of Combinatorial Chemistry*, 2000, **2**(4), 355–360.
53. P. Peluso and D. S. Wilson, *et al.*, Optimizing antibody immobilization strategies for the construction of protein microarrays, *Analytical Biochemistry*, 2003, **312**(2), 113–124.
54. G. P. Anderson and M. A. Jacoby, *et al.*, Effectiveness of protein A for antibody immobilization for a fiber optic biosensor, *Biosens. Bioelectron.*, 1997, **12**(4), 329–336.
55. J. M. Lee and H. K. Park, *et al.*, Direct immobilization of protein g variants with various numbers of cysteine residues on a gold surface, *Analytical Chemistry*, 2007, **79**(7), 2680–2687.
56. J. Ladd and C. Boozer, *et al.*, DNA-directed protein immobilization on mixed self-assembled monolayers via a streptavidin bridge, *Langmuir*, 2004, **20**(19), 8090–8095.
57. G. Jenikova and U. L. Lao, *et al.*, Elastin-calmodulin scaffold for protein microarray fabrication, *Langmuir*, 2007, **23**(5), 2277–2279.
58. T. Cha, Immobilization of oriented protein molecules on poly(ethylene glycol)-coated (Si(111)), *Proteomics*, 2004, **4**(7), 1965–1976.
59. S. Daunert and L. G. Bachas, *et al.*, Calmodulin-mediated reversible immobilization of enzymes, *Colloids Surf., B*, 2007, **58**(1), 20–27.
60. M. Schaeferling and S. Schiller, *et al.*, Application of self-assembly techniques in the design of biocompatible protein microarray surfaces, *Electrophoresis*, 2002, **23**(18), 3097–3105.
61. M. Frasconi and F. Mazzei, *et al.*, Protein immobilization at gold-thiol surfaces and potential for biosensing, *Analytical and Bioanalytical Chemistry*, 2010, **398**(4), 1545–1564.
62. V. H. Perez-Luna and M. J. O'Brien, *et al.*, Molecular recognition between genetically engineered streptavidin and surface-bound biotin, *Journal of the American Chemical Society*, 1999, **121**(27), 6469–6478.
63. A. Kondo and F. Murakami, *et al.*, Circular dichroism studies on conformational changes in protein molecules upon adsorption on ultrafine polystyrene particles, *Biotechnol. Bioeng.*, 1992, **40**(8), 889–894.

64. J. Buijs and W. Norde, *et al.*, Changes in the Secondary Structure of Adsorbed IgG and F(ab')2 Studied by FTIR Spectroscopy, *Langmuir*, 1996, **12**(6), 1605–1613.
65. W. Norde and C. E. Giacomelli, BSA structural changes during homomolecular exchange between the adsorbed and the dissolved states, *Journal of Biotechnology*, 2000, **79**(3), 259–268.
66. J. H. Lee and J. Kopecek, *et al.*, Protein-Resistant Surfaces Prepared by PEO-Containing Block Copolymer Surfactants, *Journal of Biomedical Materials Research*, 1989, **23**(3), 351–368.
67. W. R. Gombotz and W. Guanghui, *et al.*, Protein Adsorption to Poly-(Ethylene Oxide) Surfaces, *Journal of Biomedical Materials Research*, 1991, **25**(12), 1547–1562.
68. K. L. Prime and G. M. Whitesides, Self-Assembled Organic Monolayers – Model Systems for Studying Adsorption of Proteins at Surfaces, *Science*, 1991, **252**(5009), 1164–1167.
69. P. Kingshott and S. McArthur, *et al.*, Ultrasensitive probing of the protein resistance of PEG surfaces by secondary ion mass spectrometry, *Biomaterials*, 2002, **23**(24), 4775–4785.
70. L. A. Ruiz-Taylor and T. L. Martin, *et al.*, Monolayers of derivatized poly(L-lysine)-grafted poly(ethylene glycol) on metal oxides as a class of biomolecular interfaces, *Proc. Natl. Acad. Sci., U.S.A*, 2001, **98**(3), 852–857.
71. A. Larsson and C. X. Du, *et al.*, UV-patterned poly(ethylene glycol) matrix for microarray applications, *Biomacromolecules*, 2007, **8**, 3511–3518.
72. A. Wolter and R. Niessner, *et al.*, Preparation and characterization of functional poly(ethylene glycol) surfaces for the use of antibody microarrays, *Analytical Chemistry*, 2007, **79**(12), 4529–4537.
73. K. Kloth and R. Niessner, *et al.*, Development of an open stand-alone platform for regenerable automated microarrays, *Biosens Bioelectron*, 2008, **24**(7), 2106–2112.
74. H. Zhu and M. Macal, *et al.*, A miniature cytometry platform for capture and characterization of T-lymphocytes from human blood, *Analytica Chimica Acta.*, 2008, **608**(2), 186–196.
75. Y. Fang and A. G. Frutos, *et al.*, Membrane protein microarrays, *Journal of the American Chemical Society*, 2002, **124**(11), 2394–2395.
76. Y. L. Hong and B. L. Webb, *et al.*, Functional GPCR microarrays, *Journal of the American Chemical Society*, 2005, **127**(44), 15350–15351.
77. P. S. Cremer and S. G. Boxer, Formation and spreading of lipid bilayers on planar glass supports, *Journal of Physical Chemistry B*, 1999, **103**(13), 2554–2559.
78. Y. Deng and Y. Wang, *et al.*, Fluidic and air-stable supported lipid bilayer and cell-mimicking microarrays, *Journal of the American Chemical Society*, 2008, **130**(19), 6267–6271.
79. R. A. Vijayendran and D. E. Leckband, A quantitative assessment of heterogeneity for surface-immobilized proteins, *Analytical Chemistry*, 2001, **73**(3), 471–480.

80. M. Mrksich and J. R. Grunwell, *et al.*, Biospecific Adsorption of Carbonic-Anhydrase to Self-Assembled Monolayers of Alkanethiolates That Present Benzenesulfonamide Groups on Gold, *Journal of the American Chemical Society*, 1995, **117**(48), 12009–12010.
81. T. Cha and A. Guo, *et al.*, Enzymatic activity on a chip: the critical role of protein orientation, *Proteomics*, 2005, **5**(2), 416–419.
82. Y. Y. Luk and M. L. Tingey, *et al.*, Imaging the binding ability of proteins immobilized on surfaces with different orientations by using liquid crystals, *Journal of the American Chemical Society*, 2004, **126**(29), 9024–9032.
83. C. Boozer and G. Kim, *et al.*, Looking towards label-free biomolecular interaction analysis in a high-throughput format: a review of new surface plasmon resonance technologies, *Current Opinion in Biotechnology*, 2006, **17**(4), 400–405.
84. L. Z. He and A. F. Dexter, *et al.*, Biomolecular engineering at interfaces, *Chemical Engineering Science*, 2006, **61**(3), 989–1003.
85. T. S. Tsapikouni and Y. F. Missirlis, Protein-material interactions: From micro-to-nano scale, *Mat. Sci. Eng. B*, 2008, **152**(1–3), 2–7.
86. W. Norde and F. Macritchie, *et al.*, Protein Adsorption at Solid Liquid Interfaces – Reversibility and Conformation Aspects, *Journal of Colloid and Interface Science*, 1986, **112**(2), 447–456.
87. D. R. Lu and K. Park, Effect of surface hydrophobicity on the conformational changes of adsorbed fibrinogen, *Journal of Colloid and Interface Science*, 1991, **144**(1), 271–281.
88. C. E. Giacomelli and M. Bremer, *et al.*, ATR-FTIR study of IgG adsorbed on different silica surfaces, *Journal of Colloid and Interface Science*, 1999, **220**(1), 13–23.
89. B. K. Lok and Y. L. Cheng, *et al.*, Protein Adsorption on Crosslinked Polydimethylsiloxane Using Total Internal-Reflection Fluorescence, *Journal of Colloid and Interface Science*, 1983, **91**(1), 104–116.
90. J. L. Robeson and R. D. Tilton, Spontaneous reconfiguration of adsorbed lysozyme layers observed by total internal reflection fluorescence with a pH-sensitive fluorophore, *Langmuir*, 1996, **12**(25), 6104–6113.
91. C. Narasimhan and C. S. Lai, Conformational changes of plasma fibronectin detected upon adsorption to solid substrates: a spin-label study, *Biochemistry*, 1989, **28**(12), 5041–5046.
92. R. D. Tilton and C. R. Robertson, *et al.*, Manipulation of Hydrophobic Interactions in Protein Adsorption, *Langmuir*, 1991, **7**(11), 2710–2718.
93. A. Sethuraman and M. Han, *et al.*, Effect of surface wettability on the adhesion of proteins, *Langmuir*, 2004, **20**(18), 7779–7788.
94. B. Sivaraman and K. P. Fears, *et al.*, Investigation of the effects of surface chemistry and solution concentration on the conformation of adsorbed proteins using an improved Circular Dichroism method, *Langmuir*, 2009, **25**(5), 3050–3056.
95. F. Hook and M. Rodahl, *et al.*, Structural changes in hemoglobin during adsorption to solid surfaces: Effects of pH, ionic strength, and ligand binding, *Proc. Natl. Acad. Sci., U.S.A.*, 1998, **95**(21), 12271–12276.

96. O. Mermut and D. C. Phillips, *et al.*, *In situ* adsorption studies of a 14-amino acid leucine-lysine peptide onto hydrophobic polystyrene and hydrophilic silica surfaces using quartz crystal microbalance, atomic force microscopy, and sum frequency generation vibrational spectroscopy, *Journal of the American Chemical Society*, 2006, **128**(11), 3598–3607.
97. S. Noinville and M. Revault, *et al.*, Conformational changes and orientation of Humicola lanuginosa lipase on a solid hydrophobic surface: An *in situ* interface Fourier transform infrared-attenuated total reflection study, *Biophysical Journal*, 2002, **82**(5), 2709–2719.
98. D. Otzen, Differential adsorption of variants of the Thermomyces lanuginosus lipase on a hydrophobic surface suggests a role for local flexibility, *Colloids Surf., B*, 2008, **64**(2), 223–228.
99. J. Kim and G. A. Somorjai, Molecular packing of lysozyme, fibrinogen, and bovine serum albumin on hydrophilic and hydrophobic surfaces studied by infrared-visible sum frequency generation and fluorescence microscopy, *Journal of the American Chemical Society*, 2003, **125**(10), 3150–3158.
100. P. Roach and D. Farrar, *et al.*, Interpretation of protein adsorption: Surface-induced conformational changes, *Journal of the American Chemical Society*, 2005, **127**, 8168–8173.
101. S. Koutsopoulos and K. Patzsch, *et al.*, Adsorption of trypsin on hydrophilic and hydrophobic surfaces, *Langmuir*, 2007, **23**(4), 2000–2006.
102. K. L. Prime and G. M. Whitesides, Adsorption of Proteins onto Surfaces Containing End-Attached Oligo(Ethylene Oxide) - a Model System Using Self-Assembled Monolayers, *Journal of the American Chemical Society*, 1993, **115**(23), 10714–10721.
103. E. Ostuni and R. G. Chapman, *et al.*, A survey of structure-property relationships of surfaces that resist the adsorption of protein, *Langmuir*, 2001, **17**(18), 5605–5620.
104. A. L. Lewis, Phosphorylcholine-based polymers and their use in the prevention of biofouling, *Colloids Surf., B*, 2000, **18**(3–4), 261–275.
105. Z. Zhang and S. F. Chen, *et al.*, Dual-functional biomimetic materials: Nonfouling poly(carboxybetaine) with active functional groups for protein immobilization, *Biomacromolecules*, 2006, **7**(12), 3311–3315.
106. H. Kitano and A. Kawasaki, *et al.*, Resistance of zwitterionic telomers accumulated on metal surfaces against nonspecific adsorption of proteins, *Journal of Colloid and Interface Science*, 2005, **282**(2), 340–348.
107. W. K. Cho and B. Y. Kong, *et al.*, Highly efficient non-biofouling coating of zwitterionic polymers: Poly((3-(methacryloylamino)propyl)-dimethyl(3-sulfopropyl)ammonium hydroxide), *Langmuir*, 2007, **23**(10), 5678–5682.
108. M. Hederos and P. Konradsson, *et al.*, Synthesis and self-assembly of galactose-terminated alkanethiols and their ability to resist proteins, *Langmuir*, 2005, **21**(7), 2971–2980.

109. Y. Y. Luk and M. Kato, *et al.*, Self-assembled monolayers of alkanethiolates presenting mannitol groups are inert to protein adsorption and cell attachment, *Langmuir*, 2000, **16**(24), 9604–9608.
110. M. Metzke and J. Z. Bai, *et al.*, A novel carbohydrate-derived side-chain polyether with excellent protein resistance, *Journal of the American Chemical Society*, 2003, **125**(26), 7760–7761.
111. D. O. H. Teare and W. C. E. Schofield, *et al.*, Poly(N-acryloylsarcosine methyl ester) protein-resistant surfaces, *Journal of Physical Chemistry B*, 2005, **109**(44), 20923–20928.
112. R. Chelmowski and S. D. Koster, *et al.*, Peptide-based SAMs that resist the adsorption of proteins, *Journal of the American Chemical Society*, 2008, **130**(45), 14952–14953.
113. S. Herrwerth and W. Eck, *et al.*, Factors that determine the protein resistance of oligoether self-assembled monolayers - Internal hydrophilicity, terminal hydrophilicity, and lateral packing density, *Journal of the American Chemical Society*, 2003, **125**(31), 9359–9366.
114. L. Y. Li and S. F. Chen, *et al.*, Protein adsorption on oligo(ethylene glycol)-terminated alkanethiolate self-assembled monolayers: The molecular basis for nonfouling behavior, *Journal of Physical Chemistry B*, 2005, **109**(7), 2934–2941.
115. R. Michel and S. Pasche, *et al.*, Influence of PEG architecture on protein adsorption and conformation, *Langmuir*, 2005, **21**(26), 12327–12332.
116. S. Pasche and J. Voros, *et al.*, Effects of ionic strength and surface charge on protein adsorption at PEGylated surfaces, *Journal of Physical Chemistry B*, 2005, **109**(37), 17545–17552.
117. H. Chen and X. Y. Hu, *et al.*, Effect of chain density and conformation on protein adsorption at PEG-grafted polyurethane surfaces, *Colloids Surf., B*, 2008, **61**(2), 237–243.
118. G. Ladam and P. Schaaf, *et al.*, Protein adsorption onto auto-assembled polyelectrolyte films, *Langmuir*, 2001, **17**(3), 878–882.
119. A. Wittemann and B. Haupt, *et al.*, Adsorption of proteins on spherical polyelectrolyte brushes in aqueous solution, *Physical Chemistry Chemical Physics*, 2003, **5**(8), 1671–1677.
120. O. Hollmann and C. Czeslik, Characterization of a planar poly(acrylic acid) brush as a materials coating for controlled protein immobilization, *Langmuir*, 2006, **22**(7), 3300–3305.
121. W. R. Glomm and O. Halskau, *et al.*, Adsorption behavior of acidic and basic proteins onto citrate-coated Au surfaces correlated to their native fold, stability, and pI, *Journal of Physical Chemistry B*, 2007, **111**(51), 14329–14345.
122. A. Kusumo and L. Bombalski, *et al.*, High capacity, charge-selective protein uptake by polyelectrolyte brushes, *Langmuir*, 2007, **23**(8), 4448–4454.
123. K. W. Mattison and P. L. Dubin, *et al.*, Complex formation between bovine serum albumin and strong polyelectrolytes: Effect of polymer

charge density, *Journal of Physical Chemistry B*, 1998, **102**(19), 3830–3836.

124. C. Czeslik and G. Jackler, *et al.*, Salt-induced protein resistance of polyelectrolyte brushes studied using fluorescence correlation spectroscopy and neutron reflectometry, *Physical Chemistry Chemical Physics*, 2004, **6**(24), 5557–5563.
125. M. A. Andersson and R. Hatti-Kaul, Protein stabilising effect of polyethyleneimine, *Journal of Biotechnology*, 1999, **72**(1–2), 21–31.
126. F. Caruso and C. Schuler, Enzyme multilayers on colloid particles: Assembly, stability, and enzymatic activity, *Langmuir*, 2000, **16**(24), 9595–9603.
127. P. Schwinte and J. C. Voegel, *et al.*, Stabilizing effects of various polyelectrolyte multilayer films on the structure of adsorbed/embedded fibrinogen molecules: An ATR-FTIR study, *Journal of Physical Chemistry B*, 2001, **105**(47), 11906–11916.
128. A. Wittemann and M. Ballauff, Secondary structure analysis of proteins embedded in spherical polyelectrolyte brushes by FT-IR spectroscopy, *Analytical Chemistry*, 2004, **76**(10), 2813–2819.
129. C. Reichhart and C. Czeslik, Native-like Structure of Proteins at a Planar Poly(acrylic acid) Brush, *Langmuir*, 2009, **25**(2), 1047–1053.
130. B. Haupt and T. Neumann, *et al.*, Activity of enzymes immobilized in colloidal spherical polyelectrolyte brushes, *Biomacromolecules*, 2005, **6**(2), 948–955.
131. Gibson, T. D. Hulbert, J. N., Pierce, B. Webster, J. I. (1993). The stabilisation of analytical enzymes using polyelectrolytes and sugar derivatives. Stability and Stabilization of Enzymes: Proceedings of an International Symposium held in Maastricht, the Netherlands, 22_25. Amsterdam, Elsevier: 337.
132. T. D. Gibson, Protein stabilisation using additives based on multiple electrostatic interactions, *Dev Biol Stand*, 1996, **87**, 207–217.
133. C. L. Cooper and P. L. Dubin, *et al.*, Polyelectrolyte-protein complexes, *Curr. Opin. Colloid Interface Sci*, 2005, **10**(1–2), 52–78.
134. M. P. Allen, and D. J. Tildesley *Computer Simulation of Liquids*, 1987, Oxford University Press.
135. M. Karplus and G. A. Petsko, Molecular-Dynamics Simulations in Biology, *Nature*, 1990, **347**(6294), 631–639.
136. C. M. Roth and A. M. Lenhoff, Electrostatic and van der Waals contributions to protein adsorption – Comparison of theory and experiment, *Langmuir*, 1995, **11**(9), 3500–3509.
137. D. Asthagiri and A. M. Lenhoff, Influence of structural details in modeling electrostatically driven protein adsorption, *Langmuir*, 1997, **13**(25), 6761–6768.
138. Y.-J. Sheng and H. K. Tsao, *et al.*, Orientation of a Y-shaped biomolecule adsorbed on a charged surface, *Phys. Rev. E*, 2002, **66**(1 Pt 1), 011911.
139. J. Zhou and H. K. Tsao, *et al.*, Monte Carlo simulations of antibody adsorption and orientation on charged surfaces, *Journal of Chemical Physics*, 2004, **121**(2), 1050–1057.

140. M. Skepo, Model simulations of the adsorption of statherin to solid surfaces: Effects of surface charge and hydrophobicity, *Journal of Chemical Physics*, 2008, **129**(18), 185101.
141. J. Zhou and J. Zheng, *et al.*, Molecular simulation studies of the orientation and conformation of cytochrome c adsorbed on self-assembled monolayers, *Journal of Physical Chemistry B*, 2004, **108**(45), 17418–17424.
142. S. F. Chen and L. Y. Liu, *et al.*, Controlling antibody orientation on charged self-assembled monolayers, *Langmuir*, 2003, **19**(7), 2859–2864.
143. H. Wang and D. G. Castner, *et al.*, Probing the orientation of surface-immobilized immunoglobulin G by time-of-flight secondary ion mass spectrometry, *Langmuir*, 2004, **20**(5), 1877–1887.
144. T. A. t. Knotts IV and N. Rathore, *et al.*, Structure and stability of a model three-helix-bundle protein on tailored surfaces, *Proteins*, 2005, **61**(2), 385–397.
145. M. Friedel and A. Baumketner, *et al.*, Effects of surface tethering on protein folding mechanisms, *Proc. Natl. Acad. Sci., U.S.A.*, 2006, **103**(22), 8396–8401.
146. C. E. Nordgren and D. J. Tobias, *et al.*, Molecular dynamics simulations of a hydrated protein vectorially oriented on polar and nonpolar soft surfaces, *Biophysical Journal*, 2002, **83**(6), 2906–2917.
147. K. Wilson and S. J. Stuart, *et al.*, A molecular modeling study of the effect of surface chemistry on the adsorption of a fibronectin fragment spanning the 7–10th type-III repeats, *J. Biomed. Mater. Res. A*, 2004, **69A**(4), 686–698.
148. M. Agashe and V. Raut, *et al.*, Molecular simulation to characterize the adsorption behavior of a fibrinogen gamma–chain fragment, *Langmuir*, 2005, **21**(3), 1103–1117.
149. J. Feng and K. Y. Wong, *et al.*, Peptide conformations for a microarray surface-tethered epitope of the tumor suppressor p. 53, *Journal of Physical Chemistry B*, 2007, **111**(49), 13797–13806.
150. G. Raffaini and F. Ganazzoli, Understanding the performance of biomaterials through molecular modeling: crossing the bridge between their intrinsic properties and the surface adsorption of proteins, *Macromolecular Bioscience*, 2007, **7**(5), 552–566.
151. H. J. Hsu and S. Y. Sheu, *et al.*, Preferred orientation of albumin adsorption on a hydrophilic surface from molecular simulation, *Colloids Surf., B*, 2008, **67**(2), 183–191.
152. C. P. O'Brien and S. J. Stuart, *et al.*, Modeling of Peptide Adsorption Interactions with a Poly(lactic acid) Surface, *Langmuir*, 2008, **24**(24), 14115–14124.
153. J. W. Shen and T. Wu, *et al.*, Induced stepwise conformational change of human serum albumin on carbon nanotube surfaces, *Biomaterials*, 2008, **29**(28), 3847–3855.
154. B. Trzaskowski and F. Leonarski, *et al.*, Altering the orientation of proteins on self-assembled monolayers: a computational study, *Biomacromolecules*, 2008, **9**(11), 3239–3245.

155. J. Feng and K. Y. Wong, *et al.*, Transport properties of water at functionalized molecular interfaces, *Journal of Chemical Physics*, 2009, **131**, 12.
156. T. Hagiwara and T. Sakiyama, *et al.*, Molecular Simulation of Bovine beta-Lactoglobulin Adsorbed onto a Positively Charged Solid Surface, *Langmuir*, 2009, **25**(1), 226–234.
157. J. Feng and K. Y. Wong, *et al.*, Salt Effects on Surface-Tethered Peptides in Solution, *Journal of Physical Chemistry B*, 2009, **113**(28), 9472–9478.
158. S. I. Jeon and J. H. Lee, *et al.*, Protein Surface Interactions in the Presence of Polyethylene Oxide. 1. Simplified Theory, *Journal of Colloid and Interface Science*, 1991, **142**(1), 149–158.
159. R. L. C. Wang and H. J. Kreuzer, *et al.*, Molecular conformation and solvation of oligo(ethylene glycol)-terminated self-assembled monolayers and their resistance to protein adsorption, *Journal of Physical Chemistry B*, 1997, **101**(47), 9767–9773.
160. A. J. Pertsin and M. Grunze, Computer simulation of water near the surface of oligo(ethylene glycol)-terminated alkanethiol self-assembled monolayers, *Langmuir*, 2000, **16**(23), 8829–8841.
161. A. J. Pertsin and T. Hayashi, *et al.*, Grand canonical Monte Carlo Simulations of the hydration interaction between oligo(ethylene glycol)-terminated alkanethiol self-assembled monolayers, *Journal of Physical Chemistry B*, 2002, **106**(47), 12274–12281.
162. J. Zheng and L. Y. Li, *et al.*, Molecular simulation study of water interactions with oligo (ethylene glycol)-terminated alkanethiol self-assembled monolayers, *Langmuir*, 2004, **20**(20), 8931–8938.
163. J. Zheng and L. Y. Li, *et al.*, Strong repulsive forces between protein and oligo (ethylene glycol) self-assembled monolayers: A molecular simulation study, *Biophysical Journal*, 2005, **89**(1), 158–166.
164. A. E. Ismail and G. S. Grest, *et al.*, Structure and dynamics of water near the interface with oligo(ethylene oxide) self-assembled monolayers, *Langmuir*, 2007, **23**(16), 8508–8514.
165. Y. He and Y. Chang, *et al.*, Origin of repulsive force and structure/dynamics of interfacial water in OEG-protein interactions: a molecular simulation study, *Physical Chemistry Chemical Physics*, 2008, **10**(36), 5539–5544.
166. C. L. Ren and R. J. Nap, *et al.*, The Role of Hydrogen Bonding in Tethered Polymer Layers, *Journal of Physical Chemistry B*, 2008, **112**(50), 16238–16248.
167. E. A. Vogler, Structure and reactivity of water at biomaterial surfaces, *Advances in Colloid and Interface Science*, 1998, **74**, 69–117.
168. E. A. Vogler, Water and the acute biological response to surfaces, *Journal of Biomaterials Science-Polymer Edition*, 1999, **10**(10), 1015–1045.
169. R. Kimmich and T. Gneiting, *et al.*, Fluctuations, Exchange Processes, and Water Diffusion in Aqueous Protein Systems – a Study of Bovine

Serum-Albumin by Diverse Nmr Techniques, *Biophysical Journal*, 1990, **58**(5), 1183–1197.
170. V. A. Makarov and M. Feig, *et al.*, Diffusion of solvent around biomolecular solutes: A molecular dynamics simulation study, *Biophysical Journal*, 1998, **75**(1), 150–158.
171. B. Bagchi, Water dynamics in the hydration layer around proteins and micelles, *Chemical Reviews*, 2005, **105**(9), 3197–3219.
172. L. A. Pozhar and K. E. Gubbins, Transport theory of dense, strongly inhomogeneous fluids, *Journal of Chemical Physics*, 1993, **99**(11), 8970–8996.
173. L. A. Pozhar, Structure and dynamics of nanofluids: Theory and simulations to calculate viscosity, *Phys. Rev. E*, 2000, **61**(2), 1432–1446.
174. R. S. Kane and A. D. Stroock, Nanobiotechnology: protein-nanomaterial interactions, *Biotechnology Progress*, 2007, **23**(2), 316–319.

CHAPTER 13

Multiscale Design: From Theory to Practice

J. FISH,*[a] V. FILONOVA[a] AND Z. YUAN[b]

[a] Department of Civil Engineering and Engineering Mechanics, Columbia University, West 120th Street, 10027, New York, USA; [b] Multiscale Design Systems, LLC Multiscale, Burton Lane, 12211, Loudonville, NY, USA
*Email: fishj@columbia.edu

13.1 Introduction

Two approaches, often labeled as "practical" and "rigorous", are used to describe the nature of the method employed to analyze heterogeneous materials and structures. The so-called "practical" category of methods ranges from the semi-analytical approximation methods, such as the rule of mixtures dating back to the Renaissance era to various effective medium models of Eshelby,[1] Hashin,[2] Mori and Tanaka,[3] and the self-consistent approaches of Hill[4] and Christensen.[5] Methods often coined as "rigorous" are those based on the computational homogenization methods pioneered by Suquet,[6] and Guedes and Kikuchi[7] and later enhanced by several investigators[8–22] in an attempt to improve accuracy and computational efficiency. From a theoretical point of view, the latter category of methods is based on either the Hill-Mandel macrohomogeneity condition[23] or on the multiple scale asymptotic expansion methods pioneered by Babuska,[24] Benssousan[25] and Sanchez-Palencia.[26]

The question that is often posed is whether the semi-analytical-like methods are really "practical," while the computational homogenization methods are really "rigorous." Clearly, if "practicality" is measured in terms of

RSC Biomolecular Sciences No. 23
Innovations in Biomolecular Modeling and Simulations: Volume 1
Edited by Tamar Schlick

Published by the Royal Society of Chemistry, www.rsc.org

computational cost of solving a multiple scale problem, while "rigor" is defined in terms of the ability of the method to account for microstructural details in the component analysis, then the two labels are justified. Yet, there is an ambiguity in these definitions. Consider, for instance, the Airbus A380 Superjumbo consisting of hundreds of thousands of structural components. Just in the fuselage alone, 30% of which are made of composites, there are more than 750 000 holes and cutouts. A full scale finite element model of A380 has over 10 million of elements. Furthermore, to account for uncertainty, multiple loading conditions and to perform structural optimization would require numerous reanalyzes of already enormous finite element model. In this case semi-analytical-like methods would be a natural choice that provides reasonable accuracy for linear problems, but would fall short in predicting failure and component lifing with any confidence.

This chapter presents a new multiscale framework that possesses the ingredients of "rigor" and "practicality." The "rigor" aspect of this framework is provided by a recently developed computational continua (C^2) formulation,[27] which is endowed with fine-scale details, introduces no scale separation, makes no assumption about infinitesimality of the fine-scale structure, does not require higher order continuity, introduces no new degrees-of-freedom and is free of higher order boundary conditions. The "practicality" aspect of the proposed method is inherited from the reduced order homogenization approach,[28–32] which constructs residual free-fields that eliminate the bottleneck of satisfying fine-scale equilibrium equations and by characterizing the overall properties of microconstituents by solving the inverse problems that minimizes the error between multiscale predictions and observations.

Blending of the two methods into a single cohesive computational framework, hereafter to be referred to as the Reduced order Computational Continua or simply RC^2, that inherits the underlying characteristics of its two ingredients, is the main objective of the present chapter. The basic RC^2 formulation is presented in Section 13.2, which includes the development of the residual-free influence functions incorporating the gradient effects, formulation of the reduced order unit cell problem and coarse-scale discrete equilibrium equations. Numerical verification of the method against the direct numerical simulation (DNS) and comparison to O(1) computational homogenization is given in Section 13.3. There, we considered a bending problem of a beam with periodic cylindrical microstructures. Spherical microstructures also can be considered.[34] These examples potentially can be used to various multiscale applications in engineering and mechanics, *e.g.* to model a bending deformations of composite materials reinforced by filaments or composite polymers manufactured from hard and soft granules.

Multiscale characterization of model parameters is discussed in Section 13.4. The architecture of multiscale design system is presented in Section 13.5.

We consider a heterogeneous body formed by a repetition of a fine structure (unit cells) occupying an open, bounded domain $\Omega^{\zeta} \subset \mathbb{R}^3$. The unit cell domain denoted by $\Theta \subset \mathbb{R}^3$ is assumed to be finite unlike in the homogenization theories in which it is not infinitesimally small compared to the coarse-scale domain.

The following equations governing $\boldsymbol{x} \in \Omega^{\zeta}$ are stated at the fine-scale of interest

$$
\begin{aligned}
&\sigma^{\zeta}_{ij,j}(\boldsymbol{x}) + b^{\zeta}_i(\boldsymbol{x}) = 0 \quad \text{on} \quad \Omega^{\zeta} \\
&\sigma^{\zeta}_{ij}(\boldsymbol{x}) = L^{\zeta}_{ijkl}(\boldsymbol{x})\left(\varepsilon^{\zeta}_{kl}(\boldsymbol{x}) - \mu^{\zeta}_{kl}(\boldsymbol{x})\right) \quad \text{on} \quad \Omega^{\zeta} \\
&\varepsilon^{\zeta}_{ij}(\boldsymbol{x}) = u^{\zeta}_{(i,j)}(\boldsymbol{x}) \equiv \frac{1}{2}\left(u^{\zeta}_{i,j} + u^{\zeta}_{j,i}\right) \quad \text{on} \quad \Omega^{\zeta} \\
&u^{\zeta}_i(\boldsymbol{x}) = \bar{u}^{\zeta}_i(\boldsymbol{x}) \quad \text{on} \quad \partial\Omega^{u\zeta} \\
&\sigma^{\zeta}_{ij}(\boldsymbol{x}) n^{\zeta}_j(\boldsymbol{x}) = \bar{t}^{\zeta}_i(\boldsymbol{x}) \quad \text{on} \quad \partial\Omega^{t\zeta}
\end{aligned}
\tag{13.1}
$$

where u^{ζ}_i denotes displacements; σ^{ζ}_{ij} – the Cauchy stress; ε^{ζ}_{ij} – the total small strain; μ^{ζ}_{ij} – the eigenstrain arising from inelastic deformation, thermal changes, moisture effects or phase transformation; b^{ζ}_i – the body force, L^{ζ}_{ijkl} – the elastic material properties. The superscript ζ denotes existence of fine-scale features. $\partial\Omega^{u\zeta}$ and $\partial\Omega^{t\zeta}$ denote essential and natural boundaries such that $\partial\Omega^{\zeta} = \partial\Omega^{u\zeta} \cup \partial\Omega^{t\zeta}$ and $\partial\Omega^{u\zeta} \cap \partial\Omega^{t\zeta} = \varnothing$. The boundary-value problem Eq. (13.1) comprises an equilibrium equation, constitutive equation (13.1)b, kinematic (strain-displacement) relation for infinitesimal deformations, and essential (displacement) and natural (traction) boundary conditions, respectively. Strain and stress tensors are assumed to be symmetric. The strain field is additively decomposed into elastic strain and eigenstrain fields. For simplicity we assume that the body force $b^{\zeta}_i(\boldsymbol{x})$ is a smooth function, i.e. it may vary linearly over the unit cell domain.

13.2 RC^2 Formulation

13.2.1 Computational Continua

Following Fish and Kuznetsov paper[27] we define the *computational continua domain* Ω^c as a *disjoint union*[33] (sometimes called direct sum or free union) of *Computational Unit Cell* (CUC) domains $\Theta_{\hat{x}_I}$

$$
\Omega^c = \coprod_{I=1}^{\hat{N}} \Theta_{\hat{x}_I} \tag{13.2}
$$

where $\hat{\boldsymbol{x}}_I$ denotes the coordinates of centroid of the CUC domain $\Theta_{\hat{x}_I}$ and $\hat{N}$ the number of computational unit cells. Note that if $\Theta_{\hat{x}_I} \cap \Theta_{\hat{x}_J} = 0$, $\forall I \neq J$ then the disjoint union reduces to a regular union. The position of the CUC centroid, $\hat{\boldsymbol{x}}_I$, and quadrature weights are calculated so that the integral over the composite domain Ω^{ζ} would coincide with the integral over the computational continua domain Ω^c. This is accomplished *via* the so-called nonlocal quadrature[27] defined as

$$\int_{\Omega^\zeta} f^\zeta(\boldsymbol{x})d\Omega = \sum_{I=1}^{\hat{N}} \int_{\Theta_{\hat{x}_I}} \hat{\omega}(\hat{\boldsymbol{x}}_I,\chi) f^\zeta(\hat{\boldsymbol{x}}_I,\chi)d\Theta\,; \quad \hat{\omega}(\hat{\boldsymbol{x}}_I,\chi) = \frac{1}{|\Theta_{\hat{x}_I}|} J^e(\hat{\boldsymbol{x}}_I,\chi)\hat{W}_I \quad (13.3)$$

where $J^e(\hat{\boldsymbol{x}}_I,\chi)$ is the Jacobian that maps a coarse-scale element into a bi-unit cube; $\hat{W}_I$ denotes the nonlocal quadrature weight. The nonlocal quadrature leads to the dependence of the position of the unit cell centroid $\hat{\boldsymbol{x}}_I$ on the computational unit cell size relative to the coarse-scale finite element size.[27]

For each CUC domain, $\Theta_{\hat{x}_I}$, we introduce a local coordinate system with respect to the unit cell centroid defined as $\boldsymbol{\chi} = \boldsymbol{x} - \hat{\boldsymbol{x}}_I$. Unlike in the classical homogenization theory there is no scale separation, *i.e.* $\boldsymbol{\chi}$ and $\boldsymbol{x}$ are of the same order.

We further assume an additive decomposition of displacements $\boldsymbol{u}^\zeta$ into a smooth coarse-scale function $\boldsymbol{u}^c$ and an oscillatory weakly periodic function $\boldsymbol{u}^{(1)}$ normalized as

$$\int_{\Theta_{\hat{x}_I}} \boldsymbol{u}^{(1)}(\hat{\boldsymbol{x}}_I,\boldsymbol{\chi})d\Theta = \boldsymbol{0} \quad (13.4)$$

The total strain $\boldsymbol{\varepsilon}^\zeta(\hat{\boldsymbol{x}}_I,\boldsymbol{\chi})$ is additively decomposed into coarse-scale strain $\boldsymbol{\varepsilon}^c$ and fine-scale perturbation $\boldsymbol{\varepsilon}^*$ as

$$\varepsilon_{ij}^\zeta(\hat{\boldsymbol{x}}_I,\boldsymbol{\chi}) = \varepsilon_{ij}^c(\hat{\boldsymbol{x}}_I,\boldsymbol{\chi}) + \varepsilon_{ij}^*(\hat{\boldsymbol{x}}_I,\boldsymbol{\chi}) \quad (13.5)$$

defined as symmetric spatial derivative of displacements

$$\boldsymbol{\varepsilon}_{ij}^c(\hat{\boldsymbol{x}}_I,\boldsymbol{\chi}) = u_{(i,j)}^c(\hat{\boldsymbol{x}}_I,\boldsymbol{\chi}); \quad \varepsilon_{ij}^*(\hat{\boldsymbol{x}}_I,\boldsymbol{\chi}) = u_{(i,j)}^{(1)}(\hat{\boldsymbol{x}}_I,\boldsymbol{\chi}) \quad (13.6)$$

The coarse-scale strain field $\boldsymbol{\varepsilon}^c$ is approximated by a linear function over the computational unit cell domain consisting of an average strain $\bar{\varepsilon}_{ij}$, and an average strain gradient $\bar{\varepsilon}_{ij,m}$ defined as

$$\begin{aligned} &\varepsilon_{ij}^c(\hat{\boldsymbol{x}}_I,\boldsymbol{\chi}) = \bar{\varepsilon}_{ij}(\hat{\boldsymbol{x}}_I) + \bar{\varepsilon}_{ij,m}(\hat{\boldsymbol{x}}_I)\chi_m \\ &\bar{\varepsilon}_{ij}(\hat{\boldsymbol{x}}_I) \equiv \frac{1}{|\Theta_{\hat{x}_I}|}\int_{\Theta_{\hat{x}_I}} \varepsilon_{ij}^\zeta(\hat{\boldsymbol{x}}_I,\boldsymbol{\chi})d\Theta; \quad \bar{\varepsilon}_{ij,m}(\hat{\boldsymbol{x}}_I) \equiv \frac{1}{|\Theta_{\hat{x}_I}|}\int_{\Theta_{\hat{x}_I}} \frac{\partial \varepsilon_{ij}^\zeta}{\partial \chi_m}(\hat{\boldsymbol{x}}_I,\boldsymbol{\chi})d\Theta \end{aligned} \quad (13.7)$$

13.2.2 Residual-free Computational Unit Cell Problem

The bottleneck of multiscale computations is in fine-scale computations of complex microstructures. To illustrate the computational complexity involved, consider a coarse-scale problem with N_{cells} quadrature points, n load increments at the coarse scale, and I_{coarse} and I_{fine} average iterations at the two scales, respectively. The total number of linear solves of a fine-scale problem is thus $N_{\text{cells}} \cdot n \cdot I_{\text{coarse}} \cdot I_{\text{fine}}$ – a formidable computational complexity for large number of complex unit cells. This bottleneck can be successfully removed by constructing residual-free fields that *a priori* satisfy fine-scale equilibrium equations.[28–32]

For large unit cells over which the coarse-scale fields are no longer constant the residual-free expansion proposed in previous works[28–32] has to be modified to include higher order residual-free coarse-scale fields. This is accomplished by introducing the following decomposition of fine-scale displacements

$$u_i^{(1)}(\hat{\boldsymbol{x}}_I,\boldsymbol{\chi}) = H_i^{mn}(\hat{\boldsymbol{x}}_I,\boldsymbol{\chi})\bar{\varepsilon}_{mn}(\hat{\boldsymbol{x}}_I) + H_i^{mnp}(\hat{\boldsymbol{x}}_I,\boldsymbol{\chi})\bar{\varepsilon}_{mn,p}(\hat{\boldsymbol{x}}_I) + \int_{\Theta_{\hat{x}_I}} \tilde{h}_i^{mn}(\hat{\boldsymbol{x}}_I,\boldsymbol{\chi},\tilde{\boldsymbol{\chi}})\mu_{mn}^{\zeta}(\hat{\boldsymbol{x}}_I,\tilde{\boldsymbol{\chi}})d\tilde{\Theta} + \int_{S_{\hat{x}_I}} \breve{h}_i^{\breve{m}}\left(\hat{\boldsymbol{x}}_I,\boldsymbol{\chi},\breve{\boldsymbol{\chi}}\right)\delta_{\breve{m}}^{\zeta}\left(\hat{\boldsymbol{x}}_I,\breve{\boldsymbol{\chi}}\right)d\breve{S} \tag{13.8}$$

where H_i^{mn}, H_i^{mnp}, $\tilde{h}_i^{mn}$, $\breve{h}_i^{\breve{m}}$ are strain, strain gradient, eigenstrain, eigenseparation influence functions, respectively; $\delta_{\breve{m}}^{\zeta}$ the eigenseparation; the subscript $\breve{m}$ indicates components in the local Cartesian coordinate system of the CUC interface, denoted by $S_{\hat{x}_I}$. Assume that all influence functions are $\boldsymbol{\chi}$-periodic over the CUC domain (it yields $\boldsymbol{\chi}$-periodicity of $\boldsymbol{u}^{(1)}$) and satisfy normalization (13.4). Note that Eq. (13.7) can be also satisfied for weakly periodic boundary conditions,[27] which are not considered in the present book chapter. Then integral expressions (13.7) and (13.5) are satisfied.

A weak form of the CUC problem is constructed by multiplying (13.1)b with the test function $\boldsymbol{w}^{(1)} \in W_{\hat{x}_I}^{\Theta}$ and then integrating over the CUC domain, which gives

$$\int_{\Theta_{\hat{x}_I}} w_i^{(1)}(\hat{\boldsymbol{x}}_I,\boldsymbol{\chi})\frac{\partial}{\partial\chi_j}\left\{L_{ijkl}(\boldsymbol{\chi})\left(\begin{array}{l} E_{kl}^{mn}(\hat{\boldsymbol{x}}_I,\boldsymbol{\chi})\bar{\varepsilon}_{mn}(\hat{\boldsymbol{x}}_I) + E_{kl}^{mnp}(\hat{\boldsymbol{x}}_I,\boldsymbol{\chi})\bar{\varepsilon}_{mn,p}(\hat{\boldsymbol{x}}_I) \\ + \displaystyle\int_{S_{\hat{x}_I}} \breve{h}_{(k,l)}^{\breve{m}}\left(\hat{\boldsymbol{x}}_I,\boldsymbol{\chi},\breve{\boldsymbol{\chi}}\right)\delta_{\breve{m}}^{\zeta}\left(\hat{\boldsymbol{x}}_I,\breve{\boldsymbol{\chi}}\right)d\breve{S} \\ + \displaystyle\int_{\Theta_{\hat{x}_I}} \tilde{h}_{(k,l)}^{mn}(\hat{\boldsymbol{x}}_I,\boldsymbol{\chi},\tilde{\boldsymbol{\chi}})\mu_{mn}^{\zeta}(\hat{\boldsymbol{x}}_I,\tilde{\boldsymbol{\chi}})d\tilde{\Theta} - \mu_{kl}^{\zeta}(\hat{\boldsymbol{x}}_I,\boldsymbol{\chi}) \end{array}\right)\right\}d\Theta + \int_{\Theta_{\hat{x}_I}} w_i^{(1)}(\hat{\boldsymbol{x}}_I,\boldsymbol{\chi})b_i^{\zeta}(\hat{\boldsymbol{x}}_I,\boldsymbol{\chi})d\Theta = 0 \quad \forall\, \boldsymbol{w}^{(1)} \in W_{\hat{x}_I}^{\Theta} \tag{13.9}$$

where

$$E_{kl}^{mn}(\hat{\boldsymbol{x}}_I,\boldsymbol{\chi}) \equiv I_{klmn} + H_{(k,l)}^{mn}(\hat{\boldsymbol{x}}_I,\boldsymbol{\chi}); \quad E_{kl}^{mnp}(\hat{\boldsymbol{x}}_I,\boldsymbol{\chi}) \equiv I_{klmn}\chi_p + H_{(k,l)}^{mnp}(\hat{\boldsymbol{x}}_I,\boldsymbol{\chi}) \tag{13.10}$$

and I_{ijkl} denotes the components of the fourth order identity tensor.

For the influence functions problems to be independent of body forces, it is convenient to assume the body force $b_i^\zeta(\hat{\boldsymbol{x}}_I, \boldsymbol{\chi})$ to be a smooth polynomial function such that

$$\int_{\Theta_{\hat{x}_I}} w_i^{(1)} \chi_i^p d\Theta = 0 \quad p = 0, 1 \tag{13.11}$$

Given the smoothness condition (13.11) and solution periodicity (or weak periodicity) assumption, the space of the test functions is defined as

$$W_{\hat{x}_I}^{\Theta} = \left\{ \boldsymbol{w}^{(1)}(\hat{\boldsymbol{x}}_I, \boldsymbol{\chi}) \text{ on } \Theta_{\hat{x}_I},\ C^0(\Theta_{\hat{x}_I}),\ \boldsymbol{\chi} - \text{periodic},\ \int_{\Theta_{\hat{x}_I}} w_i^{(1)} \chi_i^p d\Theta = 0 \quad p = 0, 1 \right\} \tag{13.12}$$

13.2.2.1 Model Reduction

The primary objective of the model reduction is to reduce the computational complexity of solving a sequence of unit cell problems. This is accomplished by discretizing eigenstrains by $C^{-1}(\Theta_{\hat{x}_I})$ piecewise constant function in $\Theta_{\hat{x}_I}$

$$\begin{aligned}
&\mu_{kl}^\zeta(\hat{\boldsymbol{x}}_I, \boldsymbol{\chi}) = \sum_{\alpha=1}^{n} \tilde{N}^{(\alpha)}\left(\hat{\boldsymbol{x}}_I, \boldsymbol{\chi}^{(\alpha)}\right) \mu_{kl}^{(\alpha)}\left(\hat{\boldsymbol{x}}_I, \boldsymbol{\chi}^{(\alpha)}\right) \\
&\mu_{ij}^{(\alpha)}\left(\hat{\boldsymbol{x}}_I, \boldsymbol{\chi}^{(\alpha)}\right) \equiv \bar{\mu}_{ij}^{(\alpha)}(\hat{\boldsymbol{x}}_I) + \bar{\mu}_{ij,m}^{(\alpha)}(\hat{\boldsymbol{x}}_I)\boldsymbol{\chi}_m^{(\alpha)} \\
&\tilde{N}^{(\alpha)}\left(\hat{\boldsymbol{x}}_I, \boldsymbol{\chi}^{(\alpha)}\right) \equiv \begin{cases} 1 & \boldsymbol{\chi}^{(\alpha)} \in \Theta_{\hat{x}_I}^{(\alpha)} \\ 0 & \boldsymbol{\chi}^{(\alpha)} \notin \Theta_{\hat{x}_I}^{(\alpha)} \end{cases}; \quad \Theta_{\hat{x}_I} = \bigcup_{\alpha=1}^{n} \Theta_{\hat{x}_I}^{(\alpha)}; \quad \varnothing = \bigcap_{\alpha=1}^{n} \Theta_{\hat{x}_I}^{(\alpha)}
\end{aligned} \tag{13.13}$$

and eigenseparations by $C^0(\Theta_{\hat{x}_I})$ piecewise linear function

$$\begin{aligned}
&\delta_{\underset{\smile}{q}}^\zeta(\hat{\boldsymbol{x}}_I, \boldsymbol{\chi}) = \sum_{\xi=1}^{m} \breve{N}^{(\xi)}(\hat{\boldsymbol{x}}_I, \boldsymbol{\chi}) \delta_{\underset{\smile}{q}}^{(\xi)}(\hat{\boldsymbol{x}}_I) \\
&\breve{N}^{(\xi)}(\hat{\boldsymbol{x}}_I, \boldsymbol{\chi}) \equiv \begin{cases} \sum_{r \in S_{\hat{x}_I}^{(\xi)}} N_r(\hat{\boldsymbol{x}}_I, \boldsymbol{\chi}) & \boldsymbol{\chi} \in S_{\hat{x}_I}^{(\xi)} \\ 0 & \boldsymbol{\chi} \notin S_{\hat{x}_I}^{(\xi)} \end{cases}
\end{aligned} \tag{13.14}$$

where $\bar{\mu}_{ij}^{(\alpha)}(\hat{\boldsymbol{x}}_I)$ and $\bar{\mu}_{ij,m}^{(\alpha)}(\hat{\boldsymbol{x}}_I)$ are the average and linear variation of the eigenstrain, respectively; $\boldsymbol{\chi}^{(\alpha)} = \boldsymbol{\chi} - \hat{\boldsymbol{\chi}}^{(\alpha)}$ denotes the local coordinate system positioned at the centroid $\hat{\boldsymbol{\chi}}^{(\alpha)}$ of phase partition α; $\Theta_{\hat{x}_I}^{(\alpha)}$ is the volume of partition α in the CUC domain $\Theta_{\hat{x}_I}$ and n is a number of such non-overlapping partitions; $S_{\hat{x}_I}^{(\xi)}$ denotes ξ surface partition in $S_{\hat{x}_I}$ and m is the total number of surface partitions; $\breve{N}^{(\xi)}$ is chosen to be $C^0(\Theta_{\hat{x}_I})$ to ensure smoothness of interface separations constructed by a sum of finite element shape functions N_r defined over the interface partition $S_{\hat{x}_I}^{(\xi)}$.

The influence functions, H_i^{mn}, H_i^{mnp}, $\tilde{h}_i^{mn}$, $\breve{h}_i^{\breve{m}}$, are constructed to satisfy the integral equation (13.9) for arbitrary $\bar{\varepsilon}_{mn}$, $\bar{\varepsilon}_{mn,p}$, $\bar{\mu}_{ij}^{(\alpha)}(\hat{\boldsymbol{x}}_I)$, $\bar{\mu}_{ij,m}^{(\alpha)}(\hat{\boldsymbol{x}}_I)$ and $\delta_{\breve{m}}^{\zeta}$. Integrating (13.9) by parts and exploiting χ-periodicity of $L_{ijkl}(\boldsymbol{\chi})$ yields the following influence function problems in the weak form[34]

$$
\begin{aligned}
&\int_{\Theta_{\hat{x}_I}} w_{(i,j)}^{(1)}(\hat{\boldsymbol{x}}_I,\boldsymbol{\chi})L_{ijkl}(\boldsymbol{\chi})E_{kl}^{mn}(\hat{\boldsymbol{x}}_I,\boldsymbol{\chi})d\Theta=0\\
&\int_{\Theta_{\hat{x}_I}} w_{(i,j)}^{(1)}(\hat{\boldsymbol{x}}_I,\boldsymbol{\chi})L_{ijkl}(\boldsymbol{\chi})E_{kl}^{mnp}(\hat{\boldsymbol{x}}_I,\boldsymbol{\chi})d\Theta=0\\
&\int_{\Theta_{\hat{x}_I}} w_{(i,j)}^{(1)}(\hat{\boldsymbol{x}}_I,\boldsymbol{\chi})L_{ijkl}(\boldsymbol{\chi})\Lambda_{kl}^{mn(\alpha)}(\hat{\boldsymbol{x}}_I,\boldsymbol{\chi})d\Theta=0 \qquad \forall\, w^{(1)}\in W_{\hat{x}_I}^{\Theta}\\
&\int_{\Theta_{\hat{x}_I}} w_{(i,j)}^{(1)}(\hat{\boldsymbol{x}}_I,\boldsymbol{\chi})L_{ijkl}(\boldsymbol{\chi})\Lambda_{kl}^{mnr(\alpha)}(\hat{\boldsymbol{x}}_I,\boldsymbol{\chi})d\Theta=0\\
&\int_{\Theta_{\hat{x}_I}} w_{(i,j)}^{(1)}(\hat{\boldsymbol{x}}_I,\boldsymbol{\chi})L_{ijkl}(\boldsymbol{\chi})Q_{kl}^{\breve{m}(\xi)}(\hat{\boldsymbol{x}}_I,\boldsymbol{\chi})d\Theta=0
\end{aligned}
\tag{13.15}
$$

where

$$
\begin{aligned}
&\Lambda_{kl}^{mn(\alpha)}(\hat{\boldsymbol{x}}_I,\boldsymbol{\chi})\equiv P_{kl}^{mn(\alpha)}(\hat{\boldsymbol{x}}_I,\boldsymbol{\chi})-I_{klmn}^{(\alpha)}(\hat{\boldsymbol{x}}_I,\boldsymbol{\chi})\\
&\Lambda_{kl}^{mnr(\alpha)}(\hat{\boldsymbol{x}}_I,\boldsymbol{\chi})\equiv P_{kl}^{mnr(\alpha)}(\hat{\boldsymbol{x}}_I,\boldsymbol{\chi})-I_{klmn}^{(\alpha)}(\hat{\boldsymbol{x}}_I,\boldsymbol{\chi})\chi_r^{(\alpha)}\\
&P_{kl}^{mn(\alpha)}(\hat{\boldsymbol{x}}_I,\boldsymbol{\chi})\equiv\int_{\Theta_{\hat{x}_I}}\tilde{N}^{(\alpha)}(\hat{\boldsymbol{x}}_I,\tilde{\boldsymbol{\chi}})\tilde{h}_{(k,l)}^{mn}(\hat{\boldsymbol{x}}_I,\boldsymbol{\chi},\tilde{\boldsymbol{\chi}})d\tilde{\Theta}\\
&P_{kl}^{mnr(\alpha)}(\hat{\boldsymbol{x}}_I,\boldsymbol{\chi})\equiv\int_{\Theta_{\hat{x}_I}}\tilde{N}^{(\alpha)}(\hat{\boldsymbol{x}}_I,\tilde{\boldsymbol{\chi}})\tilde{h}_{(k,l)}^{mn}(\hat{\boldsymbol{x}}_I,\boldsymbol{\chi},\tilde{\boldsymbol{\chi}})\tilde{\chi}_r^{(\alpha)}d\tilde{\Theta}\\
&Q_{kl}^{\breve{m}(\xi)}(\hat{\boldsymbol{x}}_I,\boldsymbol{\chi})\equiv\int_{S_{\hat{x}_I}}\breve{h}_{(k,l)}^{\breve{m}}\left(\hat{\boldsymbol{x}}_I,\boldsymbol{\chi},\breve{\boldsymbol{\chi}}\right)\breve{N}^{(\xi)}\left(\hat{\boldsymbol{x}}_I,\breve{\boldsymbol{\chi}}\right)d\breve{S}\\
&I_{klmn}^{(\alpha)}\left(\hat{\boldsymbol{x}}_I,\boldsymbol{\chi}^{(\alpha)}\right)\equiv I_{klmn}\tilde{N}^{(\alpha)}\left(\hat{\boldsymbol{x}}_I,\boldsymbol{\chi}^{(\alpha)}\right)=\begin{cases}I_{klmn} & \boldsymbol{\chi}^{(\alpha)}\in\Theta_{\hat{x}_I}^{(\alpha)}\\ 0 & \boldsymbol{\chi}^{(\alpha)}\notin\Theta_{\hat{x}_I}^{(\alpha)}\end{cases}
\end{aligned}
\tag{13.16}
$$

Eqs. (13.15) define the weak form of the boundary value problems for periodic (or weakly periodic) influence functions, H_i^{mn}, H_i^{mnp}, $\tilde{h}_i^{mn}$, $\breve{h}_i^{\breve{m}}$, respectively. If weak periodicity is assumed, the influence function will have to satisfy additional constraints.[27]

The test $w_i^{(1)}$ and trial functions, H_i^{mn}, H_i^{mnp}, $\tilde{h}_i^{mn}$, $\breve{h}_i^{\breve{m}}$, are discretized using Galerkin finite element method. The resulting discrete influence function problems, which are algebraic linear system of equations, are solved prior nonlinear analysis.

The residual-free reduced order strain (see Eq. (13.8)) and stress fields (13.1)b are given as

$$
\begin{aligned}
\varepsilon_{kl}^{\zeta}(\hat{\boldsymbol{x}}_I,\boldsymbol{\chi}) = {} & E_{kl}^{mn}(\hat{\boldsymbol{x}}_I,\boldsymbol{\chi})\bar{\varepsilon}_{mn}(\hat{\mathbf{x}}_I) + E_{kl}^{mnp}(\hat{\boldsymbol{x}}_I,\boldsymbol{\chi})\bar{\varepsilon}_{mn,p}(\hat{\boldsymbol{x}}_I) \\
& + \sum_{\alpha=1}^{n} P_{kl}^{mn(\alpha)}(\hat{\boldsymbol{x}}_I,\boldsymbol{\chi})\bar{\mu}_{mn}^{(\alpha)}(\hat{\boldsymbol{x}}_I) + \sum_{\alpha=1}^{n} P_{kl}^{mnr(\alpha)}(\hat{\boldsymbol{x}}_I,\boldsymbol{\chi})\bar{\mu}_{mn,r}^{(\alpha)}(\hat{\boldsymbol{x}}_I) \\
& + \sum_{\xi=1}^{m} Q_{kl}^{\breve{m}(\xi)}(\hat{\boldsymbol{x}}_I,\boldsymbol{\chi})\delta_{\breve{m}}^{(\xi)}(\hat{\boldsymbol{x}}_I)
\end{aligned}
$$

$$
\sigma_{ij}^{\zeta}(\hat{\boldsymbol{x}}_I,\boldsymbol{\chi}) = L_{ijkl}(\boldsymbol{\chi})\left(\begin{aligned}
& E_{kl}^{mn}(\hat{\boldsymbol{x}}_I,\boldsymbol{\chi})\bar{\varepsilon}_{mn}(\hat{\mathbf{x}}_I) + E_{kl}^{mnp}(\hat{\boldsymbol{x}}_I,\boldsymbol{\chi})\bar{\varepsilon}_{mn,p}(\hat{\boldsymbol{x}}_I) \\
& + \sum_{\alpha=1}^{n} \Lambda_{kl}^{mn(\alpha)}(\hat{\boldsymbol{x}}_I,\boldsymbol{\chi})\bar{\mu}_{mn}^{(\alpha)}(\hat{\boldsymbol{x}}_I) + \sum_{\alpha=1}^{n} \Lambda_{kl}^{mnr(\alpha)}(\hat{\boldsymbol{x}}_I,\boldsymbol{\chi})\bar{\mu}_{mn,r}^{(\alpha)}(\hat{\boldsymbol{x}}_I) \\
& + \sum_{\xi=1}^{m} Q_{kl}^{\breve{m}(\xi)}(\hat{\boldsymbol{x}}_I,\boldsymbol{\chi})\delta_{\breve{m}}^{(\xi)}(\hat{\boldsymbol{x}}_I)
\end{aligned}\right)
\tag{13.17}
$$

The tractions along the unit cell interfaces are defined by

$$
t_{\breve{q}}\left(\hat{\boldsymbol{x}}_I,\breve{\boldsymbol{\chi}}\right) \equiv a_{\breve{q}i}\left(\hat{\boldsymbol{x}}_I,\breve{\boldsymbol{\chi}}\right)\sigma_{ij}^{\zeta}\left(\hat{\boldsymbol{x}}_I,\breve{\boldsymbol{\chi}}\right)\breve{n}_j^{S}\left(\hat{\boldsymbol{x}}_I,\breve{\boldsymbol{\chi}}\right) \tag{13.18}
$$

where $a_{\breve{q}i}\left(\hat{\boldsymbol{x}}_I,\breve{\boldsymbol{\chi}}\right)$ is transformation matrix from the global coordinates system to the local interface coordinate system; $\breve{\boldsymbol{n}}^{S}\left(\hat{\boldsymbol{x}}_I,\breve{\boldsymbol{\chi}}\right)$ is a unit normal to the interface $S_{\hat{\boldsymbol{x}}_I}$.

13.2.2.2 Reduced Order Computational Unit Cell Problem

Similarly to eigenstrains, strains at each partition of the CUC domain are assumed to have a linear variation

$$
\begin{aligned}
\varepsilon_{ij}^{\zeta}\left(\hat{\boldsymbol{x}}_I,\boldsymbol{\chi}^{(\beta)}\right) &= \sum_{\beta=1}^{n} \tilde{N}^{(\beta)}\left(\hat{\boldsymbol{x}}_I,\boldsymbol{\chi}^{(\beta)}\right)\varepsilon_{ij}^{(\beta)}(\hat{\boldsymbol{x}}_I,\boldsymbol{\chi}) \\
\varepsilon_{ij}^{(\beta)}\left(\hat{\boldsymbol{x}}_I,\boldsymbol{\chi}^{(\beta)}\right) &\equiv \bar{\varepsilon}_{ij}^{(\beta)}(\hat{\boldsymbol{x}}_I) + \bar{\varepsilon}_{ij,m}^{(\beta)}(\hat{\boldsymbol{x}}_I)\chi_m^{(\beta)}
\end{aligned}
\tag{13.19}
$$

Integrating Eq. (13.17) over the partition of the CUC and also integrating the same expression but multiplied by $\chi_p^{(\beta)} \in \Theta_{\hat{\boldsymbol{x}}_I}^{(\beta)}$ yields the following residual-free reduced order equations

$$\bar{\varepsilon}_{kl}^{(\beta)} = E_{kl}^{mn(\beta)}\bar{\varepsilon}_{mn} + E_{kl}^{mnp(\beta)}\bar{\varepsilon}_{mn,p} + \sum_{\alpha=1}^{n} P_{kl}^{mn(\beta\alpha)}\bar{\mu}_{mn}^{(\alpha)} + \sum_{\alpha=1}^{n} P_{kl}^{mnr(\beta\alpha)}\bar{\mu}_{mn,r}^{(\alpha)} + \sum_{\xi=1}^{m} Q_{kl}^{\breve{m}(\beta\xi)}\delta_{\breve{m}}^{(\xi)}$$

$$\begin{aligned}\bar{\varepsilon}_{kl,m}^{(\beta)}\lambda_{mp}^{(\beta)} = {} & \hat{E}_{klp}^{mn(\beta)}\bar{\varepsilon}_{mn} + \hat{E}_{klp}^{mnr(\beta)}\bar{\varepsilon}_{mn,r} \\ & + \sum_{\alpha=1}^{n} \hat{P}_{klp}^{mn(\beta\alpha)}\bar{\mu}_{mn}^{(\alpha)} + \sum_{\alpha=1}^{n} \hat{P}_{klp}^{mnr(\beta\alpha)}\bar{\mu}_{mn,r}^{(\alpha)} + \sum_{\xi=1}^{m} \hat{Q}_{klp}^{\breve{m}(\beta\xi)}\delta_{\breve{m}}^{(\xi)}\end{aligned} \tag{13.20}$$

where the partitioned strain coefficients are given as

$$\begin{aligned}
& E_{kl}^{mn(\beta)}(\hat{\boldsymbol{x}}_I) \equiv \frac{1}{\left|\Theta_{\hat{\boldsymbol{x}}_I}^{(\beta)}\right|}\int_{\Theta_{\hat{\boldsymbol{x}}_I}^{(\beta)}} E_{kl}^{mn}(\hat{\boldsymbol{x}}_I,\boldsymbol{\chi})d\Theta^{(\beta)}; \; E_{kl}^{mnp(\beta)}(\hat{\boldsymbol{x}}_I) \equiv \frac{1}{\left|\Theta_{\hat{\boldsymbol{x}}_I}^{(\beta)}\right|}\int_{\Theta_{\hat{\boldsymbol{x}}_I}^{(\beta)}} E_{kl}^{mnp}(\hat{\boldsymbol{x}}_I,\boldsymbol{\chi})d\Theta^{(\beta)} \\
& P_{klmn}^{(\beta\alpha)}(\hat{\boldsymbol{x}}_I) \equiv \frac{1}{\left|\Theta_{\hat{\boldsymbol{x}}_I}^{(\beta)}\right|}\int_{\Theta_{\hat{\boldsymbol{x}}_I}^{(\beta)}} P_{klmn}^{(\alpha)}(\hat{\boldsymbol{x}}_I,\boldsymbol{\chi})d\Theta^{(\beta)}; \; P_{kl}^{mnr(\beta\alpha)}(\hat{\boldsymbol{x}}_I) \equiv \frac{1}{\left|\Theta_{\hat{\boldsymbol{x}}_I}^{(\beta)}\right|}\int_{\Theta_{\hat{\boldsymbol{x}}_I}^{(\beta)}} P_{kl}^{mnr(\alpha)}(\hat{\boldsymbol{x}}_I,\boldsymbol{\chi})d\Theta^{(\beta)} \\
& Q_{kl}^{\breve{m}(\beta\xi)}(\hat{\boldsymbol{x}}_I) \equiv \frac{1}{\left|\Theta_{\hat{\boldsymbol{x}}_I}^{(\beta)}\right|}\int_{\Theta_{\hat{\boldsymbol{x}}_I}^{(\beta)}} Q_{kl}^{\breve{m}(\xi)}(\hat{\boldsymbol{x}}_I,\boldsymbol{\chi})d\Theta^{(\beta)}
\end{aligned} \tag{13.21}$$

$$\begin{aligned}
& \hat{E}_{klp}^{mn(\beta)}(\hat{\boldsymbol{x}}_I) \equiv \frac{1}{\left|\Theta_{\hat{\boldsymbol{x}}_I}^{(\beta)}\right|}\int_{\Theta_{\hat{\boldsymbol{x}}_I}^{(\beta)}} \hat{E}_{kl}^{mn}\chi_p^{(\beta)}d\Theta^{(\beta)}; \; \hat{E}_{klp}^{mnr(\beta)}(\hat{\boldsymbol{x}}_I) \equiv \frac{1}{\left|\Theta_{\hat{\boldsymbol{x}}_I}^{(\beta)}\right|}\int_{\Theta_{\hat{\boldsymbol{x}}_I}^{(\beta)}} \hat{E}_{kl}^{mnr}\chi_p^{(\beta)}d\Theta^{(\beta)} \\
& \hat{P}_{klp}^{mn(\beta\alpha)}(\hat{\boldsymbol{x}}_I) \equiv \frac{1}{\left|\Theta_{\hat{\boldsymbol{x}}_I}^{(\beta)}\right|}\int_{\Theta_{\hat{\boldsymbol{x}}_I}^{(\beta)}} P_{kl}^{mn(\alpha)}\chi_p^{(\beta)}d\Theta^{(\beta)}; \; \hat{P}_{klp}^{mnr(\beta\alpha)}(\hat{\boldsymbol{x}}_I) \equiv \frac{1}{\left|\Theta_{\hat{\boldsymbol{x}}_I}^{(\beta)}\right|}\int_{\Theta_{\hat{\boldsymbol{x}}_I}^{(\beta)}} P_{kl}^{mnr(\alpha)}\chi_p^{(\beta)}d\Theta^{(\beta)} \\
& \hat{Q}_{klp}^{\breve{m}(\beta\xi)}(\hat{\boldsymbol{x}}_I) \equiv \frac{1}{\left|\Theta_{\hat{\boldsymbol{x}}_I}^{(\beta)}\right|}\int_{\Theta_{\hat{\boldsymbol{x}}_I}^{(\beta)}} Q_{kl}^{\breve{m}(\xi)}\chi_p^{(\beta)}d\Theta^{(\beta)}; \; \lambda_{mp}^{(\beta)}(\hat{\boldsymbol{x}}_I) \equiv \frac{1}{\left|\Theta_{\hat{\boldsymbol{x}}_I}^{(\beta)}\right|}\int_{\Theta_{\hat{\boldsymbol{x}}_I}^{(\beta)}} \chi_m^{(\beta)}\chi_p^{(\beta)}d\Theta^{(\beta)}
\end{aligned} \tag{13.22}$$

Similarly, we define the partitioned traction as an average traction over the interface partition $S_{\hat{\boldsymbol{x}}_I}^{(\eta)}$

$$t_{\breve{q}}^{(\eta)}(\hat{\boldsymbol{x}}_I) \equiv \frac{1}{\left|S_{\hat{\boldsymbol{x}}_I}^{(\eta)}\right|}\int_{S_{\hat{\boldsymbol{x}}_I}^{(\eta)}} t_{\breve{q}}\left(\hat{\boldsymbol{x}}_I,\breve{\boldsymbol{\chi}}\right)d\breve{S}^{(\eta)} \tag{13.23}$$

which, using (13.17)b and (13.18), yields

$$t_{\underset{\smile}{q}}^{(\eta)} = T_{\underset{\smile}{q}}^{mn(\eta)}\bar{\varepsilon}_{mn} + T_{\underset{\smile}{q}}^{mnp(\eta)}\bar{\varepsilon}_{mn,p} + \sum_{\alpha=1}^{n} C_{\underset{\smile}{q}}^{mn(\eta\alpha)}\bar{\mu}_{mn}^{(\alpha)} + \sum_{\alpha=1}^{n} C_{\underset{\smile}{q}}^{mnr(\eta\alpha)}\bar{\mu}_{mn,r}^{(\alpha)} + \sum_{\xi=1}^{m} D_{\underset{\smile}{q}}^{\underset{\smile}{m}(\eta\xi)}\delta_{\underset{\smile}{m}}^{(\xi)} \tag{13.24}$$

and partitioned traction tensor coefficients are given as

$$\begin{aligned}
T_{\underset{\smile}{q}}^{mn(\eta)}(\hat{\boldsymbol{x}}_I) &\equiv \frac{1}{\left|S_{\hat{\boldsymbol{x}}_I}^{(\eta)}\right|}\int_{S_{\hat{\boldsymbol{x}}_I}^{(\eta)}} a_{\underset{\smile}{q}i}\left(\hat{\boldsymbol{x}}_I,\underset{\smile}{\boldsymbol{\chi}}\right)L_{ijkl}\left(\underset{\smile}{\boldsymbol{\chi}}\right)E_{kl}^{mn}\left(\hat{\boldsymbol{x}}_I,\underset{\smile}{\boldsymbol{\chi}}\right)\underset{\smile}{n}_j^S\left(\hat{\boldsymbol{x}}_I,\underset{\smile}{\boldsymbol{\chi}}\right)d\underset{\smile}{S}^{(\eta)} \\
T_{\underset{\smile}{q}}^{mnp(\eta)}(\hat{\boldsymbol{x}}_I) &\equiv \frac{1}{\left|S_{\hat{\boldsymbol{x}}_I}^{(\eta)}\right|}\int_{S_{\hat{\boldsymbol{x}}_I}^{(\eta)}} a_{\underset{\smile}{q}i}\left(\hat{\boldsymbol{x}}_I,\underset{\smile}{\boldsymbol{\chi}}\right)L_{ijkl}\left(\underset{\smile}{\boldsymbol{\chi}}\right)E_{kl}^{mnp}\left(\hat{\boldsymbol{x}}_I,\underset{\smile}{\boldsymbol{\chi}}\right)\underset{\smile}{n}_j^S\left(\hat{\boldsymbol{x}}_I,\underset{\smile}{\boldsymbol{\chi}}\right)d\underset{\smile}{S}^{(\eta)} \\
C_{\underset{\smile}{q}}^{mn(\eta\alpha)}(\hat{\boldsymbol{x}}_I) &\equiv \frac{1}{\left|S_{\hat{\boldsymbol{x}}_I}^{(\eta)}\right|}\int_{S_{\hat{\boldsymbol{x}}_I}^{(\eta)}} a_{\underset{\smile}{q}i}\left(\hat{\boldsymbol{x}}_I,\underset{\smile}{\boldsymbol{\chi}}\right)L_{ijkl}\left(\underset{\smile}{\boldsymbol{\chi}}\right)\Lambda_{kl}^{mn(\alpha)}(\hat{\boldsymbol{x}}_I,\boldsymbol{\chi})\underset{\smile}{n}_j^S\left(\hat{\boldsymbol{x}}_I,\underset{\smile}{\boldsymbol{\chi}}\right)d\underset{\smile}{S}^{(\eta)} \\
C_{\underset{\smile}{q}}^{mnr(\eta\alpha)}(\hat{\boldsymbol{x}}_I) &\equiv \frac{1}{\left|S_{\hat{\boldsymbol{x}}_I}^{(\eta)}\right|}\int_{S_{\hat{\boldsymbol{x}}_I}^{(\eta)}} a_{\underset{\smile}{q}i}\left(\hat{\boldsymbol{x}}_I,\underset{\smile}{\boldsymbol{\chi}}\right)L_{ijkl}\left(\underset{\smile}{\boldsymbol{\chi}}\right)\Lambda_{kl}^{mnr(\alpha)}(\hat{\boldsymbol{x}}_I,\boldsymbol{\chi})\underset{\smile}{n}_j^S\left(\hat{\boldsymbol{x}}_I,\underset{\smile}{\boldsymbol{\chi}}\right)d\underset{\smile}{S}^{(\eta)} \\
D_{\underset{\smile}{q}\underset{\smile}{m}}^{(\eta\xi)}(\hat{\boldsymbol{x}}_I) &\equiv \frac{1}{\left|S_{\hat{\boldsymbol{x}}_I}^{(\eta)}\right|}\int_{S_{\hat{\boldsymbol{x}}_I}^{(\eta)}} a_{\underset{\smile}{q}i}\left(\hat{\boldsymbol{x}}_I,\underset{\smile}{\boldsymbol{\chi}}\right)L_{ijkl}\left(\underset{\smile}{\boldsymbol{\chi}}\right)Q_{kl\underset{\smile}{m}}^{(\xi)}\left(\hat{\boldsymbol{x}}_I,\underset{\smile}{\boldsymbol{\chi}}\right)\underset{\smile}{n}_j^S\left(\hat{\boldsymbol{x}}_I,\underset{\smile}{\boldsymbol{\chi}}\right)d\underset{\smile}{S}^{(\eta)}
\end{aligned} \tag{13.25}$$

We further assume that cohesive law is specified, *i.e.* partitioned traction, $t_{\underset{\smile}{q}}^{(\eta)}$, along the interface partition is prescribed as a function of eigenseparation. Also suppose that partitioned average eigenstrain, $\bar{\mu}_{ij}^{(\alpha)}$, and its gradient, $\bar{\mu}_{ij,k}^{(\alpha)}$, can be obtained from the constitutive law defined in the volume partition as described in manuscripts.[30,34] Schematically, this is denoted as

$$t_{\underset{\smile}{p}}^{(\xi)} = g\left(\delta_{\underset{\smile}{q}}^{(\xi)}\right); \quad \bar{\mu}_{mn}^{(\alpha)} = f_1\left(\bar{\varepsilon}_{kl}^{(\alpha)}, \bar{\varepsilon}_{kl,p}^{(\alpha)}\right); \quad \bar{\mu}_{mn,r}^{(\alpha)} = f_2\left(\bar{\varepsilon}_{kl}^{(\alpha)}, \bar{\varepsilon}_{kl,p}^{(\alpha)}\right) \tag{13.26}$$

Eqs (13.20) and (13.24) comprise the reduced order nonlinear system of equations for the unknowns $\bar{\varepsilon}_{kl}^{(\alpha)}, \bar{\varepsilon}_{kl,p}^{(\alpha)}, \delta_{\underset{\smile}{q}}^{(\xi)}$. This nonlinear system of equations is solved by Newton method, which is nested with the coarse-scale Newton iteration.[34]

13.2.3 The Coarse-scale Weak Form

The coarse-scale weak form is obtained by multiplying the strong form (13.1)a by the coarse-scale test function $w_i^c(\boldsymbol{x}) = w_i^c(\hat{\boldsymbol{x}}_I, \boldsymbol{\chi}) \in W_\Omega^c$, integrating by parts and applying the nonlocal quadrature scheme (13.3), which yields

$$\sum_{I=1}^{\hat{N}} \int_{\Theta_{\hat{x}_I}} \hat{\omega}(\hat{\boldsymbol{x}}_I, \boldsymbol{\chi}) w_{i,j}^c(\hat{\boldsymbol{x}}_I, \boldsymbol{\chi}) \sigma_{ij}^{\zeta}(\hat{\boldsymbol{x}}_I, \boldsymbol{\chi}) d\Theta = \int_{\partial\Omega^{t\zeta}} w_i^c(\boldsymbol{x}) \bar{t}_i^{\zeta}(\boldsymbol{x}) d\Gamma + \int_{\Omega} w_i^c(\boldsymbol{x}) b_i^{\zeta}(\boldsymbol{x}) d\Omega$$
$$\forall w_i^c \in W_{\Omega}^c = \left\{ w_i^c \in C^0(\Omega) \middle| w_i^c = 0 \text{ on } \partial\Omega^{u\zeta} \right\} \tag{13.27}$$

Following computational continua paper,[27] stresses are decomposed analogously to strains (see Eq. (13.5))

$$\begin{aligned} \sigma_{ij}^{\zeta}(\hat{\boldsymbol{x}}_I, \boldsymbol{\chi}) &= \sigma_{ij}^c(\hat{\boldsymbol{x}}_I, \boldsymbol{\chi}) + \sigma_{ij}^*(\hat{\boldsymbol{x}}_I, \boldsymbol{\chi}) \\ \sigma_{ij}^c(\hat{\boldsymbol{x}}_I, \boldsymbol{\chi}) &= \bar{\sigma}_{ij}(\hat{\boldsymbol{x}}_I) + q_{ijm}(\hat{\boldsymbol{x}}_I)\chi_m \end{aligned} \tag{13.28}$$

where σ_{ij}^c is a coarse-scale stress; σ_{ij}^* the stress perturbation; $\bar{\sigma}_{ij}$ the constant part of the coarse-scale stress; and q_{ijm} denotes the coefficient of linear variation of the coarse-scale stress. Components $\bar{\sigma}_{ij}$ and q_{ijm} are evaluated from the condition that σ_{ij}^{ζ} in (13.27) could be replaced by σ_{ij}^c

$$\sum_{I=1}^{\hat{N}} \int_{\Theta_{\hat{x}_I}} \hat{\omega}(\hat{\boldsymbol{x}}_I, \boldsymbol{\chi}) w_{i,j}^c(\hat{\boldsymbol{x}}_I, \boldsymbol{\chi}) \sigma_{ij}^c(\hat{\boldsymbol{x}}_I, \boldsymbol{\chi}) d\Theta = \int_{\partial\Omega^{t\zeta}} w_i^c(\boldsymbol{x}) \bar{t}_i^{\zeta}(\boldsymbol{x}) d\Gamma + \int_{\Omega} w_i^c(\boldsymbol{x}) b_i^{\zeta}(\boldsymbol{x}) d\Omega$$
$$\forall \boldsymbol{w}^c \in W_{\Omega}^c \tag{13.29}$$

which yields

$$\int_{\Theta_{\hat{x}_I}} \hat{\omega}(\hat{\boldsymbol{x}}_I, \boldsymbol{\chi}) w_{i,j}^c(\hat{\boldsymbol{x}}_I, \boldsymbol{\chi}) \sigma_{ij}^*(\hat{\boldsymbol{x}}_I, \boldsymbol{\chi}) d\Theta = 0 \tag{13.30}$$

To calculate the coefficients of the coarse-scale stress, $\bar{\sigma}_{ij}$ and q_{ijk}, $\hat{\omega} w_{i,j}^c$ is approximated by a linear function over the CUC domain (for more general case see manuscript)[27] which yields

$$\bar{\sigma}_{ij}(\hat{\boldsymbol{x}}_I) = \frac{1}{|\Theta_{\hat{x}_I}|} \int_{\Theta_{\hat{x}_I}} \sigma_{ij}^{\zeta}(\hat{\boldsymbol{x}}_I, \boldsymbol{\chi}) d\Theta; \; \bar{q}_{ijk}(\hat{\boldsymbol{x}}_I) = \frac{1}{|\Theta_{\hat{x}_I}|} \int_{\Theta_{\hat{x}_I}} \sigma_{ij}^{\zeta}(\hat{\boldsymbol{x}}_I, \boldsymbol{\chi}) \chi_k d\Theta$$
$$q_{ijk}(\hat{\boldsymbol{x}}_I) = \frac{12}{l_{\underline{k}}^2} \bar{q}_{ij\underline{k}}(\hat{\boldsymbol{x}}_I), \quad \text{no summution over } \underline{k} \tag{13.31}$$

where l_k is the unit cell length in k^{th}-direction, and (13.31)c is calculated here for a brick volume which is usual choice of the unit cell. The resulting coarse-scale stress is given by[27]

$$\sigma_{ij}^c(\hat{\boldsymbol{x}}_I, \boldsymbol{\chi}) = \bar{\sigma}_{ij}(\hat{\boldsymbol{x}}_I) + \frac{12}{l_{\underline{k}}^2} \bar{q}_{ijk}(\hat{\boldsymbol{x}}_I)\, \chi_k \tag{13.32}$$

The coefficients of the coarse-scale stress, $\bar{\sigma}_{ij}$ and q_{ijk}, can be directly calculated from (13.17)b and (13.31), which yields

$$\bar{\sigma}_{ij} = \bar{L}_{ijmn}\bar{\varepsilon}_{mn} + \bar{L}_{ijmnp}\bar{\varepsilon}_{mn,p} + \sum_{\alpha=1}^{n} \bar{A}_{ij}^{mn(\alpha)}\bar{\mu}_{mn}^{(\alpha)} + \sum_{\alpha=1}^{n} \bar{A}_{ij}^{mnp(\alpha)}\bar{\mu}_{mn,p}^{(\alpha)} + \sum_{\xi=1}^{m} \bar{B}_{ij}^{\breve{m}(\xi)}\delta_{\breve{m}}^{(\xi)}$$

$$\bar{q}_{ijr} = \hat{L}_{ijmnr}\bar{\varepsilon}_{mn} + \hat{L}_{ijmnpr}\bar{\varepsilon}_{mn,p} + \sum_{\alpha=1}^{n} \hat{A}_{ijr}^{mn(\alpha)}\bar{\mu}_{mn}^{(\alpha)} + \sum_{\alpha=1}^{n} \hat{A}_{ijr}^{mnp(\alpha)}\bar{\mu}_{mn,p}^{(\alpha)} + \sum_{\xi=1}^{m} \hat{B}_{ijr}^{\breve{m}(\xi)}\delta_{\breve{m}}^{(\xi)} \tag{13.33}$$

where

$$\bar{L}_{ijmn}(\hat{\boldsymbol{x}}_I) \equiv \frac{1}{|\Theta_{\hat{x}_I}|}\int_{\Theta_{\hat{x}_I}} L_{ijkl}(\boldsymbol{\chi})E_{kl}^{mn}(\hat{\boldsymbol{x}}_I,\chi)d\Theta;\ \hat{L}_{ijmnr}(\hat{\boldsymbol{x}}_I) \equiv \frac{1}{|\Theta_{\hat{x}_I}|}\int_{\Theta_{\hat{x}_I}} L_{ijkl}E_{kl}^{mn}\chi_r d\Theta$$

$$\bar{L}_{ijmnp}(\hat{\boldsymbol{x}}_I) \equiv \frac{1}{|\Theta_{\hat{x}_I}|}\int_{\Theta_{\hat{x}_I}} L_{ijkl}(\boldsymbol{\chi})E_{kl}^{mnp}(\hat{\boldsymbol{x}}_I,\boldsymbol{\chi})d\Theta;\ \hat{L}_{ijmnpr}(\hat{\boldsymbol{x}}_I) \equiv \frac{1}{|\Theta_{\hat{x}_I}|}\int_{\Theta_{\hat{x}_I}} L_{ijkl}E_{kl}^{mnp}\boldsymbol{\chi}_r d\Theta$$

$$\bar{A}_{ij}^{mn(\alpha)}(\hat{\boldsymbol{x}}_I) \equiv \frac{1}{|\Theta_{\hat{x}_I}|}\int_{\Theta_{\hat{x}_I}} L_{ijkl}(\boldsymbol{\chi})\Lambda_{kl}^{mn(\alpha)}(\hat{\boldsymbol{x}}_I,\boldsymbol{\chi})d\Theta;$$

$$\hat{A}_{ij}^{mnr(\alpha)}(\hat{\boldsymbol{x}}_I) \equiv \frac{1}{|\Theta_{\hat{x}_I}|}\int_{\Theta_{\hat{x}_I}} L_{ijkl}\Lambda_{kl}^{mn(\alpha)}\chi_r d\Theta$$

$$\bar{A}_{ij}^{mnp(\alpha)}(\hat{\boldsymbol{x}}_I) \equiv \frac{1}{|\Theta_{\hat{x}_I}|}\int_{\Theta_{\hat{x}_I}} L_{ijkl}(\boldsymbol{\chi})\Lambda_{kl}^{mnp(\alpha)}(\hat{\boldsymbol{x}}_I,\boldsymbol{\chi})d\Theta;$$

$$\hat{A}_{ij}^{mnpr(\alpha)}(\hat{\boldsymbol{x}}_I) \equiv \frac{1}{|\Theta_{\hat{x}_I}|}\int_{\Theta_{\hat{x}_I}} L_{ijkl}\Lambda_{kl}^{mnp(\alpha)}\chi_r d\Theta$$

$$\bar{B}_{ij}^{\breve{m}(\xi)}(\hat{\boldsymbol{x}}_I) \equiv \frac{1}{|\Theta_{\hat{x}_I}|}\int_{\Theta_{\hat{x}_I}} L_{ijkl}(\boldsymbol{\chi})Q_{kl}^{\breve{m}(\xi)}(\hat{\boldsymbol{x}}_I,\boldsymbol{\chi})d\Theta;\ \hat{B}_{ijr}^{\breve{m}(\xi)}(\hat{\boldsymbol{x}}_I) \equiv \frac{1}{|\Theta_{\hat{x}_I}|}\int_{\Theta_{\hat{x}_I}} L_{ijkl}Q_{kl}^{\breve{m}(\xi)}\chi_r d\Theta \tag{13.34}$$

Note that all the influence functions and the overall coefficients in (13.34) are computed in preprocessing stage prior to nonlinear analysis.

13.2.4 Coarse-scale Discretization

The coarse-scale trial and test functions, u_i^c and w_i^c, are discretized using C^0 continuous coarse-scale shape functions $N_{i\alpha}^c(\hat{\boldsymbol{x}}_I, \boldsymbol{\chi})$ on Ω

$$u_i^c = N_{i\alpha}^c(\hat{\boldsymbol{x}}_I, \boldsymbol{\chi})d_\alpha^c, \qquad w_i^c = N_{i\alpha}^c(\hat{\boldsymbol{x}}_I, \boldsymbol{\chi})c_\alpha^c \tag{13.35}$$

where d_α^c denotes the nodal displacements and c_α^c the nodal values of the test function. Greek subscripts are reserved for finite element degrees-of-freedom and summation convention over repeated degrees-of-freedom is employed. The symmetric derivatives of coarse-scale trial and test functions are given as

$$\varepsilon_{ij}^c(\hat{\boldsymbol{x}}_I, \boldsymbol{\chi}) = N_{(i,j)\alpha}^c(\hat{\boldsymbol{x}}_I, \boldsymbol{\chi}) d_\alpha^c; \quad w_{(i,j)}^c(\hat{\boldsymbol{x}}_I, \boldsymbol{\chi}) = N_{(i,j)\alpha}^c(\hat{\boldsymbol{x}}_I, \boldsymbol{\chi}) c_\alpha^c \tag{13.36}$$

Note that for the second order RC[2] formulation considered in the present chapter the coarse-scale stress is assumed to vary linearly (see (13.28)b) over the unit cell domain. Yet, the polynomial order of the symmetric derivatives of coarse-scale trial and test functions depends on the choice of finite elements. Furthermore, for distorted elements this approximation may not be a polynomial function. Therefore, we will employ a B-bar approach[35] by which approximation (13.36) is replaced by the assumed linear fields$(\bar{\varepsilon}_{ij}^c, \bar{w}_{(i,j)}^c)$ defined as

$$\begin{aligned}
&\bar{\varepsilon}_{ij}^c(\hat{\boldsymbol{x}}_I, \boldsymbol{\chi}) = \left\{ \bar{N}_{(i,j)\alpha}(\hat{\boldsymbol{x}}_I) + D\bar{N}_{(i,j)m\alpha}(\hat{\boldsymbol{x}}_I)\chi_m \right\} d_\alpha^c; \quad \bar{\varepsilon}_{ij} = \bar{N}_{(i,j)\alpha} d_\alpha^c; \quad \bar{\varepsilon}_{ij,m} = D\bar{N}_{(i,j)m\alpha} d_\alpha^c \\
&\bar{w}_{(i,j)}^c(\hat{\boldsymbol{x}}_I, \boldsymbol{\chi}) = \left\{ \bar{N}_{(i,j)\alpha}(\hat{\boldsymbol{x}}_I) + D\bar{N}_{(i,j)m\alpha}(\hat{\boldsymbol{x}}_I)\chi_m \right\} c_\alpha^c
\end{aligned} \tag{13.37}$$

where

$$\bar{N}_{(i,j)\alpha}(\hat{\boldsymbol{x}}_I) \equiv \frac{1}{\left|\Theta_{\hat{x}_I}\right|} \int_{\Theta_{\hat{x}_I}} N_{(i,j)\alpha}^c(\hat{\boldsymbol{x}}_I, \boldsymbol{\chi}) d\Theta; \quad D\bar{N}_{(i,j)m\alpha}(\hat{\boldsymbol{x}}_I) \equiv \frac{1}{\left|\Theta_{\hat{x}_I}\right|} \int_{\Theta_{\hat{x}_I}} \frac{\partial N_{(i,j)\alpha}^c}{\partial \chi_m} d\Theta \tag{13.38}$$

Then a coarse-scale internal force vector for (13.39) is the following

$$f_\alpha^{\text{int}} = \sum_{I=1}^{\hat{N}} \int_{\Theta_{\hat{x}_I}} \hat{\omega}(\hat{\boldsymbol{x}}_I, \boldsymbol{\chi}) \left\{ \bar{N}_{(i,j)\alpha}(\hat{\boldsymbol{x}}_I) + D\bar{N}_{(i,j)m\alpha}(\hat{\boldsymbol{x}}_I)\boldsymbol{\chi}_m \right\} \sigma_{ij}^c(\hat{\boldsymbol{x}}_I, \boldsymbol{\chi}) d\Theta \tag{13.39}$$

The consistent tangent stiffness matrix with derivatives $\partial\Delta\sigma_{ij}^c(\hat{\boldsymbol{x}}_I, \boldsymbol{\chi})/\partial\Delta d_\beta$ is evaluated from reduced order residual system of nonlinear equations (13.20) and (13.24).[34]

13.3 Model Verification

For model verification we consider a problem with a beam consisting of eight coarse-scale elements. For simplicity, both the coarse-scale element and the computational unit cell are in the form of a cube. The unit cell contains a cylindrical inclusion placed at the CUC centroid as shown in Figure 13.1(a). The diameter of the inclusion cross-section is chosen to be 0.6 of the CUC length. For the reference solution, we consider direct numerical simulation (DNS), which employs sufficiently fine finite element mesh where the element size is considerably smaller than the size of the inclusion. The results of the reduced order computational continua (referred here as RC[2]) are also

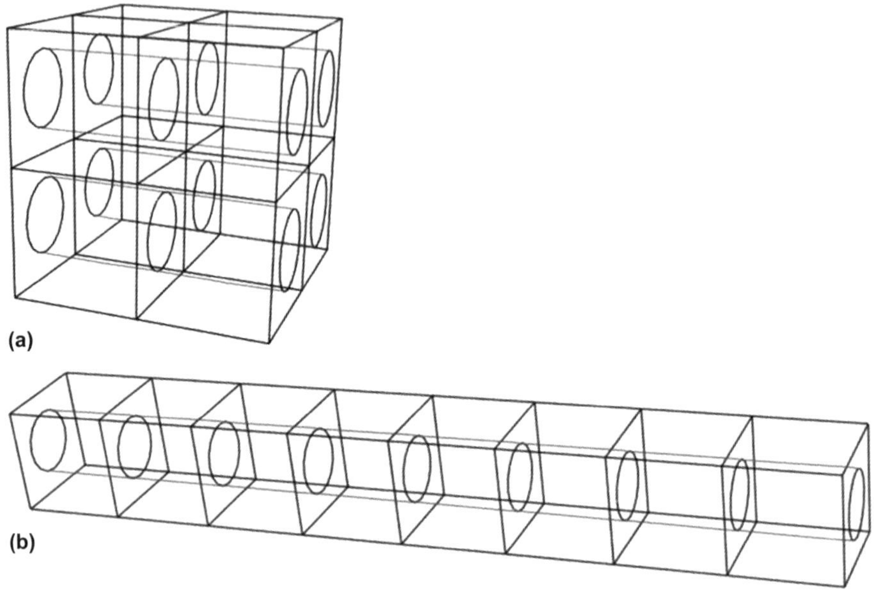

Figure 13.1 (a) A single coarse-scale finite element with cylindrical inclusion for $\zeta = 0.5$ and (b) the beam problem with eight CUCs for $\zeta = 1$.

compared to the $O(1)$ reduced order formulation[28–32] (subsequently referred as $O(1)$).

We study the dependence of the solution accuracy on CUC size characterized by dimensionless size-effect parameter ζ which is equal to a ratio of the CUC size to macro element size, Figure 13.1. For the RC^2 formulation any value of ζ in $[0, 1]$ can be considered; here we take $\zeta = 0.01, 0.25, 0.5, 0.75, 1$. However, the reference solution cannot be constructed for some choices of ζ, such as for instance $1 > \zeta > 0.5$, if the coarse-scale problem is made by spatial repetition of complete unit cells. Note that the results obtained by $O(1)$ homogenization method do not depend on the value of ζ.

We apply the following boundary condition with prescribed linear distribution of dimensionless displacements in x-direction on the right face of the beam

$$U_x = \begin{cases} 0.08t + x_R, & t \in [0, t_1) \\ 0.04y(t - t_1) + 0.08t_1 + x_R, & t \in [t_1, t_2] \end{cases}$$
$$U_y = 0; \quad U_z = 0; \quad t_1 = 1; \quad t_2 = 2$$

Here x_R denotes the coordinate of the right face of the beam, the left side is clamped, and t denotes the dimensionless quasi-time that controls the magnitude of loading. The cylindrical inclusion is placed along the beam axe in x-direction.

We consider linear isotropic hardening plasticity for matrix phase with inclusion being elastic. The hardening modulus $\bar{H}$ is assumed to be constant.

The dimensionless inelastic parameters, the yield stress and the hardening modulus for matrix phase, are $\sigma_Y = 380.e + 6$, $\bar{H} = 200.e + 8$, respectively, while elastic moduli for the two phases are: $E_{matrix} = 200.e + 9$, $\nu_{matrix} = 0.3$, $E_{inclusion} = 1000.e + 9$, $\nu_{inclusion} = 0.3$. For simplicity, we consider one volume partition for each material phase.

Implementation was conducted by using ABAQUS/Standard solver with element user subroutine (UEL) to define an element properties for reduced order computational continua approach.[34] Eight 20-node serendipity quadratic hexahedral elements are used for the coarse-scale mesh of the beam, Figure 13.2(a). Preprocessing stage for the fine-scale problem formulation is implemented separately and also based on finite element method. The unit cell finite mesh considered in both RC^2 and $O(1)$ formulations is made of linear tetrahedral elements totaling 1523 elements as shown in Figure 13.3(b). The mesh for DNS is sufficiently fine and it is consistent with the mesh of the CUC employed in the preprocessing stage (see Figure 13.2(b) and Figure 13.3(a)). Figure 13.2 depicts displacement distribution along the beam and shows good agreement between the RC^2 and the DNS models. Figure 13.3(a) depicts the Mises stress distribution at the free end of the beam.

The result for dimensionless bending moment is given in Figure 13.4(a) at $t = t_2$. It can be seen that RC^2 formulation is in excellent agreement with DNS while $O(1)$ solution results in 40% error in the moment for $\zeta = 1$. Figure 13.4(b) depicts the evolution of moment as a function of displacement at the upper corner of the beam; the coarse-scale element is chosen to be equal to the CUC, $\zeta = 1$. The results show good agreement between RC^2 and DNS during the entire loading process. When the bending initiates the O(1)

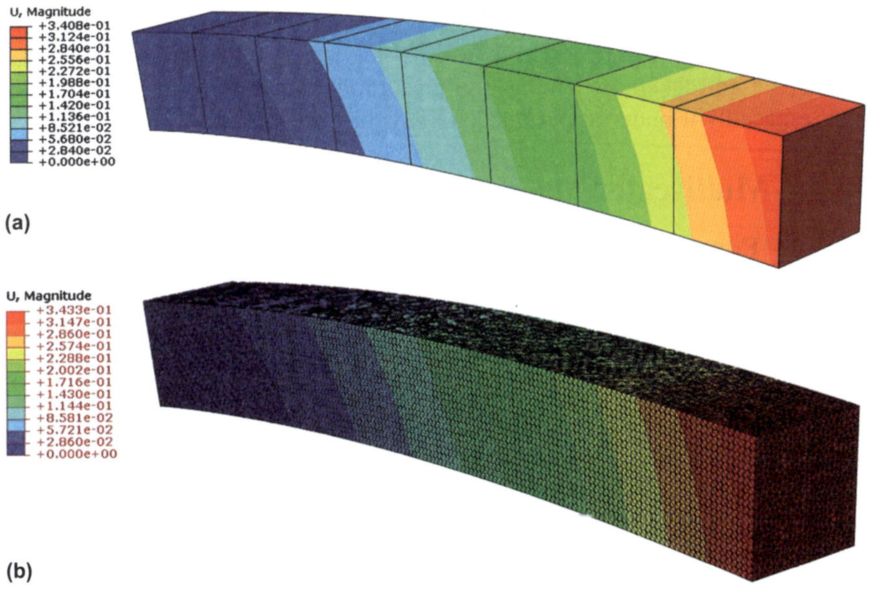

Figure 13.2 A displacement magnitude along the beam for (a) the RC^2 and (b) the DNS, $\zeta = 0.5$.

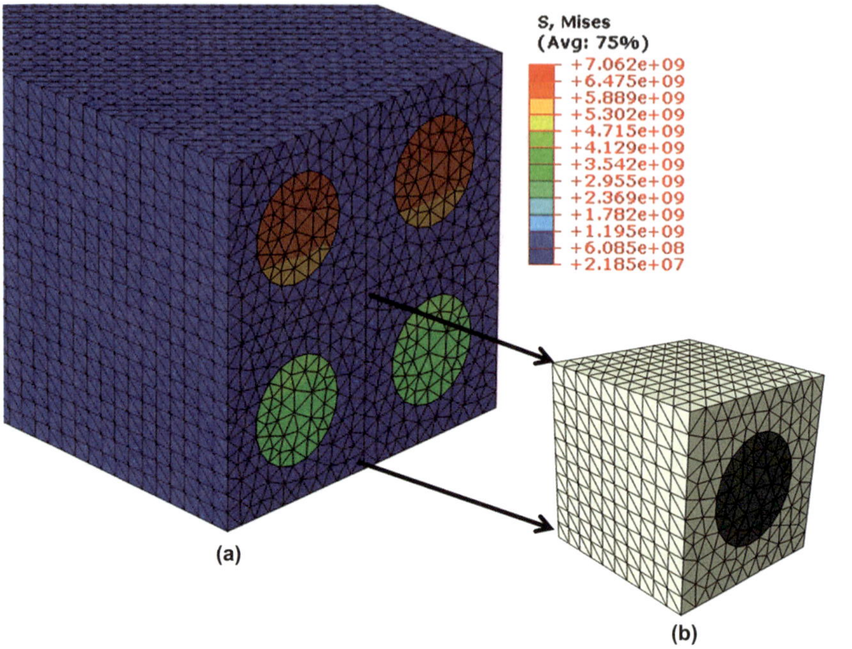

Figure 13.3 A finite element mesh for (a) the DNS (one-scale analysis) and for (b) the CUC problem (preprocessing stage).

formulation continues to correctly predict the value of reaction force[34] but errs badly in predicting the moment.

For the problems where the CUC size is small or strain gradient over the unit cell is not sufficient (uniform coarse-scale field) both reduced order formulations (RC^2 and O(1)) are similar and give good agreement with DNS simulation, especially if material parameters are calibrated properly.

13.4 Multiscale Characterization

13.4.1 Formulation of an Inverse Problem

Consider parameters defining unit cell geometry and phenomenological coefficients of fine-scale constitutive models to be characterized (identified) as components of the vector $z \in \mathbb{R}^N$. Then the inverse (or optimization) problem can be formulated as follows:

Find the vector z that minimizes the objective function

$$\Phi(z) = \frac{1}{2}\sum_{\alpha=1}^{L} \theta^{\alpha} f^{\alpha}(z) \quad l_i \leq z_i \leq u_i \quad (i = 1, \ldots, N) \tag{13.40}$$

where L is the total number of specific response quantities (denoted by α) which can be measured in the course of experiments and then

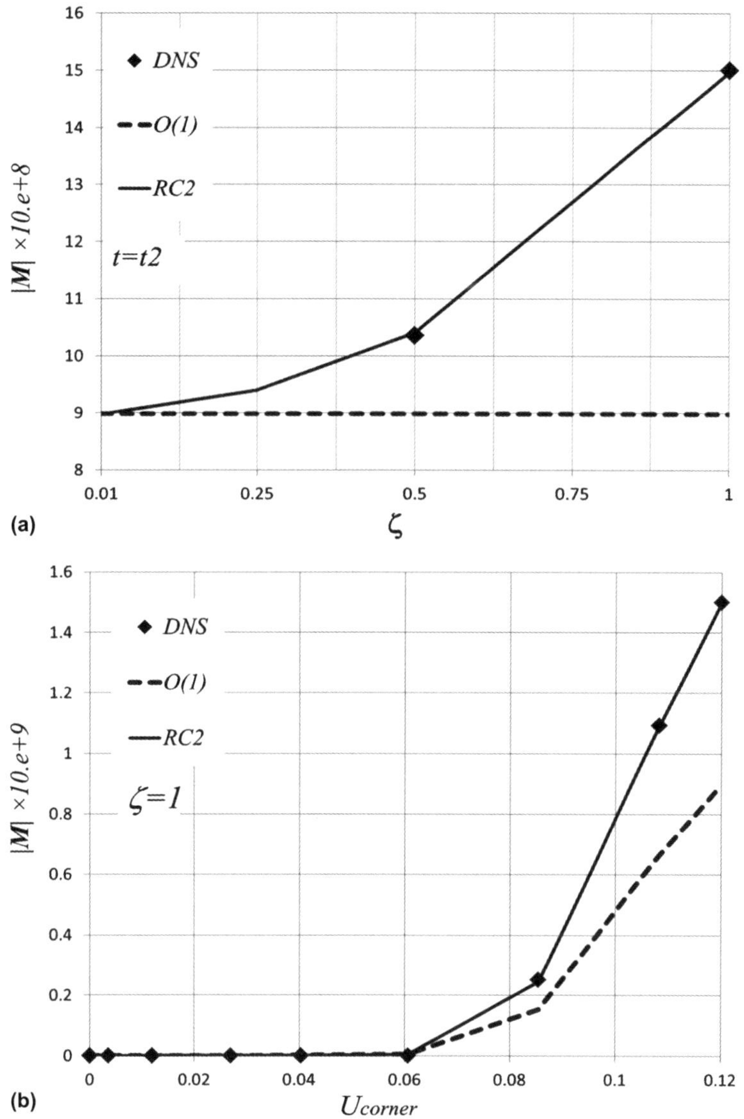

Figure 13.4 (a) Moment evolution for various CUC sizes, fixed time, and (b) moment evolution versus displacement at the right upper corner of the beam, fixed CUC size.

obtained as a result of the numerical simulation; $f^{\alpha}(z)$ is the dimensionless function

$$f^{\alpha}(z) = \frac{\sum_{s=1}^{S_{\alpha}} \left[F_s^{\alpha} - F^{\alpha}(z, \tau_s^{\alpha})\right]^2}{\sum_{s=1}^{S_{\alpha}} \left[F_s^{\alpha}\right]^2} \tag{13.41}$$

which measures the deviation between the computed αth individual response and the observed one from the experiment; τ^α is a parameter which defines the history of the process in the course of the experiment (e.g. the time or the loading parameter), and the values $\tau_s^\alpha(\alpha = 1, \ldots, L;\ s = 1, \ldots, S_\alpha)$ define the discrete set of S_α data points; F_s^α is the value of the αth measured response quantity corresponding to the value of the experiment history parameter τ_s^α; $F^\alpha(z, \tau_s^\alpha)$ is the value of the same response quantity obtained from the numerical simulation; θ^α is the weight coefficient which determines the relative contribution of information yielded by the αth set of experimental data; l_i, u_i are side constraints, stipulated by some additional physical considerations, such manufacturing, which define the search region in the space $\mathbb{R}^N$of optimization parameters.

The weight coefficient θ^α may be chosen by various criteria: (i) in a deterministic context as identity matrix or as experimental data weights which are inversely proportional to the measurement scattering; or (ii) as diagonal of covariance matrix (for uncorrelated quantities), when available from statistical processing of the measurements (experimental data are usually provided together with a standard deviation as a measure of the scattering or uncertainty intrinsically typical of the measurement devices).

It is convenient to recast the constrained minimization problem (13.40), (13.41) in the matrix form as

$$\Phi(z) = \frac{1}{2}\boldsymbol{F}(z)^T \boldsymbol{W}\boldsymbol{F}(z) \qquad l_i \le z_i \le u_i \quad (i = 1, \ldots, N) \tag{13.42}$$

where $\boldsymbol{F}(z) \in \mathbb{R}^n$ and diagonal matrix $\boldsymbol{W} \in \mathbb{R}^{n\times n}$ $\left(n = \sum_{i=1}^{L} S_i\right)$ are given by

$$\begin{gathered}\boldsymbol{F}(z)\\ = \left[(F_1^1 - F^1(z,\tau_1^1)),\ldots,\left(F_{S_1}^1 - F^1(z,\tau_{S_1}^1)\right),\ldots,(F_1^N - F^N(z,\tau_1^N)),\ldots,\left(F_{S_N}^N - F^N(z,\tau_{S_N}^N)\right)\right]\\ diag(\boldsymbol{W}) = \left[\theta^1\Big/\sum_{s=1}^{S_1}\left[F_s^1\right]^2,\ldots,\theta^1\Big/\sum_{s=1}^{S_1}\left[F_s^1\right]^2,\ldots,\theta^N\Big/\sum_{s=1}^{S_N}\left[F_s^N\right]^2,\ldots,\theta^N\Big/\sum_{s=1}^{S_N}\left[F_s^N\right]^2\right]\end{gathered} \tag{13.43}$$

The well known Gauss–Newton algorithm consists of minimization of the objective function, i.e. finding z such that $\frac{\partial\Phi}{\partial z} = 0$ and use of linear approximation of $\boldsymbol{F}(z)$, which yields

$$^{k+1}z = {}^k z - ({}^k\boldsymbol{J}^T \boldsymbol{W}\, {}^k\boldsymbol{J})^{-1}\, {}^k\boldsymbol{J}^T \boldsymbol{W}\, {}^k\boldsymbol{F} \tag{13.44}$$

where the left superscript indicates iteration count; ${}^k\boldsymbol{J} = \frac{\partial \boldsymbol{F}}{\partial z}\big|_{^k z}$ is Jacobian matrix which is a function of the sensitivity matrix ${}^kJ_{\alpha i} = \frac{\partial F^\alpha}{\partial z_i}\Big|_{^k z}$. The Gauss-Newton method makes use of the Hessian matrix approximation, $\boldsymbol{H} \approx {}^k\boldsymbol{J}^T\ \boldsymbol{W}\ {}^k\boldsymbol{J}$, instead of evaluating the exact second derivatives of the objective function $H_{ij} = \frac{\partial^2 \Phi}{\partial z_i \partial z_j}$ required by the Newton method or by approximating it in the Quasi-Newton method.

The sensitivity matrix ${}^k J_{\alpha i}$ can be determined either analytically or by using finite difference which requires computation of the observations for small variations of each of the parameter values. While the latter is time consuming since the number of necessary multiscale simulations increase linearly with the number of unknown parameters, it is a nonintrusive strategy that makes use of exiting multiscale solver as a black box.

To overcome potential singularity or near singularity Hessian matrix approximation ${}^k\boldsymbol{J}^T\, \boldsymbol{W}\, {}^k\boldsymbol{J}$ the Levenberg-Marquardt algorithm perturbs the Hessian matrix approximation by a positive definite matrix $\mu\mathbf{I}$ and guides the iterative process

$$ {}^{k+1}\hat{z} = {}^k\hat{z} - ({}^k\hat{\boldsymbol{J}}^T \hat{\boldsymbol{W}}\, {}^k\hat{J} + \mu I)^{-1}\, {}^k\hat{\boldsymbol{J}}^T \hat{\boldsymbol{W}}\, {}^k\hat{\boldsymbol{F}} \tag{13.45} $$

by active parameter strategy, which imposes simple bounds on the Levenberg-Marquardt parameter μ and on the trust region radius Δ. In (13.45) $\hat{\boldsymbol{F}}$ contains components of $\boldsymbol{F}$ for which the model parameters lie inside bounds (*i.e.* $i = 1, 2, \ldots, N_a$ such that $l_i \le z_i \le u_i$); $\hat{\boldsymbol{J}}$ is the Jacobian of $\hat{\boldsymbol{F}}$; μ is selected to ensure that iterates lie within the trust region radius Δ (i.e., $\mu = 0$ if $\|{}^k\hat{\boldsymbol{J}}^T\, \hat{\boldsymbol{W}}\, {}^k\hat{\boldsymbol{F}}\|_2 \le \Delta$ and $\mu > 0$ otherwise) where μ and Δ are computed based on a locally constrained hook step algorithm.

13.4.2 Characterization of Model Parameters in Reduced Order Computational Continua

The fine-scale model parameters in the context of reduced order homogenization include elastic constitutive model parameters of phases (such as Young's modulus, Poisson ratio), inelastic constitutive parameters of phases and interfaces (such as eigenstrain and eigenseparation evolution law parameters), and geometric unit cell parameters (such as volume fraction, parametric geometry definition of inclusions, *etc.*).

The sequence in which various model parameters are characterized is as follows:

I. Elastic constitutive model parameters of phases
The influence functions comprising material database depend on elastic properties and therefore have to be characterized (identified) first. In this process one has to consider experiments for which the unit cell geometry in known either from micrographs or based on manufacturer's specifications. Only uniform coarse-scale fields are considered.

II. Inelastic constitutive model parameters of phases and interfaces
Once the elastic properties have been characterized (identified) the inelastic constitutive parameters of phases and interfaces can be identified. As in Step I, only experiments for which the precise microstructural geometry and uniform coarse-scale fields are considered. In many situations, however, the inverse problem (13.42) for characterization of inelastic model parameters, may lead to numerically unstable solutions, *i.e.* small variations in observable response F_s^{α} may lead to large variations in inelastic model parameters. This difficulty is caused, in part because there could me many inelastic material parameters of phases and interfaces, which in some cases could be linearly dependent, or because experimental data is inadequate in the sense that some parameters are not "activated". One possibility to address this issue is to employ a multistep characterization process by which only a subset of parameters is characterized at a time against experiments that mostly affect these parameters. We will refine discussion on this point in Section 13.4.3.

III. Size effect parameter
Consideration of problems (such as beam bending) involving coarse-scale solution gradients over the unit cell domain to characterize size effect parameter ζ (see Section 13.3).

IV. Fine-scale geometric parameters
Here we can consider experiments for the precise geometry is unknown. Since coefficient tensors depend on the unit cell geometry, material database needs to be recomputed on the fly during characterization process of geometric parameters. This introduces significant computational complexity, which stems from continuous recreation of the unit cell CAD and FE models. This final step can be used for material optimization (instead of material characterization) if the precise geometry is known. The input to such a material optimization is a desired overall response, which can be stated by modifying existing experimental response.

13.4.3 Multistep Model Characterization

Consider a brittle fibrous composite material (Figure 13.5(b)) that behaves linearly until failure (Figure 13.5(a)). Elastic materials properties are as follows: matrix (E = 2050 MPa, $\nu = 0.3$), fiber (E = 410 000 MPa, $\nu = 0.3$) and fiber volume fraction: 30%. Four test have been conducted: two tension loading tests (T00 – tension in fiber direction, T90 – tension perpendicular to fibers) and two shear tests (S12 – in plane shear and S23-transverse shear) as shown in Figure 13.5. The ultimate strength properties of the composite are shown in the Table 13.1.

The initial strength fiber and matrix strength values were assumed 10 MPa and 6000 MPa, respectively.

We compare computational efficiency of one-step parameter characterization (identification) strategy by which two model parameters, fiber strength S_f and matrix strength S_m where identified using the Levenberg-Marquardt algorithm

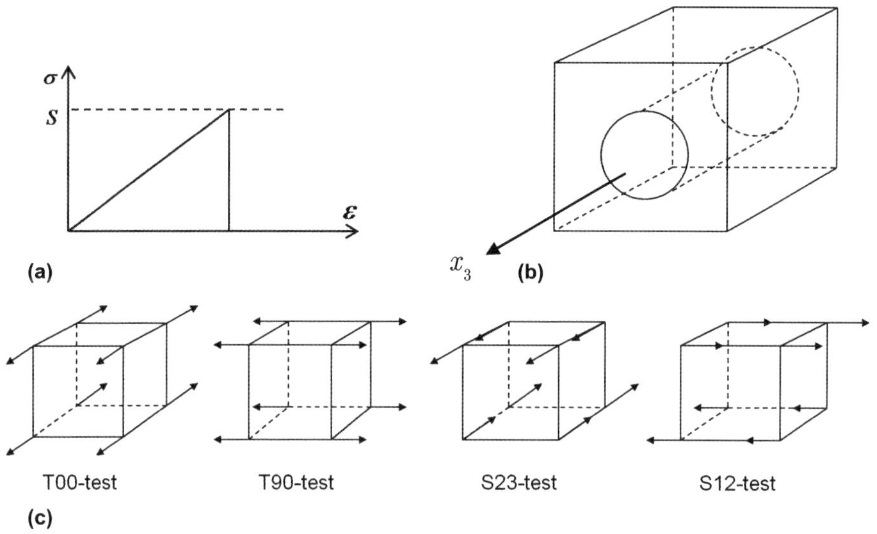

Figure 13.5 (a) Brittle failure of the composite (b) fibrous unit cell (c) tests configurations.

Table 13.1 The ultimate strength properties of the composite material.

Configuration	*T00*	*T90*	*S12*	*S23*
Strength	2112 MPa	84 MPa	15.6 MPa	19 MPa

and a three-step parameter characterization strategy. The three-step parameter characterization strategy consists of first identifying matrix strength S_m from matrix dominated tests (S12 and T90). Then keeping identified matrix shear values fixed (passive) identify fiber strength S_f based on fiber-dominated mode of deformation (S23 and T00). Finally, identify the two parameters simultaneously in single step with the initial guesses obtained from the first two steps. The one-step parameter identification strategy required 104 unit cell solutions resulting in $S_m = 22.8\ MPa$ and $S_f = 7872\ MPa$. The first two steps in a multistep approach required 12 and 20 unit cell solutions resulting in $S_m = 23\ MPa$ and $S_f = 7873\ MPa$. The final step involved additional two unit cell solutions resulting in identical identified values obtained in a multistep approach. In conclusion, the multistep approach provided a speedup of approximately factor of three compared a one-step approach.

13.5 Multiscale Design System Architecture

Based on the RC^2 formulation, a complete environment for analysis and design, subsequently to be referred as Multiscale Design System (MDS), for structural components made of heterogeneous material system can be

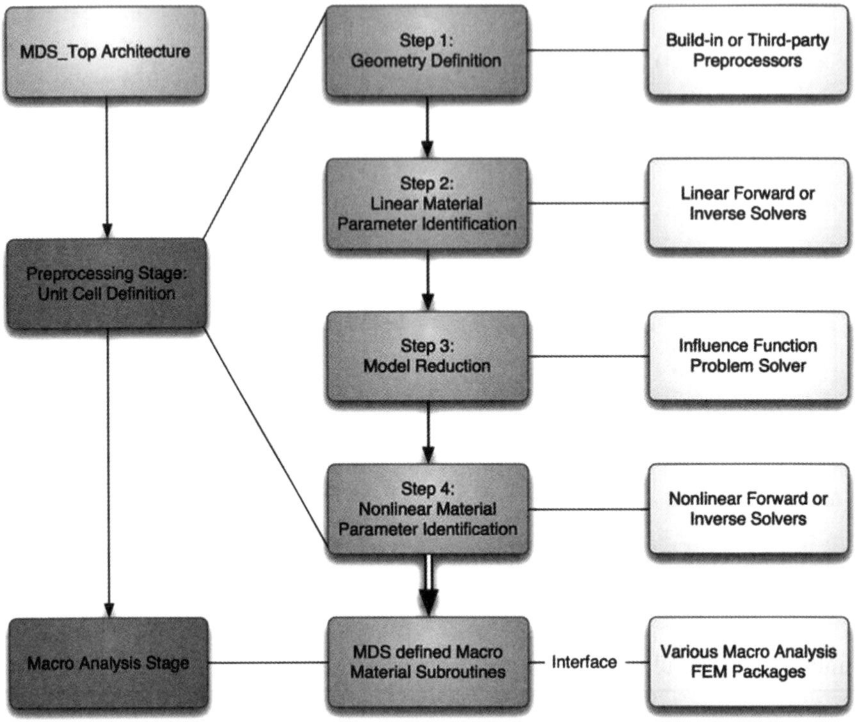

Figure 13.6 Architecture of multiscale design system.

developed (Figure 13.6). The MDS consists multiscale- solver as described in the previous sections and optimization package for model characterization (or parameter identification) and validation.

The multiscale analysis and design process consists of two analysis stages as shown in Section 13.4. The preprocessing stage includes unit cell definition, linear material parameter characterization, model reduction, and nonlinear material parameter characterization. The coarse-scale analysis stage includes coarse-scale analysis by utilizing nonlocal coarse-scale element and the unit cell database generated in the preprocessing stage.

13.6 Conclusions

The reduced order computational continua (RC^2) formulation embodies the computational efficiency of semi-analytical or phenomenological methods by constructing residual-free fields and thus avoiding costly discrete equilibrium calculations. It inherits the ability of computational continua[27] formulation to resolve microstructural details even in the regions where coarse-scale fields are no longer uniform by employing nonlocal quadrature that reproduces the exact coarse-scale variational statement. Multiscale characterization of the overall microstructural fields de facto compensates for the approximation introduced by model reduction.

Acknowledgements

This work was supported by the Office of Naval Research grant N000140310396, Rolls-Royce contract 0518502 and General Motors grant.

References

1. J. D. Eshelby, *Proc. R. Soc. London, Ser. A*, 1957, **241**, 376–396.
2. Z. Hashin, *J. Appl. Mech.*, 1962, **29**, 143–150.
3. T. Mori and K. Tanaka, *Acta Metall.*, 1973, **21**, 571–574.
4. R. Hill, *J. Mech. Phys. Solids*, 1965, **13**, 213–222.
5. R. M. Christensen and K. H. Lo, *J. Mech. Phys. Solids*, 1979, **27**, 315–330.
6. P. Suquet, *Homogenization techniques for composite media, Lecture Notes in Physics*, 1987, **272**, 193–278.
7. J. M. Guedes and N. Kikuchi, *Computer Methods in Applied Mechanics and Engineering*, 1990, **83**, 143–198.
8. K. Terada and N. Kikuchi, ASME International Mechanical Engineering Congress and Exposition, 1995.
9. F. Feyel and J. L. Chaboche, *Computer Methods in Applied Mechanics and Engineering*, 2000, **183**, 309–330.
10. J. Fish and K. Shek, *Computer Methods in Applied Mechanics and Engineering*, 1999, **172**, 145–174.
11. J. Fish, K. Shek, M. Pandheeradi and M. S. Shephard, *Computer Methods in Applied Mechanics and Engineering*, 1997, **148**, 53–73.
12. J. Fish and Q. Yu, *International Journal for Numerical Methods in Engineering*, 2001, **52**, 159–191.
13. M. G. D. Geers, V. Kouznetsova and W. A. M. Brekelmans, *Journal De Physique Iv*, 2001, **11**, 145–152.
14. S. Ghosh, K. Lee and S. Moorthy, *Computer Methods in Applied Mechanics and Engineering*, 1996, **132**, 63–116.
15. V. Kouznetsova, W. A. M. Brekelmans and F. P. T. Baaijens, *Computational Mechanics*, 2001, **27**, 37–48.
16. K. Matsui, K. Terada and K. Yuge, *Computers & Structures*, 2004, **82**, 593–606.
17. C. McVeigh, F. Vernerey, W. K. Liu and L. C. Brinson, *Computer Methods in Applied Mechanics and Engineering*, 2006, **195**, 5053–5076.
18. J. C. Michel, H. Moulinec and P. Suquet, *Computer Methods in Applied Mechanics and Engineering*, 1999, **172**, 109–143.
19. C. Miehe and A. Koch, *Archive of Applied Mechanics*, 2002, **72**, 300–317.
20. R. J. M. Smit, W. A. M. Brekelmans and H. E. H. Meijer, *Computer Methods in Applied Mechanics and Engineering*, 1998, **155**, 181–192.

21. K. Terada and N. Kikuchi, *Computer Methods in Applied Mechanics and Engineering*, 2001, **190**, 5427–5464.
22. S. Ghosh, K. H. Lee and S. Moorthy, *Int. J. Solids Struct.*, 1995, **32**, 27–62.
23. R. Hill, *J. Mech. Phys. Solids*, 1963, **11**, 357–372.
24. I. Babuska, *Numerical solution of partial differential equations, III* (Proc. Third Sympos. (SYNSPADE), Univ. Maryland, College Park, Md., 1975), 1976, 89–116.
25. A. Benssousan, J. L. Lions and G. Papanicoulau, *Asymptotic Analysis for Periodic Structures*, North-Holland, 1978.
26. E. Sanchez-Palencia, in *Lecture notes in physics*, Springer-Verlag, Berlin, 1980, vol. 127.
27. J. Fish and S. Kuznetsov, *International Journal for Numerical Methods in Engineering*, 2010, **84**, 774–802.
28. J. Fish and Z. Yuan, *Bridging the Scales in Science and Engineering*, Oxford University Press, 2008.
29. C. Oskay and J. Fish, *Computer Methods in Applied Mechanics and Engineering*, 2007, **196**, 1216–1243.
30. W. Wu, Z. Yuan, J. Fish and V. Aitharaju, *International Journal for Multiscale Computational Engineering*, 2010, **8**, 615–629.
31. Z. Yuan and J. Fish, *International Journal for Numerical Methods in Engineering*, 2009, **79**, 314–339.
32. Z. Yuan and J. Fish, *Computer Methods in Applied Mechanics and Engineering*, 2009, **198**, 2016–2038.
33. M. A. Armstrong, *Basic Topology*, Springer-Verlag, New York, 1997.
34. J. Fish, V. Filonova and Z. Yuan, in *Submitted.*
35. T. J. R. Hughes, *The Finite Element Method: Linear Static and Dynamic Finite Element Analysis*, Dover Publications, Inc., New York, 2000.

Subject Index

Note: Page numbers in **bold** refer to Volume 1 and non-bold numbers refer to Volume 2.